Civil Engineering Materials

Civil Engineering Materials

Peter A. Claisse

ELSEVIER

AMSTERDAM • BOSTON • HEIDELBERG • LONDON • NEW YORK
OXFORD • PARIS • SAN DIEGO • SAN FRANCISCO • SINGAPORE
SYDNEY • TOKYO
Butterworth-Heinemann is an imprint of Elsevier

Butterworth-Heinemann is an imprint of Elsevier
The Boulevard, Langford Lane, Kidlington, Oxford OX5 1GB, UK
225 Wyman Street, Waltham, MA 02451, USA

Notices

Knowledge and best practice in this field are constantly changing. As new research and experience broaden our understanding, changes in research methods, professional practices, or medical treatment may become necessary.

Practitioners and researchers must always rely on their own experience and knowledge in evaluating and using any information, methods, compounds, or experiments described herein. In using such information or methods they should be mindful of their own safety and the safety of others, including parties for whom they have a professional responsibility.

To the fullest extent of the law, neither the Publisher nor the authors, contributors, or editors, assume any liability for any injury and/or damage to persons or property as a matter of products liability, negligence or otherwise, or from any use or operation of any methods, products, instructions, or ideas contained in the material herein.

British Library Cataloguing-in-Publication Data
A catalogue record for this book is available from the British Library

Library of Congress Cataloging-in-Publication Data
A catalog record for this book is available from the Library of Congress

ISBN: 978-0-08-100275-9

For information on all Butterworth-Heinemann publications
visit our website at http://store.elsevier.com/

Printed in the United States of America

Transferred to Digital Printing, 2015

Working together
to grow libraries in
developing countries

www.elsevier.com • www.bookaid.org

Contents

Summary

This book covers the construction materials content for undergraduate courses in Civil Engineering and related subjects, and will also be a valuable reference for professionals working in the construction industry.

The topics are relevant to all the different stages of a course, starting with basic properties of materials and leading to more complex areas, such as the theory of concrete durability, and corrosion of steel.

The first 16 chapters cover the basic properties of materials, and how they are measured. These range from basic concepts of strength to more complex topics, such as diffusion and adsorption. The following 13 chapters cover cementitious materials. The production and use of reinforced concrete, as well as its durability, are considered in detail because this is the most ubiquitous material that is used in construction, and its premature failure is a massive problem worldwide. The subsequent chapters consider the other materials used in construction, including metals, timber, masonry, plastics, glass, and bitumens. The final chapters discuss examples of composites, and adhesives and sealants, and then cover a number of potentially important technologies that are currently being developed.

The book is intended for use both in the United Kingdom and the United States, as well as in other countries, so both metric and imperial units are discussed.

Abbreviations

ac	Alternating current
ASR	Alkali silica reaction
ASTM	American Society for Testing and Materials
BS	British Standard
BTU	British thermal unit
CE	Conformité Européenne
cgs	Centimetre gram second
CSF	Condensed silica fume
CSH	Calcium-silicate hydrate
dc	Direct current
DEF	Delayed ettringite formation
DIN	Deutsches Institut für Normung (German standards)
dpc	Damp proof course
EN	EuroNorm
ENV	Draft EuroNorm
GGBS	Ground granulated blast furnace slag
GRP	Glass reinforced polyester
HAC	High-alumina cement
HC	Hydrated cement
HDPE	High density polyethylene
ISAT	Initial surface absorption test
ISO	International Standards Organization
kip	Thousands of pounds (lb)
ksi	Thousands of pounds per square inch
Ln or Log_e	Natural logarithm
Log or Log_{10}	Logarithm to base 10
LPRM	Linear polarization resistance measurement
MKS SI	Meter kilograms seconds, Systéme International
OPC	Ordinary Portland cement
PFA	Pulverised fuel ash
psf	Pounds per square foot
psi	Pounds per square inch
QA	Quality assurance
RH	Relative humidity
SCC	Self-compacting concrete
SIP	Structural insulated panel
VFA	Voids filled with binder (for bituminous mix)
VMA	Viscosity modifying admixture (for concrete)
VMA	Voids in the mineral aggregate (for bituminous mix)
VTM	Voids in the total mix (for bituminous mix)
w/c	Water to cement ratio

Introduction

This book covers the construction materials content for undergraduate courses in Civil Engineering, and related subjects. The aim of the method or presentation is to cover the basic science, before moving on to a detailed analysis of the materials. The chapters on the science include mechanical, thermal, electrical, and transport properties of materials, and discuss the basic theory, as well as the relevance to applications in construction. The book then moves on to consider in detail each of the key materials, such as concrete and steel, and to discuss their properties, with reference to the basic science from the initial chapters.

The book is written for the age of the Internet, in which facts are readily obtained from websites. It therefore concentrates on demonstrating methods to obtain, analyse, and use information from a wide variety of sources.

Improving materials offers great scope for energy saving and environmental gains. These gains should be considered at all stages of the design and specification process, so they are discussed throughout the book.

The subject of construction materials is an area where there have been some very expensive mistakes, such as the use of High-alumina cement and calcium chloride in concrete, at times when there was ample published research to show that they should not have been used. Therefore, in addition to Chapter 15 explaining how to write detailed reports on materials experiments, Chapter 16 includes details of methods of assessment of published literature.

With the wide availability of research reports on the properties of materials, clients are now asking for calculations for the design life of structures, particularly reinforced concrete, for which the durability generally depends on the transport properties. This book presents the basic theories, and equations necessary for these calculations, and gives numerical examples of how they may be applied.

Construction materials are continually being replaced with new products. This book is, therefore, intended to give guidance on the assessment of new materials, rather than simply concentrating on those that are currently available. This includes methods to assess and analyse data on physical properties such as strength, permeability, and thermal conductivity. Similarly, new standards are continually being produced, and are now immediately available for all engineers to download. This book seeks to show the principles of test methods, so new ones can be understood and applied.

Materials cause many problems on site, and are a major area for improvement. There are, however, few simple right answers. Students will find this different from, say, the study of structures, where calculations give just one correct value for the size of a structural member. When considering the correct solution for a problem with durability, there are many different possibilities that may be appropriate for different situations, and the aim of this book is to provide the basis for the choice.

It is no longer possible for engineers in Europe to treat US Customary (Imperial) units as things of the past that are only found in the United States. When looking for a material property on the Internet, the data that is found will frequently come from the United States, and will frequently be in pounds per square inch, or degrees Fahrenheit, or similar units. All engineers should be familiar with these units, so they can make use of all the mass of data that is available on the web. The different units and the methods needed to use them are discussed in Chapter 1.

Detailed suggestions for appropriate chapters to be included at each level of a course, together with solutions to tutorial questions, and suggested laboratory exercises, are included in an Instructor's Manual that accompanies this book (available online to instructors who register at textbooks.elsevier.com/9780081002759).

Students with a particular interest in transport processes in concrete and durability modelling are referred to "Transport Properties of Concrete" by the same author, also published by Elsevier (Woodhead imprint).

UNITS

1.1 INTRODUCTION

This chapter provides an overview of some mathematical notation and methods that are essential to engineering. All students should read through it, and many will be able to confirm that they have covered it all in previous studies. It is essential that any students who are not familiar with this material should study it further because, if it is not fully understood, many section of this book and many other aspects of a degree in engineering will be impossible to understand.

The main discussion uses metre–kilogramme–second (MKS) units; but the use of Imperial (US customary), and centimetre–gramme–second (CGS) units is also described. The purpose of this is not that students in Europe should read one section, and students in the United States should read the other. When searching for values for material properties on the Internet, they may be found in any units. All students should be fully familiar with all these types of units, and able to recognize them, and convert between them. Unit conversion macros are found easily online. The important concept is to recognize which system of units the data is in, and make sure that all data in an equation or calculation is in the same system.

1.2 SYMBOLS

This book is about the properties of construction materials. Where possible, these properties are measured quantitatively; this means that values are assigned to them. For example, if a large block of concrete is considered which has one side 10-m long, this may be represented by equation (1.1):

$$L = 10\,\text{m} \tag{1.1}$$

where L is the variable we are using for the length of that side.

If the length of the block in other directions is considered and these are 8 and 12 m, this may be represented by equation (1.2):

$$L_1 = 10\,\text{m} \quad L_2 = 8\,\text{m} \quad L_3 = 12\,\text{m} \tag{1.2}$$

To calculate the mass of the block, the density must be known. This could be given by the equation (1.3):

$$\rho = 2400\,\text{kg/m}^3 \tag{1.3}$$

where ρ is the Greek letter rho that is often used as a variable for density. Greek letters are used because there are insufficient letters in the English alphabet (correctly called the Roman alphabet) for the different properties that are commonly measured.

The Greek letters that are commonly used are in Table 1.1.

Table 1.1 Greek Letters in Common Use in Engineering

Lower Case	Upper Case	Name
α		Alpha
β		Beta
γ		Gamma
δ	Δ	Delta
ε		Epsilon
η		Eta
θ		Theta
λ		Lambda
μ		Mu
ν		Nu
π		Pi
ρ		Rho
σ	Σ	Sigma
τ		Tau
φ		Phi
ω	Ω	Omega

1.3 SCIENTIFIC NOTATION

The mass of the block is given by equation (1.4):

$$m = L_1 \times L_2 \times L_3 \times \rho = 2,304,000 \, \text{kg} \qquad (1.4)$$

The result has been given in kilogrammes. It may be seen that the number is large, and not easy to visualize or use. In order to make the number easier to use, it is expressed commonly in scientific notation as 2.304×10^6 kg.

It is important to express numbers correctly in scientific notation. The 2.304 should normally be between 1 and 10. To express this number as 0.2304×10^7 or 23.04×10^5 is mathematically the same, but should not normally be used. The number raised to a power must be 10, for example, 8^6 or 7^6 should not appear in this notation. The power must be a positive or negative integer, for example, numbers such as $10^{4.5}$ or $10^{6.3}$ should not appear, but negative powers such as 10^{-4} may be used.

Because many computer printers could not at one time print superscripts, an alternative notation is often used: 2.304×10^6 is written as 2.304E6. This notation may be found in books and papers, but it is not recommended for use in formal reports. It is, however, a convenient notation, and is used in spreadsheets. The E is represented by the "EXP" key on some calculators. Note that, for example, 10^8 is entered into a calculator as 1 EXP 8 not 10 EXP 8. 10 EXP 8 is 10×10^8, which is 10^9.

1.4 UNIT PREFIXES

The alternative way of making the number easier to use is to change the units. For all metric units, the prefixes in Table 1.2 are used.

Thus 2,304,000 kg = 2,304 Mg (Megagramme). It would be technically correct, but very unusual to express it as 2.304 Gg. One Mg is equal to a metric tonne, so the mass would commonly be expressed as 2304 tonnes.

Table 1.2 Metric Prefixes		
P	Peta	10^{15}
T	Tera	10^{12}
G	Giga	10^{9}
M	Mega	10^{6}
k	Kilo	10^{3}
m	milli	10^{-3}
μ	micro	10^{-6}
n	nano	10^{-9}
p	pico	10^{-12}

1.5 LOGS

Another method of expressing large numbers is the use of logarithms, called commonly logs. These are particularly useful on graphs because both small and large numbers can be represented on the same graph (see, e.g., Fig 15.1). The log function is available on many calculators, and in all spreadsheets.

Logs are always relative to a given base, and the log of a number x to a base a is written as $\log_a(x)$, and is defined from equation (1.5):

$$a^{\log_a(x)} = x \tag{1.5}$$

Some useful relationships with logs are:

$$\log_a(xy) = \log_a(x) + \log_a(y) \tag{1.6}$$

$$\log_a\left(\frac{x}{y}\right) = \log_a(x) - \log_a(y) \tag{1.7}$$

$$\log_a(x^y) = y \times \log_a(x) \tag{1.8}$$

1.5.1 LOGS TO BASE 10

Logs to base are the logs that were used for calculations before calculators were invented. The procedure for multiplying two numbers together was to obtain the log of each, and then add these together (as in equation (1.6)), and obtain the inverse log (shown as 10^x on calculators) of the result.

It may be seen that:

$$\log_{10}(10) = 1, \ \log_{10}(100) = 2, \text{ etc.} \tag{1.9}$$

1.5.2 LOGS TO BASE e

e is a constant (= 2.718), and logs to base e are called natural logs. $\log_e(x)$ is written as $\ln(x)$ (while $\log_{10}(x)$ is written as $\log(x)$) in spreadsheets, and most calculators. It may be seen that if:

$$y = \log_e(x) \tag{1.10}$$

then

$$x = e^y \tag{1.11}$$

The function e^y is, therefore, the inverse function to the natural log, and can be used to obtain the original number from a natural log. It is often called the exponential function and written as EXP(y) in spreadsheets.

1.6 ACCURACY

Depending on the accuracy known or required, the number 2.304×10^6 might be expressed as: 2.3×10^6 kg or even 2×10^6 kg. This may be correct, but numbers should never be rounded in this way until the end of a calculation. If numbers are rounded, and then used for further calculations very large errors may be caused.

1.7 UNIT ANALYSIS

This is a method that can be used to check equations by reducing the terms on each side to base units (it is also often used without units and called dimensional analysis, but they are used in the method presented here). Thus, for example, equation (1.4) has the base units:

$$kg = m \times m \times m \times kg/m^3 \tag{1.12}$$

that may be seen to be the same on each side.

All equations should check in this way. If the two sides of the equation are different, then it is incorrect.

1.8 MKS SI UNITS

The calculation in equation (1.12) has been carried out in MKS SI (metre–kilogramme–second Systéme International) units in which masses are measured in kilogramme, and lengths in metre. The MKS SI system was developed specifically to be consistent, easy to use, and is the legally recognised system of units in Europe, and has no local variations.

A system of units consists of base units and derived units. The base units are fixed (e.g., the metre was defined as the length of a piece of steel that was kept in Paris) and the derived units are obtained from them. The main base units in the MKS system are given in Table 1.3.

Derived units are obtained from the equations that define them, for example:

Force = mass × acceleration, thus 1 Newton of force = $1 \, kgm/s^2$
Pressure = force per unit area, thus 1 Pascal of pressure = $1 \, N/m^2 = 1 \, kg/ms^2$

Within the MKS SI system, the only units that are used are based on the metric prefixes in Table 1.2 that increase or decrease by factors of 1000. Thus the metre, the kilometre, and the millimetre are used, but the centimetre is not.

1.9 US CUSTOMARY UNITS

US customary units are also known as a type of British Imperial units. In the US customary system, force is measured generally in the same units as mass (pounds).

Some of the US customary units (such as the gallon, pint and ton) are not the same as the Imperial units still used informally in the United Kingdom. 1 US liquid pint = 473 mL but 1 UK pint = 568 mL.

Table 1.3 Base Units in the MKS System		
	Name	**Symbol**
Length	metre	m
Mass	kilogramme	kg
Time	second	s
Electric charge	Coulomb	C
Temperature	degree Celsius or Kelvin	°C or °K

Table 1.4 Some US Customary Units

Quantity	MKS Unit	US Customary Unit	Conversion
Length	m	Inch, foot, thousanths of an inch (thou or mil)	1 m = 39.4 in. = 3.28 ft., 1 mm = 39.4 mil
Mass	kg	Pound (lb), thousands of pounds (kip)	1 kg = 2.205 lb = 2.205 × 10^{-3} kip
Force	N	Pound (lb)	1 N = 0.225 pounds
Stress	Pa (=N/m^2)	Pounds per square inch (psi) or thousand pounds per square inch (ksi)	1 Mpa = 145 psi = 0.145 ksi = 1 N/mm^2, 1 GPa = 145 ksi
Energy	J	British Thermal Unit (BTU)	1 kJ = 0.948 BTU
Power	W	BTU/h	1 W = 3.41 BTU/h
Volume	m^3, L	ft.^{3} or yd^3	1 m^3 = 1.31 yd^3 = 35.32 ft.^{3} = 1000 L
Density	kg/m^3	lb/yd^3	1 kg/m^3 = 1.69 lb/yd^3
Thermal conductivity	W/m°C	BTU/h ft.°F	1 W/m°C = 0.578 BTU/h ft.°F
Specific heat	J/kg°C	BTU/lb°F	1 J/kg°C = 2.39 × 10^{-4} BTU/lb°F
Temperature	°C or °K	°F	°C = (°F - 32) × 0.556 = °K - 273.12

Similarly 1 US short ton = 0.907 metric tonnes but 1 UK ton = 1.016 tonnes. For this reason these imperial units are not used in this book.

Table 1.4 shows the commonly used US customary units. Definitions of these properties such as thermal conductivity are given in subsequent chapters of this book.

Conversions to US customary units are given in appropriate chapters of this book.

1.10 CGS UNITS

CGS (centimetre–gramme–second) units were used in parts of Europe before the MKS SI system was adopted. As with the Imperial system, there are local variations within it. In addition to the metric prefixes listed in Table 1.2 the "centi" prefix (10^{-2}) is commonly used. Some CGS units that may be encountered are given in Table 1.5.

1.11 PROPERTIES OF WATER IN DIFFERENT UNITS

Some properties of water in different units are given in Table 1.6.

Table 1.5 Some CGS Units

Quantity	MKS Unit	CGS Unit	Conversion
Length	Meter	Centimeter	1 m = 100 cm
Force	Newton	Dyne	1 N = 10^5 Dynes
Viscosity	Pascal second	Poise	1 Pas = 10 Poise
Kinematic viscosity = viscosity/density	Square meter per second	Stokes (St)	1 m²/s = 10^4 St
Energy	Joule	Calorie	1 Joule = 0.24 calories

The kilocalorie may be also referred to as a Calorie (with an uppercase C) causing some confusion.

Table 1.6 Properties of Water

	MKS	US Customary	CGS
Density	1000 kg/m³	62.43 lb/ft.³ = 1686 lb/yd³	1 g/cm³
Specific heat	4186 J/kg°C	1 BTU/lb°F	1 calorie/g°C
Freezing point	0°C, 273°K	32°F	
Boiling point	100°C, 373°K	212°F	

1.12 SUMMARY

- Scientific notation and unit prefixes should be used for clarity with large or small numbers.
- 10^8 is entered into a calculator as 1 EXP 8 not 10 EXP 8. 10 EXP 8 is 10×10^8, which is 10^9.
- To maintain accuracy, numbers should never be rounded until the end of a calculation.
- Unit analysis should be used to check equations.
- MKS, CGS, and US customary are three different systems of units, and should never be mixed together in the same equation.
- Some US customary units differ from Imperial units previously used in the United Kingdom.

NOTATION

e   Mathematical constant = 2.718
L Length (m)
m Mass (kg)
ρ Density (kg/m³)

STRENGTH OF MATERIALS

2

CHAPTER OUTLINE

2.1 INTRODUCTION

The strength of a material is almost always the first property that the engineer needs to know about. If the strength is not adequate, then the material cannot be used and other properties are not even considered. The next property to be considered is often the "stiffness," or elastic modulus because this determines how far a structure will deflect under load. In this chapter, the basic concepts of force, stress, strength, strain, and elastic modulus are introduced.

2.2 MASS AND GRAVITY

In the MKS SI system, the mass of an object is defined from its acceleration when a force is applied, for example, from equation (2.1)

$$f = ma \tag{2.1}$$

where f is the force in Newton, m is the mass in kg, a is the acceleration in m/s^2.

Gravity is normally the largest force acting on a structure. On the earth's surface, the gravitational force on a mass m is given by equation (2.2)

$$f = mg \qquad (2.2)$$

where g is the gravitational constant = 9.81 m/s^2 (32.2 ft./s^2).

The gravitational force on an object is called its weight. Thus, an object will have a weight of 9.81 N/kg of mass. An approximate value of 10 is often used for g to give the commonly used value of 10 kN weight for a mass of 1 tonne (=1000 kg). In the US customary system of units, force is generally measured as a weight in pounds and, if this is done, a constant term for $g = 32.2$ ft./s^2 must be included in equation (2.1).

2.3 STRESS AND STRENGTH

2.3.1 TYPES OF LOADING

In engineering, the term strength is always defined by type, and is probably one of the following (see Fig. 2.1), depending on the method of loading.

- Compressive strength
- Tensile strength
- Flexural strength
- Shear strength

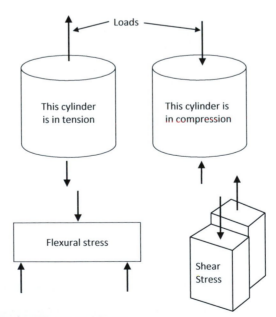

FIGURE 2.1 Compression, Tension, Flexure and Shear

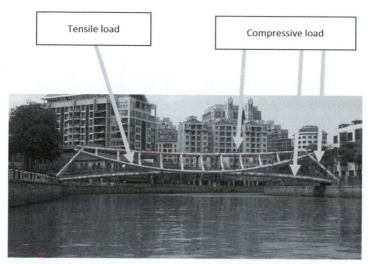

FIGURE 2.2 A Complex Bridge on the Singapore River

A force acting on an object becomes a load on the object, so force and load have the same units. In some structures, the compressive and tensile forces are not immediately apparent (Fig. 2.2).

2.3.2 TENSILE AND COMPRESSIVE STRESS

In order to define strength, it is necessary to define stress. This is a measure of the internal resistance in a material to an externally applied load. For direct compressive or tensile loading, the stress is designated σ, and is defined in equation (2.3), and measured in Newtons per square metre (Pascals) or pounds per square inch.

$$\text{Stress}, \sigma = \frac{\text{Load } W}{\text{Area } A} \tag{2.3}$$

See Fig. 2.3.

2.3.3 SHEAR STRESS

Similarly, in shear the shear stress τ is a measure of the internal resistance of a material to an externally applied shear load. The shear stress is defined in equation (2.4) (see Fig. 2.4):

$$\text{Shear stress}, \tau = \frac{\text{Load } W}{\text{Area resisting shear } A} \tag{2.4}$$

2.3.4 STRENGTH

The strength of a material is a measure of the stress that it can take when in use. The ultimate strength is the measured stress at failure, but this is not normally used for design because safety factors are required.

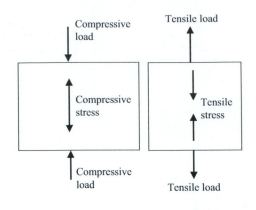

FIGURE 2.3 Load and Stress

2.4 STRAIN

In engineering, strain is not a measure of force, but is a measure of the deformation produced by the influence of stress. For tensile and compressive loads:

$$\text{Strain} = \frac{\text{Increase in length } x}{\text{Original length } L} \tag{2.5}$$

Strain is dimensionless, so it is not measured in metres, kilogrammes, etc. The commonly used unit is microstrain (μstrain), which is a strain of one part per million.

For shear loads, the strain is defined as the angle γ (see Fig. 2.4). This is measured in radians, and thus for small strains:

$$\text{Shear strain} \approx \frac{\text{Shear displacement } x}{\text{Width } B} \tag{2.6}$$

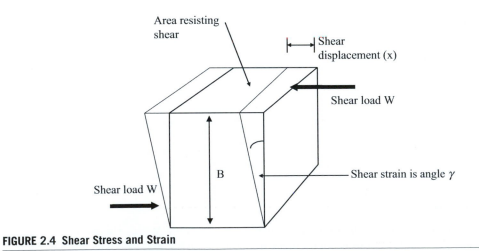

FIGURE 2.4 Shear Stress and Strain

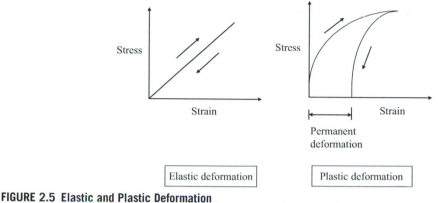

FIGURE 2.5 **Elastic and Plastic Deformation**

The arrows show the sequence of loading and unloading during a test.

2.5 **DEFORMATION AND STRENGTH**

Strain may be elastic or plastic. Figure 2.5 shows the stress on an object, and the resulting strain as it is loaded and then unloaded. If the strain is elastic, the sample returns exactly to its initial shape when unloaded. If plastic strain occurs, there is permanent deformation.

If the material exhibits plastic deformation (yields), and does not return to its original shape when unloaded, this is clearly unacceptable for most construction applications. Figure 2.6 shows a stress–strain curve for a typical metal. As the load is applied, the graph is initially linear (the stress is proportional to the strain), until it reaches a yield point. If the load is removed after yield, the sample will not return to its original shape, and is left with final residual strain.

For a brittle material (such as concrete), strength is defined from the stress at fracture, but for a ductile material (e.g., some steels) that yields a long way before failure, strength is often defined from limits to the residual strain, after loading and unloading. This concept of "proof stress" is discussed in Chapter 30.

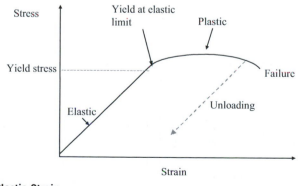

FIGURE 2.6 **Elastic and Plastic Strain**

2.6 MODULUS OF ELASTICITY

If the strain is "elastic," that is, on the linear part of a graph of stress versus strain, Hooke's law may be used to define Young's modulus as the gradient:

$$\text{Young's modulus}, E = \frac{\text{Stress}}{\text{Strain}} \tag{2.7}$$

Thus, from equations (2.4), (2.5) and (2.7)

$$E = \frac{W}{x} \times \frac{L}{A} \tag{2.8}$$

where W/x may be the gradient of a graph of load versus displacement obtained from an experiment.

The Young's modulus is also called the modulus of elasticity or stiffness, and is a measure of how much strain occurs due to a given stress. Because strain is dimensionless, Young's modulus has the units of stress or pressure.

Some typical values are given in Table 2.1.

In reality, no part of a stress–strain curve obtained from an experiment is ever perfectly linear. Thus the modulus must be obtained from a tangent or a secant. The difference between an initial tangent and secant modulus is shown in Fig. 2.7.

2.7 POISSON'S RATIO

This is a measure of the amount by which a solid "spreads out sideways" under the action of a load from above. It is defined from equation (2.9). A material like timber which has a "grain direction" will have a number of different Poisson's ratios corresponding to loading, and deformation in different directions.

$$\text{Poisson's ratio} = \frac{\text{Lateral strain}}{\text{Strain in direction of load}} \tag{2.9}$$

Table 2.1 Typical Values of Strength, Failure Strain and Modulus

	Typical Values		
	Strength	Failure/Yield Strain	Young's Modulus
	MPa (ksi)	(μstrain)	GPa (ksi)
Mild steel (tensile/compressive)	300 (43)	1,500	200 (29,000)
Concrete (compressive)	40 (6)	2,000	20 (2,900)

Notes: 1 MPa = 1 N/mm²; 1 GPa = 1 kN/mm²

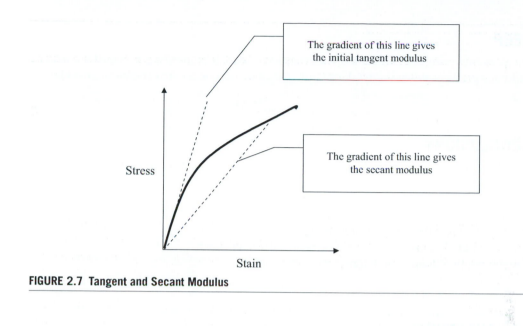

The gradient of this line gives the initial tangent modulus

The gradient of this line gives the secant modulus

Stress

Stain

FIGURE 2.7 Tangent and Secant Modulus

2.8 FATIGUE STRENGTH

If a material is continually loaded and unloaded (e.g., the springs in a car), the permanent strain from each cycle slowly decreases. This may be seen from Fig. 2.8. Eventually, the sample will fail, and the number of cycles it takes to fail will depend on the maximum stress that is being applied. The fatigue of steel is discussed in Section 30.6.4.

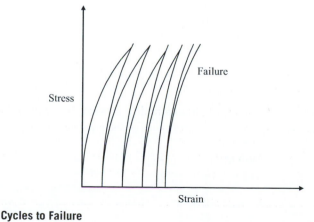

Failure

Stress

Strain

FIGURE 2.8 Fatigue Cycles to Failure

2.9 CREEP

Creep is the slow irreversible deformation of materials under load. It is surprisingly large for concrete, and tall buildings get measurably shorter during use (the guide rails on the lifts sometimes buckle).

2.10 CONCLUSIONS

- In the MKS system, force is defined from mass and acceleration, and is measured in Newtons.
- Stresses may be compressive, tensile, flexural, or shear.
- The strength of a material is the stress at failure.
- Strain is a measure of the deformation produced by a stress.
- The elastic modulus is the ratio of stress/strain.
- The Poisson's ratio is a measure of strain perpendicular to the load.
- Creep is a measure of long-term deformation under load.

TUTORIAL QUESTIONS

Assume $g = 10$ m/s^2 for all questions

1. **a.** A round steel bar with an initial diameter of 20 mm and length of 2 m is placed in tension, supporting a load of 2000 kg. If the Young's modulus of the bar is 200 GPa what is the length of the bar when supporting the load (assuming it does not yield)?
   b. If the Poisson's ratio of the steel is 0.4, what will the diameter of the bar be when supporting the load?

   Solution:

   a. Cross-sectional area $= 3.14 \times 10^{-4}$ m^2
   Load $= 2000$ kg $\times g = 2 \times 10^4$ N   (2.2)
   Stress $= 2 \times 10^4/3.14 \times 10^{-4} = 6.4 \times 10^7$ Pa   (2.3)
   Strain $= 6.4 \times 10^7/200 \times 10^9 = 3.2 \times 10^{-4}$ (2.7)
   New length $= (3.2 \times 10^{-4} \times 2) + 2 = 2.00064$ m   (2.5)

   b. Lateral strain $= 3.2 \times 10^{-4} \times 0.4 = 1.28 \times 10^{-4}$ (2.9)
   Change in diameter $= 1.28 \times 10^{-4} \times 0.02 = 2.56 \times 10^{-6}$ m   (2.5)
   New diameter $= 0.02 - 2.56 \times 10^{-6} = 1.999744 \times 10^{-2}$ m $= 19.99744$ mm

2. A 2 m long tie in a steel frame is made of circular hollow section steel with an outer diameter of 100 mm and a wall thickness of 5 mm. The properties of the steel are as follows:
   Young's modulus: 200 GPa
   Yield stress: 300 MPa
   Poisson's ratio: 0.15

a. What is the extension of the tie when the tensile load in it is 200 kN?
b. What is the reduction in wall thickness when this load of 200 kN is applied?

Solution:

a. Area $= \pi(0.05^2 - 0.045^2) = 1.491 \times 10^{-3}$ m^2
  Stress $= 200 \times 10^3/1.491 \times 10^{-3} = 1.34 \times 10^8$ Pa   (2.3)
  This stress is 134 MPa, which is well below the yield stress.
  Strain $= 1.34 \times 10^8/2 \times 10^{11} = 6.7 \times 10^{-4}$ (2.7)
  Extension $= 2 \times 6.7 \times 10^{-4} = 0.00134$ m $= 1.34$ mm   (2.5)
b. Strain $= 6.7 \times 10^{-4} \times 0.15 = 1.005 \times 10^{-4}$ (2.9)
  Reduction $= 1.005 \times 10^{-4} \times 0.005 = 5.025 \times 10^{-7}$ m $= 0.5$ μm   (2.5)

3. A water tank is supported by four identical timber posts that all carry an equal load. Each post measures 50 mm by 75 mm in cross-section, and is 1.0 m long. When 0.8 m^3 of water is pumped into the tank, the posts get 0.07 mm shorter.

a. What is the Young's modulus of the timber in the direction of loading?
b. If the cross-sections measure 75.002 mm by 50.00015 mm, after loading, what are the relevant Poisson's ratios?
c. If 300 L of water are now pumped out of the tank, what are the new dimensions of the post?

(Assume that the strain remains elastic.)

Solution:

Normally, all calculations should be carried out in base units, but these calculations of strain are an example of where it is clearly easier to work in mm.

a. 0.8 m^3 of water has a mass of 800 kg and thus a weight of $800 \times g = 8000$ N
  Stress $= 8000/(4 \times 0.075 \times 0.05) = 5.33 \times 10^5$ Pa   (2.3)
  Strain $= 0.07/1000 = 7 \times 10^{-5}$ (2.5)
  Modulus $= 5.33 \times 10^5/7 \times 10^{-5} = 7.6 \times 10^9$ Pa   (2.7)
b. Change in width on long side $= 75.002 - 75 = 0.002$ mm
  Strain on long side $= 0.002/75 = 2.66 \times 10^{-5}$ (2.5)
  Thus, Poisson ratio $= 2.66 \times 10^{-5}/7 \times 10^{-5} = 0.38$ (2.9)
  Strain on short side $= 0.00015/50 = 3 \times 10^{-6}$ (2.5)
  Thus, Poisson ratio $= 3 \times 10^{-6}/7 \times 10^{-5} = 0.043$ (2.9)
c. 300 L of water has a mass of 300 kg.
  Thus, the mass of water now in the tank is $800 - 300 = 500$ kg
  The changes in dimensions are proportional to the load, and thus they are $500/800 = 0.625$ of those in part (b).
  $0.07 \times 0.625 = 0.044$, thus length $= 1000 - 0.044 = 999.956$ mm
  $0.002 \times 0.625 = 0.00125$, thus long side $= 75.00125$ mm
  $0.00015 \times 0.625 = 0.000094$, thus short side $= 50.000094$ mm

4. The below figure shows observations that were made when a 100 mm length of 10 mm diameter steel bar was loaded in tension. The equation given is for a trendline that has been fitted to the straight portion of the graph. Calculate the following:

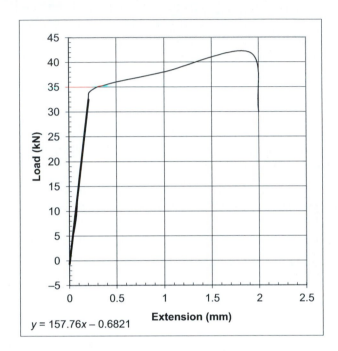

$y = 157.76x - 0.6821$

a. the Young's modulus,
b. the estimated yield stress,
c. the ultimate stress.

Solution:

Cross-section area = $\pi \times 0.01^2/4 = 7.85 \times 10^{-5} \text{ m}^2$

a. The equation given on the graph has been generated by the Excel compute package, and is a best fit to the linear part of the data (this is real experimental data). This technique is discussed in Chapter 15.
   The gradient taken from the equation is 157.76 kN/mm = 1.5776×10^8 N/m
   $E = 1.5776 \times 10^8 \times 0.1/7.85 \times 10^{-5} = 2.01 \times 10^{11}$ Pa = 201 GPa   (2.8)
b. Load = 33 kN (estimated from the end of the linear section on the graph)
   Thus, stress = 420 MPa   (2.3)
c. Load = 42 kN at failure (the highest point on the graph)
   Thus, stress = 535 MPa   (2.3)

5. **a.** A 100 mm diameter concrete cylinder 200 mm long is loaded on the end with 6 tonnes. What is the stress in it?
   b. The Young's modulus of the cylinder is 25 GPa. What is its height after loading?
   c. The Poisson's ratio of the cylinder is 0.17. What is its diameter after loading?
   d. A column 2 m high, measuring 300 mm by 500 mm on plan, is made with the same concrete as in the cylinder and supports a bridge. In order to prevent damage to the deck, the maximum permitted compression of the column is 0.3 mm. Assuming that the column is not reinforced, calculate the compression of the column when it is supporting the full weight of a 40 tonne lorry, and state whether the deck will fail.

Solution:

Area $= \pi \times 0.05^2 = 0.0078$ m^2

a. Load $= 6$ tonne $= 6000$ kg
   Weight $= 6,000 \times g = 60,000$ N   (2.2)
   Stress $= 60,000/0.0078 = 7.64 \times 10^6$ Pa   (2.3)
b. Strain $= 7.64 \times 10^6/25 \times 10^9 = 3.06 \times 10^{-4}$ (2.7)
   New height $= 0.2 - (0.2 \times 3.06 \times 10^{-4}) = 0.19994$ m   (2.5)
c. Lateral strain $= 3.06 \times 10^{-4} \times 0.17 = 5.2 \times 10^{-5}$ (2.9)
   $0.1 + (0.1 \times 5.2 \times 10^{-5}) = 0.1000052$ m   (2.5)
d. Mass of lorry $= 40$ tonne, thus Weight $= 4 \times 10^5$ N   (2.2)
   Stress $= 4 \times 10^5/(0.3 \times 0.5) = 2.66 \times 10^6$ Pa   (2.3)
   Compression $= 2.66 \times 10^6 \times 2/(25 \times 10^9) = 2 \times 10^{-4}$ m $= 0.2$ mm   (2.5) and (2.7)
   Deck will not fail.

6. **a.** How many Poisson's ratios does timber have?
   b. A timber sample measuring 50 mm $\times$ 50 mm on plan is supporting a mass of 2 tonnes. If the relevant Young's modulus and Poisson's ratio are 1.3 GPa and 0.4, what is the lateral expansion caused by the load?

Solution:

a. It has 6 Poisson's ratios (see Section 33.5.2):
   longitudinal–tangential
   tangential–longitudinal
   radial–tangential
   tangential–radial
   radial–longitudinal
   longitudinal–radial
b. Load $= 2$ tonnes $= 2000$ kg
   Weight $= 2,000 \times g = 20,000$ N   (2.2)
   Stress $= 20,000/(0.05 \times 0.05) = 8 \times 10^6$ Pa   (2.3)
   Longitudinal strain $= 8 \times 10^6/1.3 \times 10^9 = 6.15 \times 10^{-3}$ (2.7)
   Lateral strain $= 0.4 \times 6.15 \times 10^{-3} = 2.46 \times 10^{-3}$ (2.9)
   Expansion $= 2.46 \times 10^{-3} \times 0.05 = 1.23 \times 10^{-4}$ m $= 0.123$ mm   (2.5)

7. A steel specimen measuring 1 in. by 0.5 in. in section is found to have yielded at a load of 38 kips, and failed at 50 kips. Assuming that the modulus of elasticity is 29,000 ksi, and the Poisson's ratio is 0.3, calculate:

a. The tensile stress at yield and failure.
b. The strain when the stress is half the yield stress.
c. The dimensions of the section at half the yield stress.

Solution:

a. Yield stress = 38,000/(1 × 0.5) = 76,000 psi   (2.3)
   Failure stress = 50,000/(1 × 0.5) = 10^5 psi   (2.3)
b. Longitudinal strain = (0.5 × 76,000)/(2.9 × 10^7) = 1.31 × 10^{-3} (2.7)
c. Lateral strain = 1.31 × 10^{-3} × 0.3 = 3.4 × 10^{-4} (2.9)
   New width = 1 − (1 × 3.4 × 10^{-4}) = 0.99966 in.   (2.5)
   New thickness = 0.5 − (0.5 × 3.4 × 10^{-4}) = 0.49983 in.   (2.5)

8. a. A round steel bar with an initial diameter of ¾ in. and length of 6 ft. is placed in tension, supporting a load of 4,400 lb. If the Young's modulus of the bar is 29,000 ksi, what is the length of the bar when supporting the load (assuming it does not yield)?
b. If the Poisson's ratio of the steel is 0.4, what will the diameter of the bar be when supporting the load?

Solution:

a. Cross-sectional area = 0.442 in.^{2}
   Stress = 4,400/0.442 = 9,954 psi   (2.3)
   Strain = 9,954/29,000 × 10^3 = 3.43 × 10^{-4} (2.7)
   New length = (3.43 × 10^{-4} × 6) + 6 = 6.002 ft.   (2.5)
b. Lateral strain = 3.43 × 10^{-4} × 0.4 = 1.37 × 10^{-4} (2.9)
   Change in diameter = 1.37 × 10^{-4} × 0.75 = 1.027 × 10^{-4} in.   (2.5)
   New diameter = 0.75 − 1.027 × 10^{-4} = 0.749897 in.

9. A 6 ft. long tie in a steel frame is made of circular hollow section steel with an outer diameter of 4 in., and a wall thickness of 0.2 in. The properties of the steel are as follows:

Young's modulus: 29,000 ksi
Yield stress: 43 ksi
Poisson's ratio: 0.15

a. What is the extension of the tie when the tensile load in it is 44 kips?
b. What is the reduction in wall thickness when this load of 44 kips is applied?

Solution:

a. Area = $\pi(2^2 − 1.8^2)$ = 2.387 in.^{2}
   Stress = 44 × 10^3/2.387 = 18.43 ksi   (2.3)
   This stress is well below the yield stress.
   Strain = 18.43 × 10^3/2.9 × 10^7 = 6.35 × 10^{-4} (2.7)
   Extension = 6 × 12 × 6.35 × 10^{-4} = 0.0457 in.   (2.5)

b. Strain $= 6.35 \times 10^{-4} \times 0.15 = 9.5 \times 10^{-5}$ (2.9)
Reduction $= 9.5 \times 10^{-5} \times 0.2 = 1.9 \times 10^{-5}$ in.   (2.5)

10. The figure below shows observations which were made when a 4 in. length of 3/8 in. diameter steel bar was loaded in tension. The equation given is for a trendline that has been fitted to the straight portion of the graph. Calculate the following:

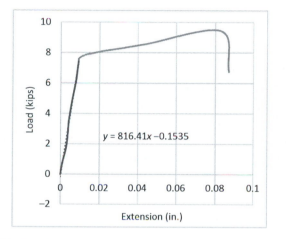

a. the Young's modulus,
b. the estimated yield stress,
c. the ultimate stress.

Solution:

Cross-section area $= \pi \times 0.375^2/4 = 0.11$ in.^{2}
a. The equation given on the graph has been generated by the Excel compute package, and is a best fit to the linear part of the data. This technique is discussed in Chapter 15.
The gradient taken from the equation is 816.4 kips/in.
$E = 816.4 \times 4/0.11 = 29,687$ ksi   (2.8)
b. Load $= 7.4$ kips (estimated from the end of the linear section on the graph)
Thus stress $= 67.3$ ksi from (2.3)
c. Load $= 8.5$ kips at failure (the highest point on the graph)
Thus stress $= 77.3$ ksi from (2.3)

11. a. A 4 in. diameter concrete cylinder 8 in. long is loaded on the end with 13440 lb. What is the stress in it?
b. The Young's modulus of the cylinder is 3600 ksi. What is its height after loading?
c. The Poisson's ratio of the cylinder is 0.17. What is its diameter after loading?

d. A column 6 ft. high, measuring 12 in. by 20 in. on plan, is made with the same concrete as in the cylinder and supports a bridge. In order to prevent damage to the deck, the maximum permitted compression of the column is 0.01 in. Assuming that the column is not reinforced, calculate the compression of the column when it is supporting the full weight of a 89600 lb. truck and state whether the deck will fail.

Solution:

Area = $\pi \times 2^2$ = 12.57 in.^{2}
- **a.** Stress = 13,440/12.57 = 1,069 psi   (2.3)
- **b.** Strain = 1069/3600 × 103 = 2.97 × 10^{-4} (2.7)
  New height = 8 − (8 × 2.97 × 10^{-4}) = 7.9976 in.   (2.5)
- **c.** Lateral strain = 2.97 × 10^{-4} × 0.17 = 5.05 × 10^{-5} (2.9)
  4 + (4 × 5.05 × 10^{-5}) = 4.0002 in.   (2.5)
- **d.** Stress = 89,600/(12 × 20) = 373 psi   (2.3)
  Compression = 373 × 6 × 12/(3,600 × 10^3) = 7.46 × 10^{-3} in.   (2.5) and (2.7)
  Deck will not fail.

NOTATION

a Acceleration (m/s^2)
A Area (m^2)
B Width (m)
E Young's modulus (Pa)
f Force (N)
g Gravitational constant (=9.81 m/s^2)
L Length (m)
m Mass (kg)
W Load (N)
x Displacement (m)
γ Shear angle (radians) or shear strain
σ Stress (Pa)
τ Shear stress (Pa)

CHAPTER

FAILURE OF REAL CONSTRUCTION MATERIALS

3.1 INTRODUCTION

In this chapter, real results from testing materials are considered. The machines used to test them are described, and then the results are examined in detail, and theories are presented to explain them. A steel sample, a concrete sample, and two timber samples tested with the grain in different directions are discussed. The failure mechanisms are found to be quite different. The steel is ductile, due to the movement of lattice dislocations, the concrete is brittle, and the timber failure depends on the direction of the grain.

3.2 THE STEEL SAMPLE
3.2.1 METHOD OF LOADING

Figure 3.1 shows the stress–strain graph for a 350 mm (14 in.) length of 6 mm (0.24 in.) diameter steel that was tested in tension. A mechanical test machine (Fig. 3.2) was used, that applies the load by rotating two large screw threads in the columns on either side of the sample. The load is measured with a load cell above the sample. The extension of the sample was recorded by the displacement transducer fixed to the bar, as shown in Fig. 3.2 (Chapter 6 explains how load cells and displacement transducers work). The output from the load cell and the displacement transducer were converted into digital

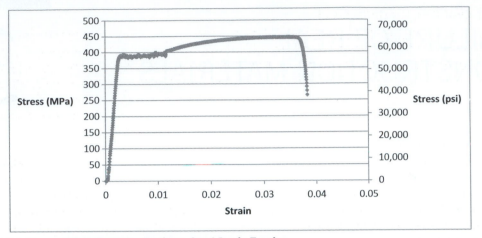

FIGURE 3.1 Stress–Strain Graph for Testing a Steel Bar in Tension

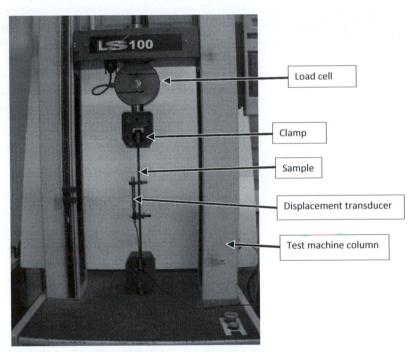

FIGURE 3.2 Arrangement of Steel Test Sample

signals for recording on the computer, and then equations (2.3) and (2.5) were used in a spreadsheet to convert them into stress, and strain.

The sample was held in clamps. During the test, it slips slightly as the jaws of the clamp dig into it, but this does not affect the displacement readings because the transducer is fixed directly to it.

3.2.2 FAILURE MECHANISMS

On the stress–strain, graph the stress increases in the elastic region to around 380 MPa (55,000 psi), as the structure is deformed. If the load had been removed in the elastic region, the sample would have recovered back to its original shape. The model for what is happening shown in Fig. 3.3. All materials are made up of atoms, and in solids, they are bonded together (see Chapter 7 for more detail of this). Under normal circumstances, the equilibrium distance between two atoms is controlled by a balance between attractive and repulsive forces. If the material is squeezed, then the atoms move together, and a net repulsion acts between them. If the material is stretched, then the atoms move apart, and a net attraction acts between them.

The next part of the curve is the horizontal section up to a strain of about 0.013. In this region, there is plastic (i.e., irreversible) deformation, with no increase in stress. This can occur in steel because the atoms in it are arranged in grains (crystals) in a regular matrix. The mechanism of deformation of these crystals is the movement of dislocations, as shown in Fig. 3.4. As atoms move to the left, the dislocation moves to the right. This enables the crystal to deform without breaking. A typical paper clip will have 10^{14} dislocations in it.

The structure of steel may be seen with an optical microscope, if a surface is polished and then chemically etched. The crystal structure only extends across small grains (typically <20 μm diameter), and is discontinuous at the grain boundaries. The dislocations cannot cross grain boundaries. New dislocations must form on the other side of the boundary. Thus, as the dislocations are stopped at the grain boundaries, the stress increases. This may be seen in Fig. 3.1, as the stress rises before failure.

The shape of the bar after failure is shown schematically in Fig. 3.5, and some failed samples are shown in Fig. 3.6. It may be seen that in the region of the failure, the steel has experienced ductile flow as the diameter has decreased (necking). This occurs immediately before failure, while the load is

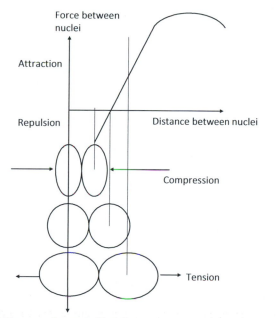

FIGURE 3.3 Effect of Applied Stress on Atoms in Material

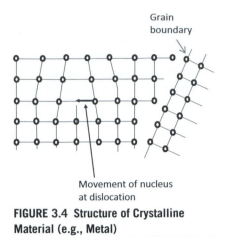

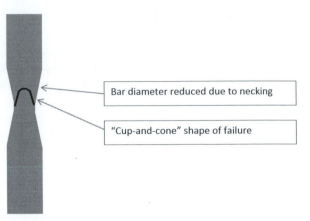

FIGURE 3.4 Structure of Crystalline Material (e.g., Metal)

FIGURE 3.5 Typical Detail of Failure of Steel Bar

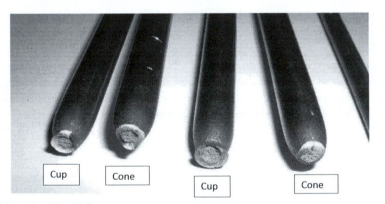

FIGURE 3.6 Steel Samples After Failure

decreasing. The other key observation is that the shape of the failure is not as predicted by the theory given in Chapter 2 that would predict flat ends to the two parts of the failed bar. One part has a "cup" shape, and one has a "cone" shape.

To explain this, it is necessary to consider a microscopic cube of the steel that is exposed to a stress σ_y (see Fig. 3.7). This is the type of approach that is used in finite element analysis, where a computer programme is used to consider the effect of a load on a large number of small elements of the sample. By making the elements progressively smaller, an increasing accurate model of the stresses and strains is obtained.

At any given angle θ to the stress, a shear force τ may be calculated as:

$$\tau = 0.5\,\sigma_y \sin 2\theta \tag{3.1}$$

where τ is the shear stress, σ_y is the direct stresses, θ is the angle of the shear plane.

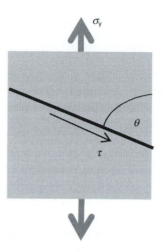

FIGURE 3.7 A Finite Element of Steel

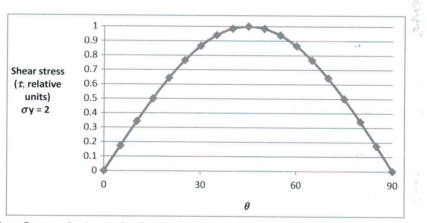

FIGURE 3.8 Shear Force at Angle θ to Applied Load

The relationship in equation (3.1) is shown in Fig. 3.8 that has been plotted for $\sigma_y = 2$. It may be seen that the maximum value of τ is 0.5 σ_y. Thus, for any material for which the shear strength is less than half the strength in direct tension, failure will occur in shear. This is the case for the steel, and the failure surface is typically at 45° to the load, fact that may be seen from Fig. 3.8 to give the highest shear stress.

3.2.3 WORK DONE DURING THE TEST

The work done during the test is defined in MKS units (Joules) as:

$$\text{Work} = \text{Force} \times \text{distance} \tag{3.2}$$

where the force is in Newtons and the distance in metres.

Thus, from equations (2.3) and (2.5):

$$\text{Work done per unit volume} = \text{Stress} \times \text{strain} \qquad (3.3)$$

The work done is thus equal to the area under the stress–strain curve, and has the unit Joules that is equivalent to Newton metres. For elastic strain with no yield, where the unloading follows exactly the same path as the loading, and there is no area between them, no work is done; all of the energy needed to deform the material is recovered when it is released. After yield, the work done will be the surface energy in the microcracks. When the failed sample is removed from the test machine, the heating in the region of the break can be felt by touching it.

Toughness is a measure of the amount of energy required before fracture under tensile stress. This is, in effect, a measure of the resistance of the material to the formation of large cracks. Ductility (i.e., deformation before failure) is an advantage for safety because it prevents sudden failure. A material is defined as being *brittle* if it is not tough, that is, it fails after a small amount of energy used (work done). This can be dangerous in construction materials. The area under the curve in Fig. 3.1 is high, indicating that steel is ductile. Concrete is normally reinforced with steel that gives it ductility, so it will resist sudden failure.

Numerous theories have been developed to predict when a system will actually fail due to complex applied stresses. The most accurate predictor of failure for use in finite element analysis has often been found to be the shear strain energy per unit volume (the Von Mises criterion).

3.3 THE CONCRETE SAMPLE
3.3.1 METHOD OF LOADING

A 100 mm (4 in.) cylinder of a typical concrete was tested in compression (Fig. 3.9). Because the failure load was much higher than the steel (42 tonnes, as opposed to 1 tonne) a hydraulic testing machine was used, rather than mechanical. This consists of a computer-controlled hydraulic pump that delivers high-pressure hydraulic oil to a ram that compressed the cylinder. The load is measured from the oil pressure, and this is used to control the rate at which it is increased. To get accurate readings, the displacement is measured using displacement transducers clamped to the cylinder. Two (or often three) transducers are arranged around the sample, and the average reading is taken because samples often deflect more on one side than the other. The graph for concrete test is shown in Fig. 3.10.

3.3.2 FAILURE MECHANISMS

The first part of the curve (up to about 5 MPa, 725 psi) shows where the sample is taking initially the load, and should be ignored. The additional strain in this region will have occurred due to slight imperfections in the sample.

The next section, up to about 30 MPa (4500 psi), is the elastic region. The mechanism for this is the same as in the steel sample, and is shown in Fig. 3.3.

Above 30 MPa (4500 psi), the sample yields and starts to fail. Figure 3.11 shows the typical shape of the remains of a sample after failure. Two pyramids of solid concrete remain after the sides have all cracked away. It may be seen that the fracture has occurred at 45° to the load, as in the steel sample. The mechanism is complex because of forces related to the Poisson's ratio (see Chapter 2).

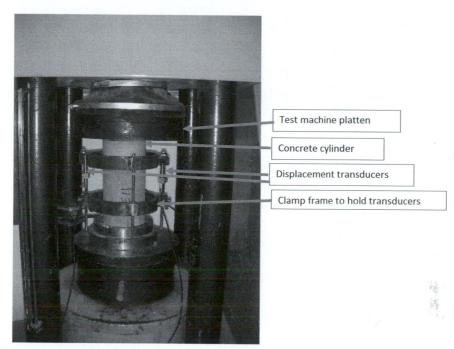

Test machine platten

Concrete cylinder

Displacement transducers

Clamp frame to hold transducers

FIGURE 3.9 Testing a Concrete Cylinder

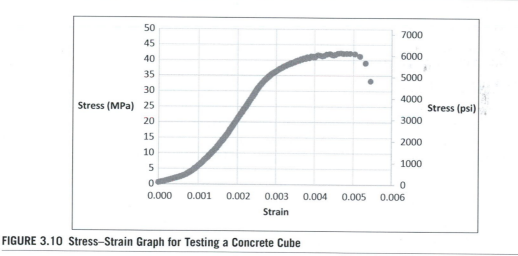

FIGURE 3.10 Stress–Strain Graph for Testing a Concrete Cube

This is further complicated (particularly for cubes) by the "platen restraint" that restrains the cube from expanding sideways at the top and bottom because it cannot slide across the machine platens that are compressing it. Nevertheless, when this standard test is used, the load is divided by the area to give a result for compressive strength. This is known as the "uniaxial" compressive strength because the load is only on one axis. There is a very specialised test in which concrete is stressed triaxially

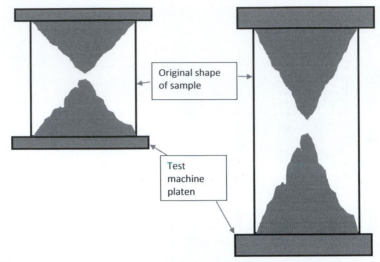

FIGURE 3.11 Typical Shape of Failed Cube (Left) and Cylinder (Right)

(on all three axes), in order to extract pore fluid for analysis. When this is done, the stresses are very much higher.

After the cube has yielded, the stress–strain curve is not linear, indicating that permanent deformation has occurred. This has happened because microcracks have started to form in the concrete. Failure occurs when these microcracks join up to form larger cracks.

3.4 THE TIMBER SAMPLES
3.4.1 METHOD OF LOADING

The test apparatus for the timber is shown in Fig. 3.12. The graphs for timber in compression are shown in Figs 3.13 and 3.14. The sample loaded parallel to the grain (Fig. 3.13) was 45 mm (1.8 in.) long by 15 mm (0.6 in.) square, and the one loaded at right angles (Fig. 3.14) was a 15 mm (0.6 in.) cube. For these very small samples, the mechanical test machine used for the steel sample had ample capacity. Because the loads were in compression, and no slippage could occur between the test machine and the sample, the displacement could be read by the test machine itself that records the rotation of the screw threads applying the load.

3.4.2 FAILURE MECHANISMS

Both graphs show a linear elastic region. For the sample loaded parallel to the grain, yield does not occur until 50 MPa (7250 psi), but at right angles to the grain yield is at just 5 MPa (725 psi).

The failure mode with the load parallel to the grain is shown schematically in Fig. 3.15, and was visible by inspection of the failed sample. The grain fibres fold over, and it may be seen from Fig. 3.13 that this leads to an immediate and progressive decrease in load. Seen at a macroscopic scale, however, this may also be considered as a shear failure at an angle to the load as in the steel.

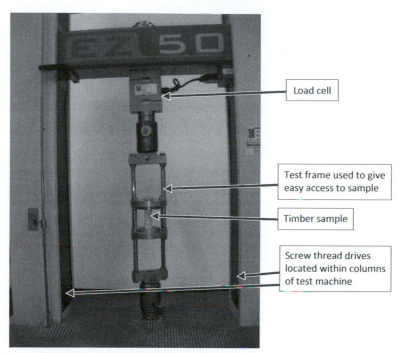

FIGURE 3.12 Testing a Timber Sample on End Grain

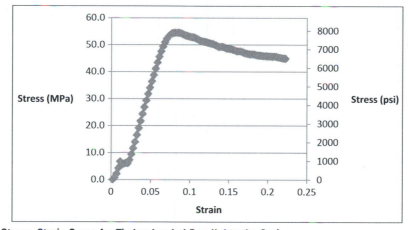

FIGURE 3.13 Stress–Strain Curve for Timber Loaded Parallel to the Grain

The failure mode with the load at right angles to the load can be visualised, as a bunch of drinking straws spreading out. After yield, they slowly spread, but the load continues to rise. This will continue as the sample gets progressively thinner, but it has effectively failed at yield, so the test machine is stopped before it overloads.

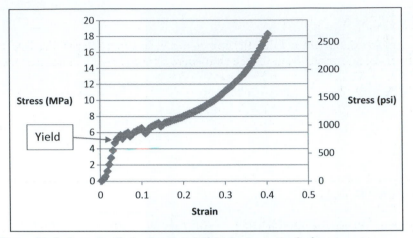

FIGURE 3.14 Stress–Strain Curve for Timber Loaded at Right Angles to the Grain

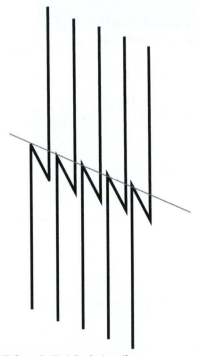

FIGURE 3.15 Schematic Diagram of Failure in End Grain Loading

3.5 SUMMARY

- The steel sample exhibited elastic deformation up to a yield stress, after which dislocations moved within the grains. Subsequently, the dislocations were stopped at grain boundaries causing an increase in stress. Failure occurred in shear.
- The concrete sample was brittle, so it failed quite rapidly after yield. Because the loading was uniaxial, it failed in lateral tension to leave a pyramid shape.
- The timber sample loaded on end grain failed, as the grain was "folded" to produce a shear plane. The timber sample loaded at right angles to the grain yielded at a low load.

TUTORIAL QUESTIONS

1. A 10 mm diameter steel bar has a tensile strength of 400 MPa, and a shear strength of 100 MPa. Calculate the maximum load that it can carry in tension.

   Solution:
   Since the shear strength is less than half the tensile strength, failure will be in shear.

   Maximum $\sigma_y = 2 \times \tau = 200$ MPa
   Area of bar $= \pi \times 0.01^2 = 3.14 \times 10^{-4}$
   Maximum tensile load $= 200 \times 10^6 \times 3.14 \times 10^{-4} = 15.7$ kN

2. A ½ in. diameter steel bar has a tensile strength of 60 ksi and a shear strength of 15 ksi. Calculate the maximum load that it can carry in tension.

   Solution:
   Since the shear strength is less than half the tensile strength, failure will be in shear.

   Maximum $\sigma_y = 2 \times \tau = 30$ ksi
   Area of bar $= \pi \times 0.5^2/4 = 0.1963$ in.^{2}
   Maximum tensile load $= 30 \times 0.1963 = 5.89$ kips

NOTATION

θ Angle of the shear plane (degrees)
σ_y Direct stresses (Pa)
τ Shear stress (Pa)

THERMAL PROPERTIES

4.1 INTRODUCTION

Thermal properties are relevant to a number of considerations of the performance of materials in complete structures, such as heat loss from buildings, and the effect of sunlight on buildings. They are also key considerations when designing concrete structures because of the heat generation during cement hydration, and the thermal cracking that it may cause.

Heat (thermal energy) may be transmitted by three main processes:

- Heat conduction through a material is a process in which the vibration of atoms (that is responsible for the temperatures that we observe) causes other nearby atoms to vibrate, and transmit the energy. This is discussed in Section 4.5
- Radiant heat, such as sunlight, or the heat felt standing in front of a fire, is electromagnetic radiation that can transmit through air, or through a vacuum. This is discussed in Section 4.9.
- Heat convection occurs when a heated fluid, such as air, moves and so the energy is moved with it. Since hot air has a lower density, it will rise from around a heater and circulate around a room.

4.2 TEMPERATURE

Temperature is a measure of the energy in a material. The common units are degrees centigrade (°C). The names Celsius or centigrade have the same meaning.

For many calculations, degrees Kelvin (°K) must be used. The temperature in °K is obtained from the temperature in °C, by adding 273.1. This unit is used because 0°K, which is −273.1°C (−459.7°F), is the temperature at which the vibration of atoms stops because they have no energy. It is actually impossible to reach it, irrespective of how good a cooling system is used.

4.3 ENERGY

Work was defined in equation (3.2). Energy has the same units as Work, and is a measure of the ability of a system to do Work.

$$\text{Power (Watts)} = \text{The rate of energy "used"} = \frac{\text{Energy (Joules)}}{\text{Time (seconds)}} \tag{4.1}$$

When energy is in the form of electricity, the units of measurement are often kilowatt-hours. One kilowatt hour is $60 \times 60 \times 1000 = 3.6$ million J; thus, the Joule is a very small unit.

In US customary units, energy is measured in British Thermal Units (BTU). The BTU is defined from the specific heat of water (see below), and 1 BTU is equal to 1.05 kJ.

4.4 SPECIFIC HEAT

The specific heat capacity (C_p) of a material is the number of J (BTU) required to raise the temperature of 1 kg (1 lb) of the material through 1°C (1°F). Thus:

$$\text{Temperature change} = T_2 - T_1 = \frac{\text{Energy}}{C_p \times \text{mass}} \tag{4.2}$$

where T_1 and T_2 are the initial and final temperatures of an object in °C (°F), C_p is the specific heat in J/kg°C (BTU/lb°F).

The BTU is defined from the specific heat of water, that is, 1 BTU/lb°F. Thus, one BTU is the energy required to heat one pound of water by 1°F. The calorie is similarly defined for the CGS system as the energy required to heat 1 g of water by 1°C (see Table 1.6).

The blocks in night storage heaters are an example of a material that has been deliberately specified with a high specific heat. These heaters use cheap electricity that is available during the night, and store it in large dense blocks inside them, for use in daytime.

Equation (4.2) assumes that the material does not change phase from solid to liquid, or liquid to gas. If a phase change takes place, the "latent heat" of the phase change will be absorbed or released. Materials that change phase (normally from liquid to solid) at given temperatures have been developed as a means of storing energy.

4.5 THERMAL CONDUCTIVITY

The thermal conductivity (k) is the measure of the ability of a material to transmit heat by conduction. The heat (Q) is measured in Watts (BTU/h). k is defined from the equation (4.3):

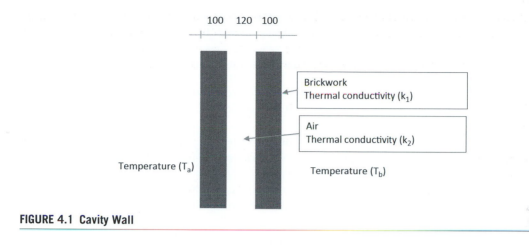

FIGURE 4.1 Cavity Wall

$$Q = \frac{kA\,(T_a - T_b)}{d}$$

(4.3)

where T_a and T_b are the temperatures either side of an element of material in °C (°F), d is the thickness in m (ft.), A is the area in m^2 (ft.^{2}), k is the thermal conductivity in W/m°C (BTU/hft.°F).

Thermal insulators are designed to have a low thermal conductivity. In fluids (liquids and gases), the main form of heat transfer may be convection, so many insulators, such as glass wool, are simply intended to prevent air movement.

The thermal properties of walls are often quoted as "U values." These are a synthesis of the k values for the actual materials, and the heat transfer coefficients at the hot and cold sides.

4.5.1 HOW TO CALCULATE HEAT LOSS THROUGH A CAVITY WALL

Consider a cavity wall with two skins of brickwork 100 mm thick and a 120 mm air cavity (Fig. 4.1).

If the conductivity of brick is k_1 and air is k_2. The air is equivalent to a very thick skin of brickwork with thickness 120 mm $\times$ k_1/k_2. Thus equation (4.3) becomes:

$$Q = \frac{k_1 A\,(T_a - T_b)}{0.2 + 0.12 \times k_1/k_2}$$

(4.4)

where 0.2 m is the thickness of the two skins of brickwork and 0.12 m is the thickness of the air cavity between them. Details of cavity wall construction are given in Chapter 34.

4.6 THERMAL CAPACITY, THERMAL DIFFUSIVITY, AND THERMAL INERTIA

Specific heat and thermal conductivity are the two basic thermal properties that determine heat flow in solids. There are, however, three other quantities, which are calculated from them:

$$\text{Thermal capacity} = \text{Specific heat} \times \text{mass}$$

(4.5)

$$\text{Thermal diffusivity} = \frac{\text{Thermal conductivity}}{\text{Density} \times \text{specific heat}} \tag{4.6}$$

$$\text{Thermal inertia} = \text{Thermal conductivity} \times \text{density} \times \text{specific heat} \tag{4.7}$$

The thermal capacity may be calculated for a whole structure, not just a material. Thus, a structure built with heavy materials with a high thermal capacity (such as brick and concrete) will tend to stay cool in the day, and warm at night.

If heat is applied to one side of a building element, the diffusivity will determine how fast the other side warms up. If it has high diffusivity, it has a low specific heat, so it cannot absorb much heat but it has a high conductivity, so it can move it away easily.

The surface of a material with a high thermal inertia will heat up more slowly when heat is applied to it because it has both the capacity to absorb the heat, and the conductivity to carry it away.

4.7 COEFFICIENT OF THERMAL EXPANSION

This is defined as the proportional length change per degree of temperature change. Whether or not thermal expansion causes cracking, will clearly depend on the degree of restraint. If an element of a structure is held at both ends, it will crack when it shrinks. If it has movement joints, the shrinkage is harmless.

If the coefficient of thermal expansion is X, the length change will be:

$$\Delta L = X \times L \times \Delta T \tag{4.8}$$

where L is the length and ΔT is the change in temperature.

4.8 HEAT GENERATION

Heat may be generated in a number of different ways. Burning fuel generates a lot of heat. For example, burning high-grade petrol (gasoline) will give off 45 MJ/kg (20,000 BTU/lb). Hydrating of cement gives off about 0.5 MJ/kg (230 BTU/lb).

4.9 HEAT ABSORPTION, REFLECTION, AND RADIATION

In the same way that a mirror reflects light, it will also tend to reflect heat, rather than absorb it. Surfaces that do not absorb heat will also not radiate heat. In general, white or reflective surfaces will not absorb or radiate heat, but black, nonreflecting ones will both absorb and radiate it. Thus, a matt black asphalt roof may get hot in strong sunlight and fail, but a thin layer of reflecting aluminum will protect it. Foil backed plasterboard has a reflecting surface that makes it a poor absorber of heat but, most importantly, a poor emitter – so it retains heat in a room. Cities have been found to be cooler if light coloured concrete road surfaces are used, rather than black bitumen (avoiding the "heat island" process).

In normal circumstances, a structure may lose or gain heat substantially through convection. If this process does not take place, and the object is black and nonreflecting, the heat loss is:

$$Q = \sigma A T^4 \tag{4.9}$$

where Q is the heat loss in Watts (J/s), σ is Stefan's constant $= 5.67 \times 10^{-8}$ W/m^2°K^4, A is the area in m^2, T is the temperature in °K.

This is the process responsible for causing ground frosts and ice on roads when the air temperature above them is still a few degrees above freezing.

The thermal radiation emitted by warm objects has a significantly longer wavelength than visible light, and is known as infrared. Thermal imaging cameras may be use to look at the infrared thermal radiation given off by warm surfaces, and identify where heat is being lost from a structure. Increasing carbon dioxide contents in the atmosphere reduce the transmission of long wavelength infrared radiation, without affecting shorter wavelengths such as visible light, and thus cause global warming.

4.10 TYPICAL VALUES

Typical values for the main thermal properties of some common construction materials are given in Table 4.1.

Water has a high specific heat, so wetting materials will raise their specific heat. Air has a low thermal conductivity but to use it as an insulator it must be stopped from moving, for example, by putting in glass fibre insulation.

4.11 SUMMARY

- Temperature is a measure of the energy in a material.
- The specific heat is a measure of the amount of energy needed to raise the temperature of a material.
- The thermal conductivity is a measure of energy transmission through a material.

Table 4.1 Typical Values

	Density	Specific Heat	Thermal Conductivity	Coefficient of Thermal Expansion
	kg/m^3 (lb/yd^3)	J/kg°C (BTU/lb°F)	W/m/°C (BTU/hft.°F)	μstrain/°C (μstrain/°F)
Steel	7800 (13180)	480 (0.115)	50 (29)	11 (6.1)
Concrete	2400 (4000)	840 (0.2)	1.4 (0.8)	11 (6.1)
Brickwork	1700 (2873)	800 (0.19)	0.9 (0.5)	8 (4.4)
Water	1000 (1690)	4186 (1)	0.6 (0.35)	
Air	1.2 (2.03)	1000 (0.24)	0.03 (0.017)	

- The thermal capacity is a measure of the energy an object can hold.
- The thermal inertia is a measure of the rate at which the temperature of a heated surface of an object will rise.
- The thermal diffusivity is a measure of the rate at which the temperature of the opposite surface of an object will rise.
- Surfaces, which absorb heat well, will also radiate it.

TUTORIAL QUESTIONS

1. Two identical water tanks are supported on timbers, which are 100 mm by 50 mm by 1 m long, and are resting on a flat, rigid floor. Each tank rests on the whole of the 50 mm wide edge of two timbers (see diagram).

a. If the tanks are at the same level, when empty, what is the difference in level when one tank has 1.5 m^3 more water in it than the other?

b. If 0.5 m^3 of hot water is added to the tank with less water in it, and the temperature of the timbers under it is raised by 30°C, what does the difference in level become?

The properties of the timbers are as follows:

- Young's modulus: 80 N/mm^2 (assumed constant for all temperatures)
- Coefficient of thermal expansion: 34 μstrain/°C

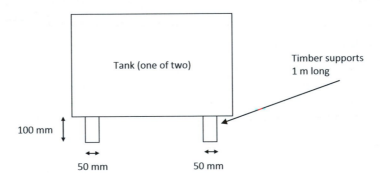

Tank (one of two)

Timber supports
1 m long

100 mm

50 mm   50 mm

Solution:

a. Area of timber under each tank = .05 × 1 × 2 = 0.1 m^2
Mass of water = 1500 kg
Load = 1,500 × 10 = 15,000 N   (2.2)
Stress from 1.5 m^3 of water = 15,000/0.1 = 150,000 N/m^2 (2.3)
Thus strain = 150,000/(80 × 10^6) = 1.9 × 10^{-3} (2.7)
Thus level difference = 100 × 1.9 × 10^{-3} = 0.19 mm   (2.5)

  b. The change in level will be proportional to the amount of water in the tank.
Thus, level change from weight of 0.5 m^3 of water $= 0.19 \times 0.5/1.5 = 0.06$ mm
Level change from heating $= 30 \times 34 \times 10^{-6} \times 100 = 0.102$ mm   (4.8)
Thus, new level difference $= 0.19 - 0.06 + 0.102 = 0.232$ mm
(Note the way these add together.)

2. The bricks in a wall have a specific heat of 800 J/kg/°C, a thermal conductivity of 0.9 W/m/°C, and a density of 1700 kg/m^3. The wall is solid, and is 215 mm thick, 2 m high, and 10 m long.

  a. If the wall is at 20°C on the inside, and 10°C on the outside, what is the rate of heat loss through it, in Watts?

  b. If the inside and outside temperatures are lowered by 10°C, what is the heat loss from the wall, in Joules?

Solution:

  a. Area of wall $= 2 \times 10 = 20$ m^2
     Heat loss $= 0.9 \times 20 \times (20 - 10)/0.215 = 837$ W   (4.3)

  b. The whole wall will be cooled by 10°C
     Mass of wall $= 20 \times 0.215 \times 1700 = 7310$ kg
     Heat energy $= 10 \times 800 \times 7310 = 5.8 \times 10^7$ J   (4.2)

3. A 200 mm thick concrete wall is cast using a mix containing 300 kg/m^3 of cement. The shuttering on each side of the wall is 20 mm plywood. If the heat of hydration is being generated at a rate of 7 W/kg of cement, and is all being lost through the shutters, what is the temperature drop across them?
(Assume the thermal conductivity of plywood $= 0.15$ W/m^2/°C.)

Solution:

Mass of cement per square metre of wall $= 0.2 \times 300 = 60$ kg
Heat per square metre $= 60 \times 7 = 420$ W
Area of shutter per square metre of wall $= 2$ m^2 (one on each side)
Temperature drop $= 420 \times 0.02/(0.15 \times 2) = 28$°C   (4.3)

4. What is the heat loss per square metre through a 320 mm cavity brick wall with a 120 mm central cavity, if the temperature difference across it is 20°C?

Solution:

Thermal conductivity of brick $= 0.9$ W/m/°C
Thermal conductivity of air $= 0.03$ W/m/°C
Heat loss $= 0.9 \times 20/\{0.2 + [0.12 \times (0.9/0.03)]\} = 4.7$ W/m^2 (4.4)
It may be seen that the 120 mm of air in the cavity is equivalent to $0.12 \times (0.9/0.03) = 3.6$ m of brick.

5. a. On a cold clear night, a road surface is at 3°C. What will be the rate of heat loss from it? (Assume a perfect nonreflecting surface.)
   b. If the surface layer is 100 mm thick, and is insulated from the layers below, and has a specific heat of 840 J/kg/°C, and a density of 2300 kg/m^3, what will its temperature change be in 1 h?

Solution:

a. Temperature $= 3 + 273.12 = 276.12°$K
Heat loss $= 5.67 \times 10^{-8} \times (276.12)^4 = 330$ W   (4.9)
b. Mass $= 0.1 \times 2300 = 230$ kg/m^2
Energy loss $= 330 \times 60 \times 60 = 1.19 \times 10^6$ J
Temperature change $= 1.19 \times 10^6/(840 \times 230) = 6°$C   (4.2)

6. The bricks in a wall have a specific heat of 700 J/Kg/°C, a thermal conductivity of 1.7 W/m/°C, a density of 1650 kg/m^3, and a coefficient of thermal expansion of 7 µstrain/°C. The wall is solid, and is 215 mm thick, 3 m high, and 8 m long.

   a. If the wall is at 23°C on the inside and 17°C on the outside, what is the rate of heat loss through it, in Watts?
   b. If the inside and outside temperatures are lowered by 8°C, what is the heat loss from the wall, in Joules?
   c. What is the length of the wall, after the temperature is lowered by the 8°C?

Solution:

a. Area of wall $= 3 \times 8 = 24$ m^2
Heat loss $= 1.7 \times (23-17) \times 24/0.215 = 1136$ W   (4.3)
b. Mass $= (3 \times 8 \times 0.215) \times 1650 = 8514$ kg
Energy $= 700 \times 8514 \times 8 = 4.76 \times 10^7$ J   (4.2)
c. Strain $= 7 \times 10^{-6} \times 8 = 5.6 \times 10^{-5}$ (4.8)
Length $= 8 - (8 \times 5.6 \times 10^{-5}) = 7.99955$ m   (2.5)

7. The four materials listed below are being considered for the cladding of a building:

Material	Proposed thickness (mm)	Thermal conductivity (W/m/°C)	Coefficient of thermal expansion (µstrain/°C)
Concrete	70	1.4	11
Steel	3.5	84	11
Aluminum	5	200	24
Brick	104	0.9	8

For each material:

   a. Calculate the heat loss in Watts through each square metre of the cladding, if the temperature difference between the interior of the building and the outside air is 10°C.

b. State the assumptions you made in your calculation in part (a), and indicate whether they are realistic.
c. Calculate the spacing of the expansion joints in the cladding, if the maximum acceptable expansion of a panel between joints is 2 mm for a 20°C temperature rise.

Solution:

a. Heat Loss $= k \times 1 \times 10/d = 10$ k/d   (4.3)
   Loss from concrete loss $= 10 \times 1.4/0.07 = 200$ W
   Similarly loss from steel $= 2.4 \times 10^5$ W; aluminum: 4×10^5 W; brick: 86 W.
b. These calculations assume that the surfaces of the cladding are at the temperature of the environment, meaning that all of the temperature drop is in the cladding. This is generally unrealistic, particularly for the steel and aluminum.
c. Coefficient $\times$ spacing $\times 20 = 0.002$ (4.8)
   Thus, spacing $= 10^{-4}$/coefficient
   Spacing for concrete $= 10^{-4}/11 \times 10^{-6} = 9$ m
   Similarly spacing for steel $= 9$ m; spacing for aluminum $= 4.1$ m; spacing for brick $= 12.5$ m.

8. A building has a 100 mm thick concrete floor slab with rooms above and below it. The slab is 6 m long and 4 m wide. The properties of the concrete are as follows:

Thermal conductivity:	1.4 W/m/°C
Density:	2300 kg/m^3
Specific heat:	840 J/kg/°C
Coefficient of thermal expansion:	11 μstrain/°C

a. What is the heat transmission through the floor, in W, if the temperature difference between the two rooms is 15°C?
b. What is the energy in J that will be absorbed by the slab, if the average temperature of the rooms rises by 10°C on a hot day?
c. What is the increase in length of the slab when the temperature is increased by 10°C?

Solution:

a. Heat transmission $= 1.4 \times (6 \times 4) \times 15/0.1 = 5040$ W   (4.3)
b. Mass $= 6 \times 4 \times 0.1 \times 2300 = 5520$ kg
   Energy $= 5520 \times 840 \times 10 = 4.63 \times 10^7$ J   (4.2)
c. Expansion $= 6 \times 11 \times 10^{-6} \times 10 = 6.6 \times 10^{-4}$ m   (4.8)

9. a. What is the energy in Joules used by a 4 kW heater, in 30 min?
b. What would the change in temperature of a 2 m^3 block of concrete with a density of 2300 kg/m^3 and a specific heat of 850 J/kg/°C be, if it was heated by a 4 kW heater for 30 min? (Assume no heat loss)
c. If the floors of a building are constructed with a material with a high specific heat, what effect is this likely to have on the temperature of the rooms?

Solution:

a. Energy $= 4000 \times 30 \times 60 = 7.2 \times 10^6$ J $\hspace{2cm}$ (4.1)

b. Mass $= 2300 \times 2 = 4600$ kg

Temperature change $= 7.2 \times 10^6/(4600 \times 850) = 1.84°C$ $\hspace{2cm}$ (4.2)

c. The temperature will be more stable – cool in the day, and warm at night.

10. Two identical water tanks are supported on timbers, which are 4 in. by 2 in. by 3 ft. long, and are resting on a flat rigid floor. Each tank rests on the whole of the 2 in. wide edge of two timbers (see diagram).

a. If the tanks are at the same level when empty, what is the difference in level when one tank has 2 yd^3 more water in it than the other?

b. If 0.5 yd^3 of hot water is added to the tank with less water in it, and the temperature of the timbers under it is raised by 55°F, what does the difference in level become?

The properties of the timbers are as follows:

- Young's modulus: 12 ksi (assumed constant for all temperatures)
- Coefficient of thermal expansion: 19 μstrain/°F

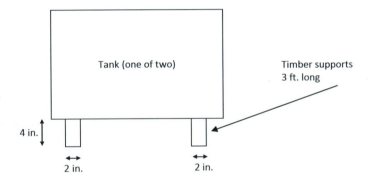

Solution:

a. Area of timber under each tank $= 2 \times 36 \times 2 = 144$ in.^{2}

Mass of water $= 1.5 \times 1686 = 2529$ lb. (see Table 1.6 for density)

Stress from 2 yd^3 of water $= 2529/144 = 17.56$ psi $\hspace{2cm}$ (2.3)

Thus strain $= 17.56/(12 \times 10^3) = 1.46 \times 10^{-3}$ $\hspace{2cm}$ (2.7)

Thus level difference $= 4 \times 1.46 \times 10^{-3} = 5.84 \times 10^{-3}$ in. $\hspace{2cm}$ (2.5)

b. The change in level will be proportional to the amount of water in the tank.

Thus, level change from weight of 0.5 yd^3 of water $= 5.84 \times 10^{-3} \times 0.5/1.5 = 1.95 \times 10^{-3}$ in.

Level change from heating $= 55 \times 19 \times 10^{-6} \times 4 = 4.18 \times 10^{-3}$ in. $\hspace{2cm}$ (4.8)

Thus, new level difference $= 5.84 \times 10^{-3} - 1.95 \times 10^{-3} + 4.18 \times 10^{-3} = 8.07 \times 10^{-3}$ in.

(Note the way these add together.)

11. The bricks in a wall have a specific heat of 0.19 BTU/lb°F, a thermal conductivity of 0.5 BTU/hft.°F, and a density of 2870 lb/yd^3. The wall is solid and is 9 in. thick, 6 ft. high, and 10 yd long.

   a. If the wall is at 65°F on the inside and 50°F on the outside, what is the rate of heat loss through it in BTU/h?

   b. If the inside and outside temperatures are lowered by 20°F, what is the heat loss from the wall, in Joules?

   Solution:

   a. Area of wall = 6 × 30 = 180 ft.^{2}
   Thickness of wall = 0.75 ft.
   Heat loss = 0.5 × 180 × (65 − 50)/0.75 = 1800 BTU/h   (4.3)

   b. The whole wall will be cooled by 20°F.
   Mass of wall = 180 × 0.75 × 2,870/27 = 14,350 lb.
   Heat energy = 20 × 0.19 × 14,350 = 54,530 BTU   (4.2)

12. What is the heat loss per yd^2 through an 11 in. cavity brick wall with a 5 in. central cavity, if the temperature difference across it is 35°F?

   Solution:

   - Thermal conductivity of brick = 0.5 BTU/hft.°F
   - Thermal conductivity of air = 0.017 BTU/hft.°F
   - Thickness of brick = 8/12 = 0.66 ft.
   - Thickness of air = 5/12 = 0.416 ft.

   Heat loss = 0.5 × 9 × 35/{0.66 + [0.416 × (0.5/0.017)]} = 12.22 BTU/h/yd^2 (4.4)

NOTATION

A Area (m^2)
C_p Specific heat (J/kg°C)
d Thickness (m)
k Thermal conductivity (W/m°C)
L Length (m)
Q Heat loss (W)
T Temperature (°C or °K)
X Coefficient of thermal expansion (°C^{-1})
σ Stefan's constant = 5.67 × 10^{-8} W/m^2°K^4

PRESSURE

CHAPTER OUTLINE

5.1 INTRODUCTION

Many aspects of construction technology relate to pressure in a fluid. These include water and air movement, either through pipes and structures, or through permeable materials (see Chapter 9). In this chapter, the basic concepts of pressure are introduced. The movement of pressure waves is then considered; and particularly ultrasound that is used to measure the elastic modulus of concrete in structures (see Chapter 27).

5.2 PRESSURE ON A FLUID

In Chapter 2, the forces on solids were considered and expressed in units of pressure (Pa or N/mm^2 etc.). In these situations, the effect of a load in a single direction, that is, "uniaxial" was considered and, with the Poisson's ratio, the movement in other directions was calculated.

In order to apply pressure to a fluid (i.e., a liquid or a gas) it must be contained in all directions, that is, "triaxially." When a fluid is under pressure:

- The pressure at any point is the same in all directions.
- The pressure exerts a force normal to the containing surface.
- The pressure increases with depth due to gravity but, with this exception, is the same throughout the volume.

Figure 5.1 is a schematic diagram of a pressure intensifier that makes use of these properties. For this system:

$$\text{Force in shaft} = \text{Pressure} \times \text{area} = p_1 A_1 = p_2 A_2 \tag{5.1}$$

47

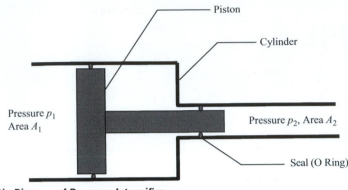

FIGURE 5.1 Schematic Diagram of Pressure Intensifier

where A_1 and A_2 are the areas of each end of the piston, and p_1 and p_2 are the pressures.
The pressure is thus increased by a factor A_1/A_2.

5.3 THE EFFECT OF GRAVITY ON PRESSURE

In MKS units, the pressure at a depth h metres below an open surface of a liquid is given by:

$$p = \rho g h \tag{5.2}$$

where ρ is the density (1000 kg/m^3 for water) and g is the gravitational constant (9.81 m/s^2).

> MKS units of pressure:
> Atmospheric pressure = 101 kPa (approx. varies with time and location) = 10.3 m water head (approx.)

In US customary units, if the pressure is measured in pounds per square foot (psf), it is not necessary to include g in equation (5.2). Atmospheric pressure (14.7 psi, 2110 psf) is approximately equivalent to 33.8 ft. of water with a density of 62.43 lb/ft.^{3}.

Figure 5.2 shows a schematic diagram of a mercury barometer that illustrates this concept. The height of the column, that is approximately 1 m (3.3 ft.), will change as atmospheric pressure on the open-end changes.

5.4 THE EFFECT OF TEMPERATURE ON GAS PRESSURE

There is an approximate equation that may be used to calculate the change in pressure in a gas, when it is heated. The "ideal gas equation" is:

$$\frac{p_1 V_1}{T_1} = \frac{p_2 V_2}{T_2} \tag{5.3}$$

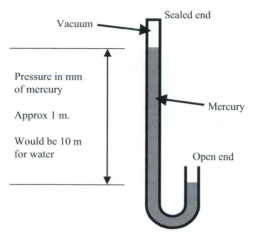

FIGURE 5.2 **Mercury Barometer**

where p_1, V_1, and T_1 are the pressure, volume, and temperature before the change, and p_2, V_2, and T_2 are the values for after the change. Note that T must be in Kelvin, not centigrade.

If several gasses are mixed together, the total pressure is the sum of the "partial pressures" due to each gas. The pressure due to each gas may be calculated as if the others were not present.

5.5 **PROPAGATION OF WAVES**

If the pressure of a fluid is suddenly increased, at a point it will then decrease rapidly, and continue decreasing beyond the initial value, and then briefly oscillate up and down. This pressure oscillation will then travel through the fluid as a pressure wave. These pressure waves are sound or, at higher frequencies, ultrasound, and can travel through most materials.

A typical pressure wave is shown in Fig. 5.3 and represented in equation (5.4).

$$p = a\sin(\omega t - kx)$$

(5.4)

If the pressure is observed at a single point, it will vary as a sine function with time, and also, if it is observed at a single time, it will vary as a sine function with distance.

The sine function of a variable repeats itself when the number is increased by 2π. The frequency is defined as the number of complete cycles per second. It may be seen from equation (5.4) that if the time t is increased by $2\pi/\omega$ then ωt will be increased by 2π and thus, this is the time taken for one cycle. If, for example, this was 1/6 of a second, then the frequency would be six cycles per second. Thus, the frequency is $\omega/2\pi$.

The wavelength is defined as the distance from a given point on one wave to the same point on the next wave (see Fig. 5.3). It may be seen from equation (5.4) that if x is increased by $2\pi/k$ then kx will be increased by 2π and thus, the wavelength is $2\pi/k$.

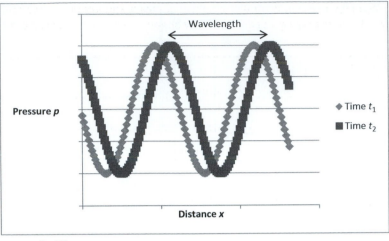

FIGURE 5.3 Movement of a Wave

Pressures at time t_1 and t_2.

If there is a point at position x_1 at time t_1 and, a short time later, the pressure wave has moved so the same pressure occurs at point x_2 at time t_2, then the velocity of the wave is distance/time:

$$v = \frac{x_2 - x_1}{t_2 - t_1} \tag{5.5}$$

But the pressure is the same, thus:

$$\omega t_1 - k x_1 = \omega t_2 - k x_2 \tag{5.6}$$

Combining equations (5.5) and (5.6) gives:

$$V = \frac{\omega}{k} \tag{5.7}$$

And thus, from the values for frequency and wavelength:

$$\text{The velocity of the wave} = \text{The frequency} \times \text{the wavelength} \tag{5.8}$$

The full solution for the velocity depends on the Poisson's ratio and is:

$$E = v^2 \rho \left[\frac{(1+v)(1-2v)}{(1-v)} \right] \tag{5.9}$$

where E = Young's modulus (Pa), v = pulse velocity (m/s), ρ = density (kg/m^3), v = Poisson's ratio.

In US customary units, if the modulus is measured in psi, an extra term for the gravitational constant g in in./s^2 must be added to the left hand side of this equation. v will then be measured in in./s and ρ in lb/in.^{3}.

The formula shows that, for a three-dimensional solid, the velocity depends on the Young's modulus, and the Poisson's ratio. Ultrasonic testing is used to measure the modulus of materials, and from it to estimate the strength (see Section 27.3.1).

Note that because the pressure is oscillating, the modulus is slightly different from the static value. It is known as the *dynamic modulus,* and is normally slightly greater than the static modulus.

5.6 THE BULK MODULUS

For pressure waves in a fluid (e.g., air) the linear definition of modulus in equation (2.7) cannot apply (if air is compressed in one direction only, it will simply flow away). Thus a volume V_1 of fluid must be considered when it is compressed to a volume V_2 by a pressure P. The *bulk modulus* B is defined as follows:

$$B = \frac{P V_1}{(V_2 - V_2)} \tag{5.10}$$

This may be seen to be exactly equivalent to the definition for E in equation (2.7) where $(V_1 - V_2)/V_1$ is the volumetric strain.

5.7 ATTENUATION OF WAVES

The ability of materials to attenuate pressure waves that are passing through them, or reflecting off them, is of critical importance in sound insulation. The basic mechanism for this is the energy of the sound being converted into a small amount of heat, due internal "friction" in the fluid. Wave attenuation will increase with increasing density; however, sound insulators are normally complex materials with substantial air voidage, and the attenuation mechanisms are also complex. In particular, they are highly dependent on the surrounding structure in which the material is used. For this reason, it is not possible to grade a material (e.g., a building block) for its sound insulation properties without knowing details of the structure in which it is to be used.

5.8 CONCLUSIONS

- Pressure in a fluid exerts an equal force in all directions.
- Pressure in a fluid will depend on depth and the fluid density.
- Pressure of a fixed volume of gas will increase with increasing temperature and decreasing volume.
- The velocity of the wave = the frequency × the wavelength.
- Dense materials are good for sound insulation.

TUTORIAL QUESTIONS

1. A 200 mm thick concrete wall is tested using ultrasonics. The measured transit time is 60 μs.
   a. Calculate the Young's modulus of the concrete assuming a density of 2400 kg/m³ and a Poisson's ratio of 0.12.

   What is the percentage error in the answer in (a) if:
   b. The actual density is 2350 kg/m³.
   c. The actual Poisson's ratio is 0.15.

   d. The actual wall thickness is 180 mm.

   e. Discuss the consequences of these percentage errors on the precautions that should be taken when carrying out ultrasonic testing.

Solution:

   a. Velocity $= 0.2/(60 \times 10^{-6}) = 3330$ m/s

      $E = 3330^2 \times 2400 \times [(1 + 0.12) \times (1 - 0.24)]/(1 - 0.12) = 2.573 \times 10^{10}$ Pa   (5.9)

   b. $E = 3330^2 \times 2350 \times [(1 + 0.12) \times (1 - 0.24)]/(1 - 0.12) = 2.519 \times 10^{10}$ Pa   (5.9)

      $(2.573 \times 10^{10} - 2.519 \times 10^{10})/(2.573 \times 10^{10}) = 2.1\%$

   c. $E = 3330^2 \times 2400 \times [(1 + 0.15) \times (1 - 0.3)]/(1 - 0.15) = 2.520 \times 10^{10}$ Pa   (5.9)

      $(2.573 \times 10^{10} - 2.520 \times 10^{10})/(2.573 \times 10^{10}) = 2\%$

   d. Velocity $= 0.18/(60 \times 10^{-6}) = 3000$ m/s

      $E = 3000^2 \times 2400 \times [(1 + 0.12) \times (1 - 0.24)]/(1 - 0.12) = 2.088 \times 10^{10}$ Pa   (5.9)

      $(2.573 \times 10^{10} - 2.088 \times 10^{10})/(2.573 \times 10^{10}) = 19\%$

   e. It is very important to measure the path length accurately.

2. a. Five litres of an ideal gas is at a pressure of 1 atm, and a temperature of 0°C. If the volume is reduced to 3 L, and the temperature rises to 10°C, what does the pressure become?

   b. A pulse of ultrasound takes 1.2×10^{-4} s to travel through a concrete wall 450 mm thick. If the Poisson's ratio of the concrete is 0.13, and the density is 2400 kg/m^3, what is the Young's modulus?

Solution:

   a. $p_1 \, V_1/T_1 = 1 \times 5/273 = p_2 \times 3/283$ (5.3)

      $p_2 = 1.72$ atm

Note that a nonstandard unit (atmosphere) has been used here. This gives the correct answer in this situation but this approach should be used with caution.

   b. $v = 0.45/1.2 \times 10^{-4} = 3750$ m/s

      $E = 3750^2 \times 2400 \times [(1 + 0.13) \times (1 - 0.26)/(1 - 0.13)] = 3.2 \times 10^{10}$ Pa   (5.9)

3. a. Air at 700 kPa and 15°C is used to power a pneumatic breaking hammer for road repairs. If the volume of the air increases by a factor of 6 in the hammer, what is its temperature when it is vented? (Assume ideal gas and no heat loss.)

   b. A pulse of ultrasound takes 1.1×10^{-4} s to travel through a concrete wall 400 mm thick. If the Poisson's ratio of the concrete is 0.15, and the density is 2400 kg/m^3, what is the Young's modulus?

Solution:

   a. The final pressure p_2 is atmospheric $= 10^5$ Pa (approximately).

      $7 \times 10^5 \times V_1/288 = 10^5 \times 6(V_1/T_2)$ (5.3)

      Thus: $T_2 = 246°K = -26°C$

   b. $v = 0.4/1.1 \times 10^{-4} = 3636$ m/s

      $E = 3636^2 \times 2400 \times (1.15 \times 0.7/0.85) = 3.005 \times 10^{10}$ Pa   (5.9)

4. Gas is leaking from a cylinder to atmosphere. The pressure in the cylinder is 10 MPa and the temperature is 20°C. If the temperature of the gas after escaping is −30°C, what is the percentage change in volume as it escapes? (Assume an ideal gas with no heat loss.)

Solution:

P_1 = 1 atm = 0.1 MPa
P_2 = 10 MPa
T_1 = 20°C = 293°K
T_2 = −30°C = 243°K
V_2/V_1 = (10 × 243)/(0.1 × 293) = 82.9 = 8290%   (5.3)

5. A pulse of ultrasound takes 1.2×10^{-4} s to travel through a concrete wall 18 in. thick. If the Poisson's ratio of the concrete is 0.13, and the density is 4050 lb/yd^3, what is the Young's modulus?

Solution:

$v = 18/1.2 \times 10^{-4}$ = 150,000 in./s
Density = 4,050/46,656 = 0.087 lb/in.^{3}
Gravitational constant = 386 in./s^2
$E = 150,000^2 \times 0.087 \times [(1 + 0.13) \times (1 - 0.26)/(1-0.13)]/386 = 4.87 \times 10^6$ psi   (5.9)

NOTATION

A Area (m^2)
B Bulk modulus (Pa)
a Amplitude (Pa)
E Young's modulus (Pa)
g Gravitational constant (= 9.81 m/s^2)
h Pressure head (m)
k 2π/wavelength (m^{-1})
p Pressure (Pa)
T Temperature (°C or °K)
t Time (s)
v Velocity (m/s)
V Volume (m^3)
x Distance (m)
v Poisson's ratio
ρ Density (kg/m^3)
ω Angular velocity (radians/s)

ELECTRICAL PROPERTIES

CHAPTER OUTLINE

6.1 INTRODUCTION

All construction works will include electrical installations, both in the permanent works, and in the plant, and temporary works required to construct them. It is important to understand the effects of electric current on the materials through that it flows. These effects can include heating that involves power loss, or corrosion that is discussed in Chapter 31. Electrical systems are also used for most methods for testing materials, and this is discussed in Section 6.10.

6.2 ELECTRIC CHARGE

All materials are made up of a mixture of positively, and negatively charged particles (this structure is discussed in more detail in Chapter 7). An object is electrically charged, if the negative charge in it is not equal to the positive charge. This situation occurs when charged particles are either removed from it, or added to it. In normal conductors, such as metals, the charged particles that move are particles called electrons. If 6.25×10^{18} electrons flow into an object (and do not flow out again), it has a negative charge of 1 Coulomb (C).

If a charged object comes near another charged object, a force acts between them. Opposite charges attract. Thus, when, for example, a positively charged particle flows out of an object and leaves it with a net negative charge, it will be subject to an attractive force to draw it back in. This is what happens when steel is corroding, and a positive metal ion flows out of it. This is discussed in detail in Chapter 31.

Like charges repel, so the electrostatic force will tend to make charges dissipate. Thus, if the electrons in an object are not uniformly distributed, the electrostatic force will tend to make them move until there is no net charge at any point.

6.3 ELECTRIC CURRENT

It is possible for charge to move easily in some materials; these are called conductors. Materials in which charge cannot move are called insulators. An example of a conductor is a copper wire.

If 1 C of charge is moving down a wire every second, the current flowing in it is defined as 1 Amp (A). Thus:

$$\text{Charge} = \text{Current} \times \text{time} \tag{6.1}$$

As with static charge, objects with current flow in them will have forces between them. These are magnetic forces. If the currents are in the same direction, the force is one of attraction between the wires. Because the force is at right angles to the flow of current, it will have no effect on the current in either wire. In order to increase the current in the wires, they must be moved. This is the basis of an electrical generator. If one of the wires is being turned around a shaft, it will move backward and forward relative to the other, and the current in it will be driven first in one direction, and then the other. This is the easiest way to generate electricity, and the current is called *alternating current* (AC), as opposed to *direct current* (DC) that is a continuous flow in one direction. The current in the main supply reverses direction 50 or 60 times per second, that is, the *frequency* is 50 or 60 cycles per second, or 50 or 60 Hz. (50 Hz is used for main supply in the United States and United Kingdom, and 60 Hz in much of the rest of Europe.)

SAFETY

If you are working with high voltage direct current systems, you should be aware that they are far more dangerous than alternating current. 50 V DC can kill, but 110 V AC (as normally used for main supply in the USA) is less dangerous. However, the UK and European mains voltage of 220 V is very dangerous.

The theory of magnetism is complex, and is not covered in detail here. Some of its consequences are:

- If a wire carrying direct current is flowing close to a conductor, opposite charges will build up on each side of the conductor. This may, for example, cause corrosion.

- If a wire carrying alternating current is flowing close to a conductor, it will induce a current in it. Fortunately, at mains frequency, this effect is not normally great, but it increases with frequency. High frequency cables must therefore be "screened" with a conductor.

6.4 VOLTAGE

Because of the electrostatic force, the current in a wire will tend to flow toward or away from a nearby charged object (depending on whether the charge is positive or negative). The voltage on the wire is a measure of this effect. It may be defined as follows, in terms of the energy required to move the charge: if the energy required to work against the electrostatic force, and move 1 C of charge along a wire, from one point to another, is 1 J, then the voltage between the points is 1 V. Thus:

$$W = QV \tag{6.2}$$

where W is the work in (J), Q is the charge in (C), V is the voltage (V).

The voltage is also known as the potential difference. Thus the potential at one point relative to another is the voltage between them.

6.5 ELECTRIC FIELD

The electric field (technically the electrostatic field) is a measure of the voltage gradient.

$$E = \frac{V}{x} \tag{6.3}$$

where E is the electric field (V/m), X is the distance through the material (m).

Thus from equations (3.2), (6.2) and (6.3):

$$f = QE \tag{6.4}$$

where f is the force in Newtons on a charged object in an electric field.

6.6 RESISTANCE

6.6.1 PROPERTIES OF CONDUCTORS

If a voltage is applied to a perfect conductor, a very large current will flow until the voltage is the same at each end. A real wire is, however, not a perfect conductor, and when current flows down it, some energy will be lost, and the voltage at each end will never be quite the same. Normally, the energy is lost in the form of heat but, for example, in an incandescent light bulb, some of it is light.

The resistance is defined as the ratio between the voltage drop along the wire and the current in it. Thus:

$$R = \frac{V}{I} \tag{6.5}$$

where R is the resistance in ohms (Ω), V is the voltage in volts and I is the current in amperes.

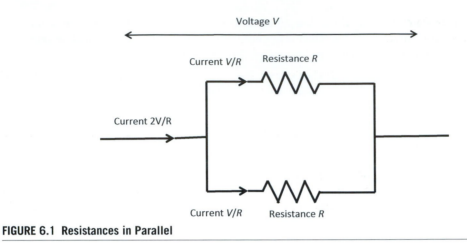

FIGURE 6.1 Resistances in Parallel

6.6.2 RESISTIVITY

Now, consider the situation if two wires are connected together at each end, and a voltage is applied across them (see Fig. 6.1). The current through each wire will be V/R, so the total current will be $2V/R$. This means that the two wires are equivalent to a single wire with resistance $R/2$. Thus, by doubling the size of the wire, we halve the resistance. In a similar way, it may be seen that doubling the length would double the resistance.

This effect is shown in an equation as:

$$R = \rho \times \frac{L}{A} \tag{6.6}$$

where R is the resistance of the wire, L is its length, A is its cross-sectional area, and ρ is the *resistivity* (in Ωm) of the material from which the wire is made. The resistivity is a material property, and does not depend on the dimensions of the wire. Values of resistivity are given in Table 6.1. By definition, materials with high resistivity are insulators, and those with low resistivity are good conductors. The resistivity of materials almost always increases as their temperature increases.

Table 6.1 Values of Resistivity	
Substance	**Resistivity (Ωm)**
Silver	1.6×10^{-8}
Aluminum	2.8×10^{-8}
Copper	1.7×10^{-8}
Lead	2.05×10^{-7}

6.6.3 CONDUCTIVITY

The *conductivity* is defined as:

$$\text{Conductivity} = \frac{1}{\text{Resistivity}} \tag{6.7}$$

6.7 CAPACITANCE

For the study of corrosion rates and circuits (see Chapter 31), it is necessary to define capacitance. Like resistance, this is a property of a component with a voltage across it but, unlike a resistor, no direct current flows through a perfect capacitor. When a voltage is applied across a capacitor, charge is stored in it. The capacitance is defined as:

$$\text{Capacitance} = \frac{\text{Stored charge}}{\text{Applied voltage}} \tag{6.8}$$

and is measured in coulomb/volt or Farad. A 1 F capacitor would be very large. Real capacitors are rated in microfarad.

A capacitor is shown schematically in Fig. 6.2.

If a (direct) current flows in from the left, it will cause a positive charge to build up on the plate on the left. This will attract a negative charge to the right hand plate that will flow along the wire to the right, giving a net positive current away from it. However, this will soon stop. The only type of current that can flow through, it is an alternating current.

6.8 POWER

Power is the rate of doing work, and is measured in joules/second or watts (see equation (4.1)). Current is the rate of flow of charge in coulombs/second or amperes. Thus, from equations (6.1) and (6.2):

$$P = VI \tag{6.9}$$

where P is the power in watts.

This gives the rate of delivery of power down a wire.

Combining equations (6.5) and (6.9) gives:

$$P = I^2 R = \frac{V^2}{R} \tag{6.10}$$

FIGURE 6.2 Schematic Diagram of a Capacitor

If the voltage in equations (6.9) and (6.10) is the voltage between the positive and negative wires, the power will be the power delivered through the wires. If the voltage is the voltage drop in the wire itself, then the power will be the power loss (through heating) of the wire itself.

Extension leads for use with mains electricity often have labels on them that give two different current ratings, for example, 4 A coiled, and 13 A uncoiled. This is important because the heat generated in the cable increases with the square of the current, and as the temperature goes up, the resistance (and thus the heat generated) increases. In a cable, the current must flow along the neutral wire, as well as the live wire, and heat will be generated in both.

6.9 ELECTRIC CURRENT IN CONCRETE

Concrete is not normally used as a conductor for electric current, but charged ions can move in the pore solution (see Chapter 7), and this has a similar effect to the movement of electrons in metals described above. The mechanism by which the charged ions move is called electromigration (see Chapter 10), and is important in the process of corrosion of steel in concrete (Chapter 31), and also in electrical tests procedures for concrete durability, such as the rapid chloride test (Chapters 22 and 27).

Because the charge carriers in concrete are in limited supply (unlike electrons in a circuit of wires), passing a direct current through a concrete sample will create areas in which they become depleted, and no current will flow. Thus, the resistivity of concrete (often measured to predict durability) is always measured with alternating current.

6.10 ELECTRICAL TEST APPARATUS
6.10.1 STRAIN GAUGES

Strain gauges are fixed to test samples with adhesive to record the strain when they deflect under load. A strain gauge is simply a fine network of wires arranged in a zigzag pattern on a membrane, so they are elongated when the sample extends (see Figs 6.3 and 6.4). As the wires become microscopically longer and thinner, due to the strain, their resistance will increase, and this can be measured.

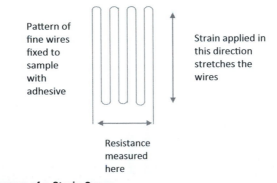

Pattern of fine wires fixed to sample with adhesive

Strain applied in this direction stretches the wires

Resistance measured here

FIGURE 6.3 Schematic Diagram of a Strain Gauge

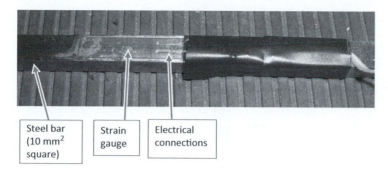

Steel bar (10 mm² square)   Strain gauge   Electrical connections

FIGURE 6.4 A Steel Bar With a Strain Gauge on it

6.10.2 **LOAD CELLS**

A load cell is a piece of metal with known properties with strain gauges on it. When a load is applied, the metal will deform, and this is recorded by the gauges. If the modulus of elasticity of the metal is known, this can be converted into a load reading.

Figure 6.5 shows load cells in use for testing a concrete beam. Figures 3.2 and 3.12 also show load cells in other configurations.

6.10.3 **DISPLACEMENT TRANSDUCERS**

Displacement transducers measure movement. In Fig. 6.6, this is shown as movement between a sample and a test frame. A constant voltage (typically 5 V) is applied to the ends of the resistive strip. When the movement occurs, the sliding connector moves along it, and the measured voltage changes. Displacement

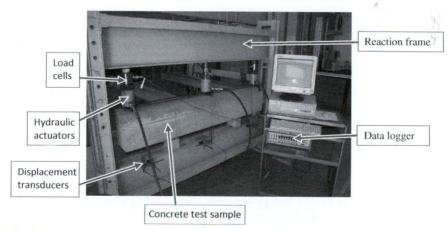

Load cells   Reaction frame   Hydraulic actuators   Data logger   Displacement transducers   Concrete test sample

FIGURE 6.5 Testing a Concrete Beam

This beam only had very light reinforcement, so deflection of the frame was not significant. For a stronger beam, the displacement transducers should not be fixed to the load-bearing part of the frame.

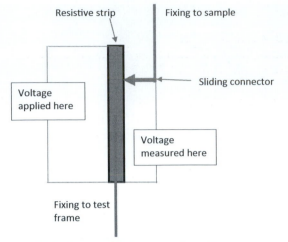

Resistive strip

Fixing to sample

Sliding connector

Voltage applied here

Voltage measured here

Fixing to test frame

FIGURE 6.6 Schematic Diagram of Potentiometer Displacement Transducer

transducers may be seen in Figs 3.9 and 6.5. The ones shown in these figures are linear variable differential transformers (LVDTs) that work on the more complex principle of variable inductance, and require an alternating current supply. They are more expensive than potentiometers, but more responsive because they do not require the sliding connector shown in Fig. 6.6.

6.11 CONCLUSIONS

- Electric charge is a measure of the surplus (or deficit) of charged particles in an object.
- Electric current is a measure of the movement of charge.
- Voltage is a measure of work done against the electrostatic force.
- Electric field is a measure of voltage gradient.
- Resistance is a measure of the voltage required to create a current.
- Capacitance is a measure of the ability to store charge.
- Electricity flows through concrete when charged ions move through it.

TUTORIAL QUESTIONS

1. A 5 mm diameter copper wire has a resistivity of 1.7×10^{-8} Ωm and is carrying a current of 100 A.
   a. What is the power loss per m length in the bar?
   b. If the bar has a specific heat of 390 J/kg°C, and no heat is lost from it, what is its temperature rise per minute? (Density of copper = 8900 kg/m^3)
   c. If the coefficient of thermal expansion of the bar is 17×10^{-6}/°C, what is the percentage expansion after 3 min?

Solution:

a. Area = $\pi \times 0.0025^2 = 1.96 \times 10^{-5}\,\text{m}^2$

   Resistance per metre = $1.7 \times 10^{-8}/1.96 \times 10^{-5} = 8.7 \times 10^{-4}\,\Omega/\text{m}$ (6.6)

   Power loss per metre = $100^2 \times 8.7 \times 10^{-4} = 8.7\,\text{W}$ (6.10)

b. Mass = $1.96 \times 10^{-5} \times 8900 = 0.174\,\text{kg}$

   Heating energy per minute = $8.7 \times 60 = 533\,\text{J}$ (4.1)

   Temperature change per minute = $533/(390 \times 0.174) = 7.7\,°\text{C}$ (4.2)

c. Expansion = $3 \times 7.7 \times 17 \times 10^{-6} = 3.9 \times 10^{-5} = 0.039\%$ (4.8)

2. A 4 mm diameter copper wire is carrying a current of 20 A and has a resistivity $1.6 \times 10^{-8}\,\Omega\text{m}$.
    a. What is the power loss per metre length in the wire?
    b. What is the voltage drop per metre length along the wire?
    c. If the wire is 30 m long, and the supply voltage is 240 V, what is the percentage power loss?

Solution:

a. Area = $1.2 \times 10^{-5}\,\text{m}^2$

   Resistance per metre = $1.6 \times 10^{-8}/1.2 \times 10^{-5} = 1.33 \times 10^{-3}\,\Omega/\text{m}$ (6.6)

   Power = $20^2 \times 1.33 \times 10^{-3} = 0.53\,\text{W/m}$ (6.10)

b. Voltage drop = $20 \times 1.33 \times 10^{-3} = 0.026\,\text{V/m}$ (6.5)

c. Power transmitted = $240 \times 20 = 4800\,\text{W}$ (6.9)

   Power loss = $0.53 \times 30 = 15.9\,\text{W}$

   $15.9/4800 = 0.33\%$

3. A current of 30 A is flowing first through a copper conductor, then through an aluminum one, and finally through a lead one. All of the conductors have a circular cross-section with a diameter of 3 mm. The properties of the conductors are as follows:

Material	Resistivity (Ωm)	Density (kg/m³)	Specific Heat (J/kg/°C)
Copper	1.7×10^{-8}	8,900	390
Aluminum	2.8×10^{-8}	2,700	880
Lead	1.8×10^{-7}	11,300	130

For each metre length of each conductor, calculate the following:
a. Resistance
b. Voltage drop
c. Power loss
d. Mass
e. Temperature rise in 10 min (assume no heat loss)

Solution:

Area $= \pi \times 0.0015^2 = 7.06 \times 10^{-6} \text{ m}^2$

		Copper	Aluminum	Lead
Resistance (Ω) = resistivity/area	Equation (6.6)	2.4×10^{-3}	3.9×10^{-3}	2.5×10^{-2}
Voltage drop (V) = IR	Equation (6.5)	0.072	0.117	0.75
Power (W) = VI	Equation (6.9)	2.1	3.51	22.5
Mass (kg) = density $\times$ area		0.063	0.019	0.079
Temperature rise (degrees) = 600 s $\times$ Power/(sp. ht. $\times$ mass)	Equation (4.2)	50	125	1314

4. The maximum acceptable voltage drop on a 32 kV transmission line is 1%/km.
    a. What cross-section area of copper wire is required for a power of 40 MW?
    b. How much would the wire between two pylons 100 m apart weigh?
    c. What would the weight be if aluminum wire was used?

Use the following data:

	Copper	Aluminum
Resistivity Ωm	1.7×10^{-8}	2.8×10^{-8}
Density kg/m^3	8900	7800

Solution:

a. Acceptable voltage drop over 1 km is 32,000/100 = 320 V
    Current = 40×10^6/32,000 = 1,250 A   (6.9)
    Resistance = 320/1,250 = 0.25 Ω (6.5)
    Area = $1.7 \times 10^{-8} \times 1,000/0.25 = 6.8 \times 10^{-5} \text{ m}^2 = 68 \text{ mm}^2$ (6.6)
b. Mass = $6.8 \times 10^{-5} \times 100 \times 8,900 = 60$ kg
c. Area = $2.8 \times 10^{-8} \times 1,000/0.25 = 1.12 \times 10^{-4} \text{ m}^2$ (6.6)
    Mass = $1.12 \times 10^{-4} \times 100 \times 7,800 = 87$ kg

NOTATION

A Area (m^2)
E Electric field (V/m)
f Force (N)
I Current (A)
L Length (m)
P Power (W)
Q Charge (C)
R Resistance (Ω)
V Voltage (V)
W Work (J)
x Distance (m)
ρ Electrical resistivity (Ωm)

CHEMISTRY OF CONSTRUCTION MATERIALS

CHAPTER OUTLINE

7.1 INTRODUCTION

By far, the most common chemical reaction in the construction process is the hydration of cement. This is considered in detail in Chapter 20. In this chapter, the basic processes of all chemical reactions are considered. Most chemical reactions fall into one of the following two categories:

- Reactions between acids and bases.
- Reactions between oxidizing agents and reducing agents.

These are considered in Sections 7.6 and 7.7, and two "chemical cycles" that occur in construction materials production are:

- The lime cycle
- The gypsum cycle

these are considered in Sections 7.9 and 7.10.

7.2 THE COMPONENTS OF THE ATOM

In Chapter 3 it was noted that all materials are formed from atoms. Atoms are made up of three different particles:

- Protons. These have a mass of 1.67×10^{-27} kg and a charge of 1.6×10^{-19} C. The number of protons in an atom (the atomic number) determines what element it is.
- Neutrons. These have the same mass as a proton, but no charge. They normally occur in approximately equal numbers to the protons in an atom. The number of neutrons determines the *isotope* of the element.
- Electrons (shown with the symbol e^-). These have a mass of 9.11×10^{-31} kg, and a charge equal, and opposite to the proton. If one of these is released from an atom, it will leave a positively charged atom, and a free electron that may carry electric current in a metal.

The protons and neutrons are in the nucleus, which is a small solid core, and the electrons orbit around them.

7.3 CHEMICAL ELEMENTS

Some of the elements that are relevant to the study of construction materials are in Table 7.1.

The atomic mass is the sum of the number of protons and the number of neutrons. It is not always a whole number because this is the average number of neutrons in different isotopes. For example,

Table 7.1 Some Elements in Construction Materials

Atomic Number	Atomic Mass	Symbol	Name
1	1	H	Hydrogen
6	12	C	Carbon
7	14	N	Nitrogen
8	16	O	Oxygen
11	23	Na	Sodium
12	24.3	Mg	Magnesium
13	27	Al	Aluminum
14	28.1	Si	Silicon
17	35.4	Cl	Chlorine
19	39.1	K	Potassium
20	40.1	Ca	Calcium
26	55.8	Fe	Iron
29	63.5	Cu	Copper
82	207.2	Pb	Lead

chlorine has two stable isotopes. The one with 18 neutrons has 75% abundance, and the one with 20 neutrons has 25% abundance. Together with the 17 protons, this averages out at a mass of 35.4.

7.4 MOLECULES

A molecule is a number of atoms, which are held together by bonds between them. For example, H_2 is a hydrogen molecule with two hydrogen atoms in it. Hydrogen atoms in hydrogen gas almost always form this molecule. The molecular mass is the sum of the atomic masses of the atoms in it and is, thus, 2 for a hydrogen molecule.

One mole (abbreviated as "mol") of a material is defined 6.02×10^{23} molecules. Thus, using the value for the mass of a proton given previously, the mass of 1 mol of a material with a molecular mass of m is:

$$m \times 1.67 \times 10^{-27} \times 6.02 \times 10^{23} \, \text{kg} = m \, \text{g}$$

The mol may be referred to as the g-mol. One kmol or kg-mol of a material will have a mass of m kg.

If the material is a gas, the following formula applies:

$$pV = nRT \tag{7.1}$$

where n is the number of mols of gas present, R is a constant (the gas constant) = 8.31 J/mol/°K, p is the pressure (Pa), V is the volume (m^3), T is the temperature (°K).

Thus, at a given temperature and pressure, one mol of any gas will occupy the same volume. This may be seen to be an extension of equation (5.3), and also assumes that the gas is ideal.

7.5 CHEMICAL REACTIONS

7.5.1 EXOTHERMIC AND ENDOTHERMIC REACTIONS

When chemicals are mixed together, they may react. There are two main types of reaction:

Exothermic reactions. These reactions proceed as soon as the components come into contact and give off energy (heat). Examples are the mixing of the components of an epoxy adhesive, or the hydration of cement.

Endothermic reactions. These reactions require heat to make them proceed. An example is the manufacture of cement in kilns, where the components would not react at all if not heated.

There are many exothermic reactions, such as the combustion of timber that require some energy input to start them, but then proceed without further energy input.

7.5.2 REACTION RATES

Reaction rates vary from very fast (explosions) to very slow (e.g., corrosion). Reaction rates are affected by the following.

- Temperature. An approximate rule is that an increase of 20°C doubles the rate of reaction.
- Pressure. High pressures will increase reaction rates.

- The energy released by the reaction.
- Physical consistency of the materials. A fine powder will react faster than a coarse one (e.g., rapid hardening cement is finely ground).
- Presence of a catalyst. This is a material, which is not used up by the reaction, but makes it proceed faster.

7.6 ACIDS AND BASES

Most acids occur in water (aqueous solutions), and the water contains free positive hydrogen ions. The equilibrium reaction that takes place may be represented by equation (7.2):

$$H_2O \leftrightarrow H^+ + OH^- \tag{7.2}$$

This is a reaction equation. Note that the number of atoms must be the same on both sides, and the $\leftrightarrow$ symbol implies that it may proceed in either direction.

The acidity is measured by the pH that is defined by equation (7.3):

$$pH = \log\left(\frac{1}{H^+}\right) \tag{7.3}$$

where H^+ is the number of hydrogen ions in kilogrammes per cubic metre (= the number of grammes per litre).

If there are 10^{-7} kg of hydrogen ions per m^3, the pH of water is 7, and is defined as neutral, and is not very reactive, and is typical for tap water. Acids have pH below 7, and alkalis (bases) have pH above 7. Concrete has a pH of 12.5 (i.e., it has $10^{-12.5}$ g of hydrogen ions per litre of pore solution), so it is highly alkaline.

In a reaction between an acid and a base, the acid acts as a "hydrogen donor." Thus, for example, the reaction between sulphuric acid (H_2SO_4) and sodium hydroxide (NaOH) is:

$$H_2SO_4 + 2NaOH \rightarrow Na_2SO_4 + 2H_2O \tag{7.4}$$

In this reaction, the acid has given up two hydrogen atoms.

7.7 OXIDIZING AGENTS AND REDUCING AGENTS

The term "oxidizing agent" was originally applied to compounds that would add oxygen to substances, but it is now generally applied to all compounds that lose electrons in reactions.

A reaction which falls within this wider definition is:

$$2Na^+ + Cl_2^- \rightarrow 2NaCl \tag{7.5}$$

in which sodium gives up an electron to chlorine to form common salt.

The tendency of a given element or compound to give electrons depends on the energy level of the "outermost" electrons in it. This energy is called the "redox potential," or "Eh." The difference in Eh between two samples may often be measured electrically by putting them in a conducting solution, and measuring the voltage between them. This concept is developed when discussing corrosion in Chapter 31.

7.8 CHEMICALS DISSOLVED IN WATER

In construction, many materials are in solution in water (e.g., in concrete pore solution) and the atoms may be present as charged ions. *Anions* are negative because they are attracted to the anode, which is positive, and similarly, *cations* are positive (see Chapter 31 for a discussion of anodes and cathodes). Thus, if salt is in solution, the following reaction will occur:

$$NaCl \rightarrow Na^+ + Cl^- \tag{7.6}$$

where the Na^+ is the cation and Cl^- is the anion. In the same way, hydrated lime (calcium hydroxide, $Ca(OH)_2$) will dissociate into Ca^+ and OH^- ions.

Tap water is usually described as "hard" or "soft," depending on its acidity, which is normally determined by its calcium carbonate content. Hard water is alkaline (pH > 7), soft water is acidic (pH < 7). Metals, such as lead from old plumbing, or aluminum and mercury from wastes, will dissolve in acidic (soft) waters. This is a major problem because most metals are highly toxic in solution. Acid rain contains acids derived from pollutants (often sulphates from coal burning), and adds to this problem.

7.9 THE LIME CYCLE

The lime cycle is one of the most important in construction materials, and also one of the oldest chemical processes used on a large scale. The Romans produced lime in large quantities.

The first process is calcination of limestone, by heating to a high temperature. This is carried out in kilns.

Calcium carbonate (limestone) → Calcium oxide (quicklime) + carbon dioxide

$$CaCO_3 \rightarrow CaO + CO_2 \tag{7.7}$$

The quicklime is highly reactive. Some of it is used directly (e.g., in concrete block production), but most of it is slaked in the hydration reaction, which is violent and gives off a lot of heat.

Quicklime + water → Hydrated (slaked) lime

$$CaO + H_2O \rightarrow Ca(OH)_2 \tag{7.8}$$

Hydrated lime is sold at builders merchants for use in bricklaying mortar. Traditional mortars set by the carbonation reaction:

Hydrated lime + Carbon dioxide → Calcium carbonate + water

$$Ca(OH)_2 + CO_2 \rightarrow CaCO_3 + H_2O \tag{7.9}$$

In modern construction, cement is used in the mortar because this reaction is too slow. This reaction also occurs with the free lime in concrete, and it is called carbonation (see Section 25.3.2). It also contributes to the removal of some of the carbon dioxide produced in the calcination process, and this is called carbon sequestration (see Section 17.9).

7.10 THE GYPSUM CYCLE

Gypsum (calcium sulphate dihydrate $CaSO_4\, 2H_2O$) is a mineral that occurs in many areas of the world. It is also produced as a by-product of flue gas desulphurisation, which is the process used at thermal power stations to reduce the amount of acid rain that they cause. Various industrial processes, such as titanium dioxide production, also produce gypsum as a by-product.

If gypsum is heated, it will calcine to produce hemihydrate (stucco) and water:

$$2CaSO_4 \cdot 2H_2O \leftrightarrow 2CaSO_4 \cdot \tfrac{1}{2}H_2O + 3H_2O \tag{7.10}$$

Stucco is the material used for plaster and plasterboard (drywall or wallboard). When mixed with water, it will revert back to gypsum, that is, the plaster will set. It can also be used for block manufacture (for internal use), and floor screeds.

Figure 7.1 shows a plasterboard production line. At the point where the picture was taken, the hemihydrate slurry has just been poured onto the paper below it. It then progresses up the factory at a walking pace, and by the time it reaches the end, it has set sufficiently for the boards to be cut to length and picked up.

Hemihydrate can be calcined further to produce anhydrite:

$$2CaSO_4 \cdot \tfrac{1}{2}H_2O \rightarrow 2CaSO_4 + H_2O \tag{7.11}$$

Anhydrite also occurs as a natural mineral in some places.

Gypsum is added to cement during manufacture to prevent flash set (see Chapter 18). Many producers aim to add a mixture of gypsum and anhydrite. It is added during the grinding process, and if this

FIGURE 7.1 Plasterboard Manufacture

runs too hot, the cement will contain hemihydrate produced from the gypsum. When this is mixed with water, it will produce a rapid "false set," which is a problem.

The calcination and hydration of gypsum is a reversible process. Thus, waste plasterboard in the form of off-cuts from construction or recovered material from demolition can be recycled. The paper is stripped from it, and it is returned to the plant to be calcined again. Problems with this recycling may occur if the board is foil backed, or if it has waterproofing additives in it.

7.11 SUMMARY

- Atoms contain positive protons, negative electrons, and neutral neutrons.
- Different chemical elements have different atomic masses.
- Molecules are groups of atoms.
- Chemical pH is determined by the number of free hydrogen ions in a solution.
- Lime is made by calcining limestone.
- Gypsum plaster is primarily calcium sulphate hemihydrate that is made from dihydrate gypsum.

TUTORIAL QUESTIONS

1. Calculate the volume of 1 mol of a gas at 0°C and 1 atm.

Solution:

1 atm = 0.1 MPa = 10^5 Pa
0°C = 273°K
V = $1 \times 8.31 \times 273/10^5$ = 0.022 m^3 = 22 L   (7.1)
This will apply to any gas.

2. Calculate the volume of 2 mols of a gas at 20°C and a pressure of 3 Bar.

Solution:

V = $2 \times 8.31 \times 293/3 \times 10^5$ = 0.016 m^3 = 16 L   (7.1)

3. Producing 1 tonne of cement releases approximately 1 tonne of carbon dioxide (CO_2) into the atmosphere. What is the volume of this gas at atmospheric pressure and 20°C?

Solution:

Molecular mass of CO_2 = 12 + 2 × 16 = 44
Thus, 1 mol of CO_2 weighs 44 g.
Thus, 1 tonne is $1000/44 \times 10^{-3}$ = 2.27×10^4 mols
V = $2.27 \times 10^4 \times 8.31 \times 293/10^5$ = 553 m^3 (7.1)

4. A lecture room is 15 m long, 10 m wide and 4 m high. CO_2 normally forms approximately 0.04% of air. Calculate the mass of CO_2 in the lecture room at 20°C.

   Solution:

   Volume of room = $15 \times 10 \times 4 = 600$ m^3
   Volume of CO_2 = $600 \times 0.04 = 0.24$ m^3
   No. of mols = $10^5 \times 0.24/(8.31 \times 293) = 9.86$ (7.1)
   Mass = $9.86 \times 44 = 433$ g = 0.43 kg

5. A solution has a pH of 3.3. How many grammes of hydrogen ions are there in it per litre?

   Solution:

   $3.3 = \log(1/H^+)$ (7.3)
   Thus, $10^{3.3} = 1995 = 1/H^+$
   Thus, $H^+ = 5 \times 10^{-4}$ g/L

NOTATION

H^+ Number of kg of hydrogen ions per cubic metre
m Mass (g)
n Number of mols
p Pressure (Pa)
R Gas constant = 8.31 (J/mol°K)
T Temperature (°K)
V Volume (m^3)

PROPERTIES OF FLUIDS IN SOLIDS

CHAPTER OUTLINE

8.1 INTRODUCTION

Fluids in construction materials are contained in the pores. Movement of fluid through materials is known as fluid transport, and is considered in Chapter 9. In this chapter, a number of properties of fluids in solids are considered, which are relevant to:

- Durability of materials
- Flow of fresh concrete
- Waterproofing buildings (because nether bricks nor concrete are fully waterproof)
- The effect of frost on building materials
- Damp in buildings
- Penetration of salt into materials
- Shrinkage and swelling, particularly in timber.

8.2 VISCOSITY

The rate at which fluids (i.e., liquids and gases) move through solids will depend on their viscosity. Thus, for example, treacle has a high viscosity, and water has a low viscosity. For the definition of viscosity see Fig. 8.1. The fluid is shown moving at different velocities in different regions: slowly (v_2) which could be near the edge of a pipe or a pore, and faster (v_1) near the centre.

The viscosity e is defined as:

$$e = \frac{\tau \delta y}{v_1 - v_2} = \frac{\text{Shear stress}}{\text{Shear strain rate}} \text{Pas} \qquad (8.1)$$

where τ is the shear stress (Pa). See Section 2.4.3 for the definition of shear stress and strain.

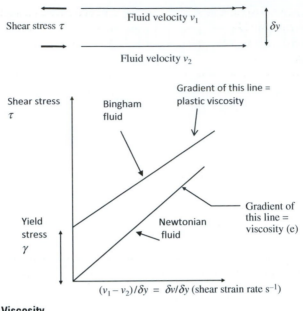

FIGURE 8.1 Definition of Viscosity

In the study of viscosity, two types of fluid are Bingham and Newtonian; the difference between them is shown in Fig. 8.1. Bingham fluids have a *yield stress* that means that when a low shear stress (below the yield) is applied, they will not move at all. Newtonian fluids will move (possibly slowly) however low the shear stress. Wet concrete is best approximated to as a Bingham fluid.

The *plastic viscosity* is the gradient of the line as shown in Fig. 8.1. For a Newtonian fluid, it is the same as the viscosity, but for a Bingham fluid, it is different.

A fluid is described as *thixotropic* if its viscosity decreases when it is stirred. Wet concrete is thixotropic.

The *kinematic viscosity* is the ratio of viscosity to density, and has units of square metres per second. The *dynamic viscosity* is a term used for the viscosity as defined in equation (8.1).

The CGS units of Poise for viscosity, and Stokes for kinematic viscosity are frequently used (see Table 1.5)

8.3 WATER AND WATER VAPOUR

At temperatures below boiling point, water partially filling a sealed container will be in equilibrium, with water vapour in the air above it. This vapour should not be confused with steam or fog that are droplets of liquid water. The water vapour will be one of many gases that make up the air, others are nitrogen, oxygen, etc. Each gas is responsible for part of the pressure in the container. If the total pressure is 1 atm the vapour will, at 20°C, account for about 2% of it (2000 Pa). (See Fig. 8.2a.)

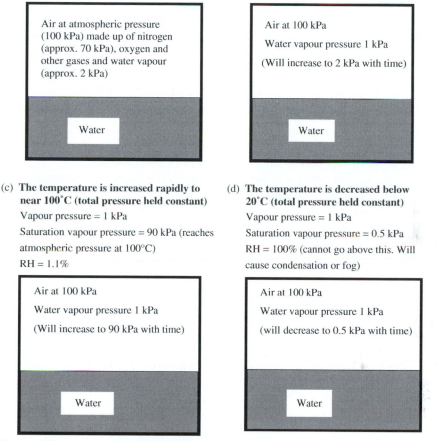

(a) Water and air in equilibrium at 20˚C
Vapour pressure = 2 kPa
Saturation vapour pressure = 2 kPa
RH = 100%

Air at atmospheric pressure (100 kPa) made up of nitrogen (approx. 70 kPa), oxygen and other gases and water vapour (approx. 2 kPa)

Water

(b) Half of the gas is replaced with dry air
Vapour pressure = 1 kPa
Saturation vapour pressure = 2 kPa
RH= 50% (typical for inside a building)

Air at 100 kPa
Water vapour pressure 1 kPa
(Will increase to 2 kPa with time)

Water

(c) The temperature is increased rapidly to near 100˚C (total pressure held constant)
Vapour pressure = 1 kPa
Saturation vapour pressure = 90 kPa (reaches atmospheric pressure at 100°C)
RH = 1.1%

Air at 100 kPa
Water vapour pressure 1 kPa
(Will increase to 90 kPa with time)

Water

(d) The temperature is decreased below 20˚C (total pressure held constant)
Vapour pressure = 1 kPa
Saturation vapour pressure = 0.5 kPa
RH = 100% (cannot go above this. Will cause condensation or fog)

Air at 100 kPa
Water vapour pressure 1 kPa
(will decrease to 0.5 kPa with time)

Water

FIGURE 8.2 Water and Water Vapour in a Sealed Container

In this condition, the humidity in the container is at 100% relative humidity (RH). The air is saturated with water vapour. The relative humidity is defined as:

$$\text{Relative humidity, RH} = \frac{\text{Vapour pressure}}{\text{Saturation vapour pressure}} \tag{8.2}$$

Now consider what happens if some dry air is blown through the container, and it is sealed again. It will take a long time for water to evaporate, and resaturate the air. If, at some time, the partial pressure of the water vapour is 1000 Pa (half of what it was initially), the humidity is 50% RH (Fig. 8.2b).

Now, consider the effect of temperature (Fig. 8.2c). If the temperature is increased, the saturation pressure will increase rapidly until 100°C, at which point, it is equal to atmospheric pressure, and the water boils. If, for example, the saturation pressure has increased to 10,000 Pa, and the partial pressure of vapour is still 1,000 Pa, increasing the temperature will have decreased the RH to 10%. Similarly, reducing the temperature will reduce the saturation pressure (Fig. 8.2d). The RH cannot go above 100% – water precipitates as condensation or fog. Typical values for RH are 50% in a building, and 70% outside.

If the pressure is raised, then the boiling point of the water will also increase. This is how autoclaved concrete is made for precast items such as kerbs. The concrete is cured rapidly in pressure vessels at temperatures well over 100°C (see Section 20.2).

8.4 POROSITY

Many materials used in construction contain pores. Concrete, brick, and wood all have considerable porosity however steel, for example, does not. These pores should not be confused with the gaps between atoms in the structure of the material; pores are big enough to contain water.

The origin of the pores is different in different materials. For example, in concrete they are generally formed by the extra water that is added to make the mix flow when it is wet.

A solid with pores has two volumes: the bulk volume, and the net or skeletal volume. The bulk volume is the skeletal volume + the volume of pores.

The porosity (ε) is defined as:

$$\varepsilon = \frac{\text{Volume of pores}}{\text{Bulk volume}} \tag{8.3}$$

The bulk volume is measured by measuring the dimensions of the object, and multiplying height × width × depth. For irregular samples, the density is most easily measured by water displacement (Fig. 8.3). The submerged weight is corrected for the uplift caused by the displacement of the cradle. The difference between the dry weight and the submerged weight will then be the mass of water displaced (principle of Archimedes), and this may be divided by the density of water to get the volume. Thus:

$$\text{Density} = \frac{\text{Dry weight} \times \text{density of water}}{\text{Dry weight} - \text{submerged weight}} \tag{8.4}$$

The skeletal volume may be measured by grinding the solid to a powder, and measuring the increase in volume of a fluid (with which it does not react) when the fluid, and powder are mixed. To ensure full saturation, samples may be saturated by placing them in a vacuum chamber, evacuating it, and then introducing water which will be drawn into the pores (this is called vacuum saturation).

The definitions are given in Table 8.1.

Thus

$$\text{Porosity} = \frac{\text{Bulk volume} - \text{net volume}}{\text{Bulk volume}}$$

$$= \frac{\text{Absolute density} - \text{dry density}}{\text{Absolute density}} \tag{8.5}$$

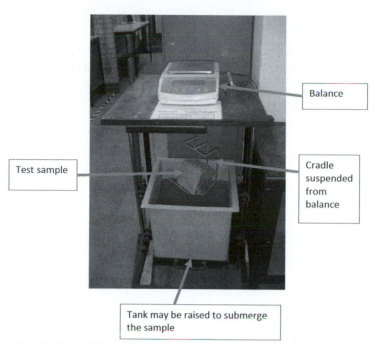

FIGURE 8.3 Balance for Measuring Mass and Density

Table 8.1 Definitions of Volume and Porosity	
Bulk volume	The overall volume defined by the outer dimensions of the sample
Skeletal volume	The volume of solids (excluding pores)
Density (or dry density)	The mass divided by the bulk volume
Absolute density	The mass divided by the skeletal volume
Specific gravity	Absolute density/density of water. When expressed in kg/litre this is equal to the absolute density because the density of water = 1 kg/L

For the definitions of density for granular materials such as aggregate the net volume is taken as the volume of the particles, and the bulk or "dry rodded" volume is taken as including the spaces among the particles.

Pores can be of several different types, the most important distinction is whether they are continuous or not, that is, whether they form a continuous pathway through the material. Even if they are continuous, the pores may form such a tortuous pathway that very little can flow through them.

8.5 CONDENSATION IN PORES

Fine pores fill with water because there is insufficient space for the molecules to move freely as a vapour (the mean free path of the molecules is greater than the pore size).

The conditions under which water will sustain a meniscus in small pores are given by the Kelvin equation:

$$r\,\mathrm{Ln(RH)} = \frac{-2M\,s}{\rho\,RT} \tag{8.6}$$

where r is the pore radius in metre, $\mathrm{Ln(RH)}$ is the natural log of the relative humidity, M is the molecular weight of water (i.e., the weight of 1 mol) = 0.018 kg (for the definition of a mol see Section 7.4), s is the surface tension of water = 0.073 N/m, ρ is the density of water = 1000 kg/m³, R is the gas constant = 8.31 J/mol/°K, T is the temperature = 293°K (at 20°C).

This equation shows that a pore of radius 3×10^{-9} m will fill with water at humidity over 70%. Thus, in a moist atmosphere, concrete will absorb a large amount of water. The effect of "ink bottle" pores is to prevent drying out of large pores with small necks (see Fig. 9.5).

8.6 WATER IN PORES

It is possible to define a relative humidity in a porous solid from the maximum pore size that is full of water (from equation (8.6)). It is measured either by weighing the solid, or by leaving small sample of, say, wood in contact with it, until they come to the same humidity, and weighing the small sample.

Most materials will swell when they get wet. The amount of swelling varies from 1000 μstrain for weak mortars, and wood across the grain, down to 100 μstrain for strong bricks. Materials with finer pores will swell more, and strong materials will swell less. Apart from causing an item, such as a door, not to fit, swelling may damage materials when it occurs unevenly. For example, if the outer surface of a solid wets and swells, but the inner core does not, it may be damaged.

8.7 DRYING OF MATERIALS

Evaporation from the surface of solids is a complex process. The rate of evaporation will depend on temperature, the relative humidity near the surface, the chemistry of the pore solution, and the exposed surface area.

The relative humidity near the surface will be controlled by air movement (wind or convection).

The chemistry of the pore solution may reduce the rate of evaporation. Materials where this happens (e.g., materials with salt in the pores) will tend to attract moisture and are described as "hygroscopic."

The exposed surface area will be the porosity × the area. Because the water is only present in the pores, it can only evaporate from them. The proportion of any cut surface that is open pores is the porosity.

8.8 SUMMARY

- Viscosity is the ratio of shear stress to shear strain rate.
- Relative humidity is the ratio of the vapour pressure to the saturation vapour pressure.

- Raising the temperature raises the saturation vapour pressure.
- Porosity is the proportion of voids by volume.
- Absolute density is the density of the solid material excluding pores.
- Small pores fill with water at lower humidities.
- Drying rates increase with porosity.

TUTORIAL QUESTIONS

1. A concrete cube has a density of 2400 kg/m³ and a porosity of 8%. What is the density of the solid material in it?

Solution:

Consider 1 m³ of the concrete.
Volume of pores = 8% of 1 m³ = 0.08 m³
Thus, volume of solid material = 1 − 0.08 = 0.92 m³
Thus, density of solid material = 2400/0.92 = 2609 kg/m³

2. What is the largest radius of pore in a brick that will be full of water at 0°C and 50% RH?

Solution:

$$R = -2 \times 0.018 \times 0.073/(\ln(0.5) \times 1000 \times 8.31 \times 273) = 1.7 \times 10^{-9} \text{ m} = 1.7 \text{ nm} \qquad (8.6)$$

3. a. A brick is fully saturated and has a density of 1800 kg/m³. When it is dried, it has a density of 1600 kg/m³. What is its porosity?
   b. If the brick is cut in half, what proportion of the cut surface will be made up of pores?

Solution:

a. Consider 1 m³ of brick.
   1800 kg total mass = 1600 kg brick + 200 kg water
   200 kg water = 200 L water
   Thus, porosity 200/1000 = 20%   (8.3)
b. 20%

4. Grout is being pumped along a pipe. All of the grout that is more than 10 mm from the wall of the pipe is moving at 0.2 m/s, and the material in contact with the pipe wall is not moving. Assuming that the grout has a viscosity of 0.12 Pas, calculate the drag on each square metre of the surface of the pipe.

Solution:

0.12 = shear × 0.1/0.2   (8.1)
Thus, shear = 2.4 N/m². This is the drag on each square metre of the surface.

5. A sample of wood is fully saturated, and has a density of 800 kg/m³. It is dried completely, and the density is reduced to 500 kg/m³. What is its porosity?

Solution:

Mass of water in 1 m^3 is 300 kg
Thus volume of water is 300 L
Porosity = 300/1000 = 30%   (8.3)

6. On a cold morning, the air in a car is at 100% RH, and the partial pressure of the water vapour in it is 0.01 bar. As the car warms up, the saturation pressure for water in the air in it is increased to 0.08 bar. Assuming that no water evaporates from surfaces in the car, and the partial pressure of the water vapour remains constant, what will the humidity be?

Solution:

At 100% RH – partial pressure = saturation pressure = 0.01 bar
When warm – partial pressure/saturation pressure = 0.01/0.08 = 12.5%

7. **a.** Describe the change in RH that occurs when the temperature inside a building is increased.
   b. Describe the effect on porous materials such as timber or masonry, when the RH is increased.

Solution:

   a. If the temperature is increased rapidly, the vapour pressure will not change, but the saturation pressure will increase, so the RH will decrease.
   b. As the humidity increases, larger pores will become saturated, and the moisture content will increase causing swelling.

8. A brick is 215 mm long by 65 mm high by 102 mm wide, and has no frog or perforations. The brick is fully saturated with water and weighs 2.5 kg. The brick is then dried and weighs 2.32 kg. What is the:

   a. Bulk dry density of the brick?
   b. Density of the solid material in it (excluding the pores)?
   c. Porosity of the brick?
   d. The brick is now placed in a high humidity environment, so that 70% of its porosity is full of water. What will its mass be?

Solution:

   a. Bulk volume of brick = 0.215 × 0.065 × 0.102 = 0.00142 m^3
       Dry density = 2.32/0.00142 = 1627 kg/m^3
   b. Volume of pores (2.5 − 2.32)/1000 = 0.00018 m^3
       Net volume = 0.00142 − 0.00018 = 0.00124 m^3
       Thus, net density = 2.32/0.00124 = 1871 kg/m^3
   c. Porosity = 0.00018/0.00142 = 12.7%   (8.3)
   d. Mass = 2.32 + (2.5 − 2.32) × 0.7 = 2.446 kg

NOTATION

e	Viscosity (Pas)
M	Molecular weight (kg)
r	Radius (m)
R	Gas constant = 8.31 (J/mol°K)
T	Temperature (°K)
s	Surface tension (N/m)
v	Velocity (m/s)
y	Distance (m)
ε	Porosity
ρ	Density (kg/m^3)
τ	Shear stress (Pa)

TRANSPORT OF FLUIDS IN SOLIDS

CHAPTER OUTLINE

9.1 INTRODUCTION

Most problems that reduce the durability of construction materials involve the transport of ions or fluids from the external environment into the material. This fluid may be water that may cause freeze–thaw damage, or the ion may be salt, which may corrode embedded steel. The salt may also be dissolved in water and move with it.

The transport properties are also relevant to a number of other aspects of materials in construction, such as waste containments barriers, where the transport of leachate must be limited, and dams, where pore pressures must be controlled.

In this chapter, the movement of fluids in solids is considered, and in Chapter 10 the movement of ions through fluids is considered.

9.2 FLOW IN A POROUS SOLID

The flow of a fluid through a porous solid, such as concrete or brick, is measured as the average speed of the fluid through the solid, the Darcy velocity v.

The volumetric flow Q is defined as the volume of fluid flowing per second:

$$Q = Av \, \mathrm{m^3/s} \tag{9.1}$$

where A is the cross-sectional area in $\mathrm{m^2}$, v is the Darcy velocity in m/s.

Now consider a porous solid with a porosity ε. The cross-sectional area available for the fluid to flow in will be $A\varepsilon$, where ε is the porosity. If the actual velocity (the seepage velocity) of the fluid in the pores is v_s

$$Av = A\varepsilon v_s \ \text{m}^3/\text{s} \tag{9.2}$$

thus

$$v = v_s \varepsilon \ \text{m/s} \tag{9.3}$$

Thus the fluid needs to flow faster to get the same volume through a limited porosity. It may be seen that when the porosity = 1 (e.g., when a fluid is flowing in a clear pipe), $v = v_s$.

9.3 PRESSURE DRIVEN FLOW

Permeable flow (i.e., flow which is controlled by the permeability) is driven by a pressure gradient (Fig. 9.1).

There are two different parameters that are used to define the permeability. The coefficient of permeability is commonly used in geotechnology, but only applies to water. The intrinsic permeability is used for the science of materials, and can be applied to any fluid because it includes a term for the viscosity.

The coefficient of permeability k (also known as the hydraulic conductivity) has the units of m/s is defined from:

$$v = \frac{k(h_1 - h_2)}{x} \ \text{m/s} \tag{9.4}$$

where the fluid is flowing through a thickness x (m) with pressure heads h_1 and h_2 (m) on each side.

The intrinsic permeability K has the units of m² and is defined from equation (9.5):

$$v = \frac{K(p_1 - p_2)}{ex} \ \text{m/s} \tag{9.5}$$

where e is the viscosity of the fluid and p_1 and p_2 are the pressures on each side in Pa.

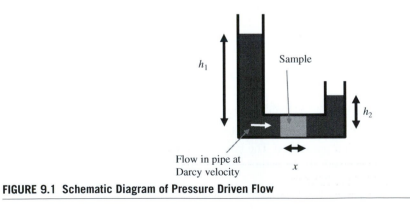

FIGURE 9.1 Schematic Diagram of Pressure Driven Flow

Note that the permeability equations should strictly be expressed in differential form:

$$v = k \frac{dh}{dx} = \frac{K}{e}\left(\frac{dp}{dx}\right)$$

The non-differential forms given may be inaccurate for thick barriers, large pressure drops, and systems that are not in a steady state. However, they are used in computer modelling where very small elements are considered, so they are accurate.

The pressure of a fluid arising from a fluid head is given by equation (5.2). Combining this with equations (9.4) and (9.5) gives

$$k = \frac{K \rho g}{e} \, \text{m/s} \tag{9.6}$$

A typical value of k for water in concrete is 10^{-12} m/s. The density of water (ρ) is 1000 kg/m^3, the gravitational constant (g) is approximately 10 m/s^2, and the viscosity of water (e) is 10^{-3} Pas, so K is approximately 10^{-19} m^2.

9.4 THERMAL GRADIENT

Water will move from hot regions to cold regions in solids. The rate at which it moves will depend on the permeability of the solid. This process is independent of, and additional to the drying process (evaporation) that will take place on exposed surfaces that are hot. In saturated concrete, or masonry, ions in hotter regions will migrate towards colder regions. The mechanism is shown in Fig. 9.2, and depends on probability. At a microscopic level, the temperature of a solid is a measure of the kinetic energy of the atoms, and molecules within it (see Section 4.2). An ion or molecule that is moving faster on the hot side, has a greater probability of crossing the sample than one on the cold side.

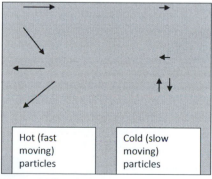

FIGURE 9.2 Schematic Diagram of Thermal Migration

The longer arrows indicate the greater movement of ions in the hot region.

9.5 CAPILLARY SUCTION

Surfaces of materials may be seen to be of two types: "wetting" and "nonwetting." On a "nonwetting" surface, such a car body after a good wax polish, droplets of water will remain almost spherical as they would be in free air. On a "wetting" surface (such as, very clean glass) the water will spread out to form a thin layer (see Fig. 9.3). These effects are caused by the relative sizes of the surface tension forces on the water/air, water/solid, and air/solid interfaces.

Capillary suction occurs in fine voids (capillaries) with wetting surfaces, and is caused by surface tension. In the experiment shown in Fig. 9.4, water rises higher up a smaller diameter glass capillary tube, and this shows how this mechanism has greatest effect in systems with fine pores. This leads to the situation that concretes with finer pore structures (normally higher grade concretes) will experience greater capillary suction pressures. Fortunately, the effect is reduced by the restriction of flow by generally lower permeabilities.

A good demonstration of the power of capillary suction in concrete can be observed by placing a cube in a tray of salt water, and simply leaving it in a dry room for several months. The water with the salt in it will be drawn up the cube by "wicking" until it is close to an exposed surface, and can evaporate. As this happens, the near-surface pores fill up with crystalline salt that will eventually achieve sufficient pressure to cause spalling. This mechanism of damage by salt crystallisation is common in climates where there is little rain to wash the salt out again (see Chapter 25).

The height of the column of water in a fully wetting capillary is:

$$h = \frac{2s}{r\rho g}\, \text{m} \tag{9.7}$$

where s is the surface tension (= 0.073 N/m for water), r is the radius of the capillary.

A typical pore in concrete has a radius of 10^{-8} m. Putting this into the equation, gives a height of 1460 m indicating that damp should rise up concrete to this height. The reason why it does not is the irregularity of the pores (see Fig. 9.5). The pressure decreases as the pore radius increases. Water will be drawn up through the neck of the pore, but will then reach the point where the radius is much larger, and the capillary suction pressure thus much lower, and will stop at that point (typically, larger pores

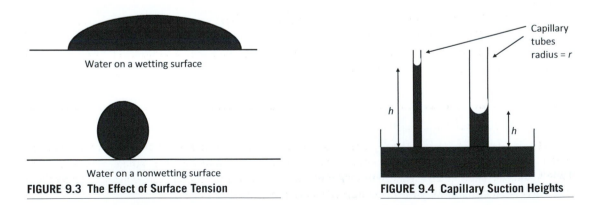

Water on a wetting surface

Water on a nonwetting surface

FIGURE 9.3 The Effect of Surface Tension

Capillary tubes radius = r

h

h

FIGURE 9.4 Capillary Suction Heights

←— Entry radius

FIGURE 9.5 "Ink Bottle" Pore

may have diameters of 5×10^{-7} m). However, if the capillary suction is driving a continuous process, such as wicking this height, may be used in equation (9.4) to calculate the flow rate.

9.6 OSMOSIS

Osmosis depends on what is called a semipermeable membrane. This is a barrier through which the water can pass, but material dissolved in it cannot pass as easily. An example is the surface layer of concrete that will permit water to enter, but restrict the movement of the lime dissolved in the pore water. The osmotic effect causes a flow of water from the weak solution to the strong solution. Thus water on the outside of concrete (almost pure, i.e., a weak solution) is drawn into the pores where there is a stronger solution. The process is illustrated in Fig. 9.6. If two solutions were placed either side of a barrier as shown, the level in one of them would rise, although in practice, this would be very difficult to observe because concrete is permeable, and the liquid would start to flow back as soon as a pressure difference developed.

If different solutions are present on each side of the sample, osmosis will still take place even if the concentrations are the same. The direction of flow will depend on their relative "osmotic coefficients." Osmosis could be a significant process for drawing chlorides and sulphates into concrete. Having entered the concrete they may move further in by diffusion (see Chapter 10).

9.7 ELECTRO-OSMOSIS

If a solid material has an electrically charged surface, the water in the pores will acquire a small opposite charge. Clay, both in soil and bricks, has a negatively charged surface, so the pore water will have a positive charge. The water will thus move towards a negatively charged plate, and away from a positively charged one (see Section 6.2). This system is used in one of the commercially available

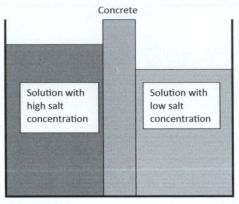

FIGURE 9.6 Schematic Diagram of Osmosis

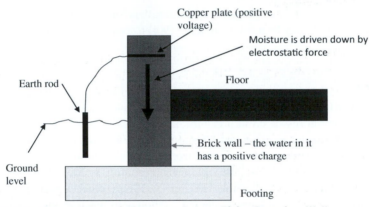

FIGURE 9.7 Arrangement of Electro-Osmosis System to Prevent Rising Damp in a Wall

systems for removing damp from existing buildings as shown in Fig. 9.7 (the more common one is resin injection to block the pores).

9.8 SUMMARY

- The coefficient of permeability is used to calculate flow rates for different heads of water.
- The intrinsic permeability is used to calculate flow rates for different pressures and viscosities.
- A thermal gradient will move fluids from a hot region to a cold one.
- Capillary suction pressures are higher in smaller capillaries.
- Electro-osmosis occurs when water molecules assume a net charge.

TUTORIAL QUESTIONS

1. **a.** A brick wall is built without a damp proof course, and the bottom of it is exposed to water. If the typical pore radius in the bricks is 0.1 μm, what is the theoretical height to which the water will rise? (Surface tension = 0.073 N/m for water.)
   b. Why will it not rise to this height in practice?

   Solution:

   a. $H = 2 \times 0.073/(10^{-7} \times 1000 \times 9.81) = 149$ m (9.7)
   b. Because the pores are not uniform.

2. Water is flowing through the base of a concrete tank at a rate of 1.5 mL/m²/day. If the base of the tank is 100 mm thick, and the depth of water in the tank is 8 m, what is the coefficient of permeability of the concrete? (Assume that there is no water pressure on the outside of the tank.)

   Solution:

   1.5 mL/m²/day $= 1.5 \times 10^{-6}/(24 \times 60 \times 60) = 1.7 \times 10^{-11}$ m³/s
   $Q = Av$ and $A = 1$, thus $v = 1.7 \times 10^{-11}$ m³/s (9.1)
   $v = k \times 8/0.1$ (9.4)
   Thus $k = 2.1 \times 10^{-13}$ m/s

3. Water is flowing through a concrete plug in a pipe. The heads of water on each side of the plug are 22 m and 5 m. The internal diameter of the pipe is 400 mm, the plug is 700 mm thick, and the flow rate is 1.5 mL/day. What is the coefficient of permeability of the concrete?

   Solution:

   1.5 mL/day $= 1.5 \times 10^{-6}/(60 \times 60 \times 24) = 1.7 \times 10^{-11}$ m³/s
   $v = 1.7 \times 10^{-11}/(\pi \times 0.2^2) = 1.37 \times 10^{-10}$ m/s (9.1)
   $k = 1.37 \times 10^{-10} \times 0.7/(22 - 5) = 5.66 \times 10^{-12}$ m/s (9.4)

4. A rectangular concrete culvert with external dimensions 3 m high and 4 m wide, and with a wall thickness of 100 m, contains water at an average pressure of 500 kPa. If there is no water pressure on the outside of the culvert, and the coefficient of permeability of the concrete is 10^{-12} m/s, what is the flow in millilitre per day through the walls of each metre length of culvert?

   Solution:

   Area of 1 m length $= (3 + 3 + 4 + 4) \times 1 = 14$ m²
   Pressure head $= 5 \times 10^5/(10^3 \times 10) = 50$ m (5.2)
   $v = 10^{-12} \times 50/0.1 = 5 \times 10^{-10}$ m/s (9.4)
   $Q = 14 \times 5 \times 10^{-10} \times 60 \times 60 \times 24 = 6.05 \times 10^{-4}$ m³/day $= 605$ mL/day (9.1)

5. **a.** A concrete wall of a tank has a head of 10 m of water against it and is 100 mm thick. What is the rate of flow of water through each square metre of it if the coefficient of permeability is 10^{-12} m/s?
   b. What is the intrinsic permeability? Use $e = 10^{-3}$ Pas for water.
   c. The tank is now filled with gas with a viscosity of 2×10^{-5} Pas. At the same pressure, what will the flow rate be?

Solution:

a. $v = 10^{-12} \times 10/0.1 = 10^{-10}$ m/s   (9.4)

 $A = 1$

 Thus $Q = 10^{-10}$ m^3/s   (9.1)

b. Pressure $= 1000 \times 10 \times 10 = 10^5$ Pa   (5.2)

 Intrinsic permeability $= 10^{-10} \times 10^{-3} \times 0.1/10^5 = 10^{-19}$ m^2 (9.5)

c. Flow $= 10^{-19} \times 10^5/(2 \times 10^{-5} \times 0.1) = 5 \times 10^{-9}$ m^3/s   (9.5)

6. a. A concrete wall 150 mm thick forms the side of a water tank with the outside of the wall open to the atmosphere. If the coefficient of permeability of the concrete is 6.5×10^{-12} m/s, and the porosity is 12%, calculate the seepage velocity of the flow of water through the wall in metre per second at a depth of 5 m below the water surface.

b. Calculate the rate of flow of water in mL/m^2/s through the wall.

Solution:

a. Darcy velocity, $v = 6.5 \times 10^{-12} \times 5/0.15 = 2.166 \times 10^{-10}$ m/s   (9.4)

 Seepage velocity $= 2.166 \times 10^{-10}/0.12 = 1.8 \times 10^{-9}$ m/s   (9.3)

b. Flow through 1 m$^2 = 2.166 \times 10^{-10} \times 1$ m^3/m^2/s $= 2.166 \times 10^{-4}$ mL/m^2/s   (9.1)

NOTATION

A Area (m^2)
e Viscosity (Pas)
g Gravitational constant (=9.81 m/s^2)
h Pressure head (m)
k Coefficient of permeability (m/s)
K Intrinsic permeability (m^2)
p Pressure (Pa)
Q Volumetric flux (m^3/s)
r Radius (m)
s Surface tension (N/m)
v Velocity (m/s)
x Distance (m)
vs Seepage velocity (m/s)
ε Porosity
ρ Density (kg/m^3)

TRANSPORT OF IONS IN FLUIDS

10.1 INTRODUCTION

In this chapter, the processes that move ions into materials without any net flow of the pore fluid are discussed. These processes are driven either by a difference in chemical concentration (diffusion), or by an electrical voltage (electromigration). The situation is complicated by adsorption in which ions become bound in the matrix (i.e., the structure of the material), and cannot move.

10.2 IONS IN SOLUTION

It was noted in Section 7.8 that many molecules would dissociate into separate parts (ions) when they are in solution. Thus, although the whole molecule is dissolved in the water, we often consider the movement of individual ions through it. In particular, when considering electromigration in Section 10.6, two ions such as Na^+ and Cl^- from common salt will have opposite charges, so they will move in opposite directions.

10.3 FLOW RATES

Diffusion is normally defined in terms of flux (F) that is the flow in kg per second per unit cross-sectional area of the porous material, and is measured in $kg/m^2/s$.

This is slightly different from the volumetric flow Q given in equation (9.1) for pressure driven flow. F and Q may be related as follows:

$$\text{Total mass flow} = F = \frac{CQ}{A} = Cv \, \text{kg/m}^2/\text{s} \tag{10.1}$$

where F is the flux in kg/m^2/s, C is the concentration in kg/m^3, v is the Darcy velocity in m/s, Q is the volumetric flow in m^3/s, A is the cross-sectional area in m^2.

10.4 DIFFUSION IN A NONADSORBING SYSTEM

In order to understand the basic mechanism of diffusion, we first consider a nonadsorbing system. That is one in which ions in solution are free to move at all times, and are not being bound to the matrix.

Diffusion is driven by concentration gradient. If a strong solution is in contact with a weak solution, they will both tend towards the same concentration. Thus, for example, if a pile of salt is placed in one corner of a container full of water, diffusion will be the process that ensures that when the salt has dissolved, it will assume a uniform concentration throughout the water. Thus, if the system in Fig. 10.1 is left for a long time, the salt will ultimately have a uniform concentration throughout the tank, and also in the pore solution in the concrete. It is not necessary for the water to move for this to take place.

The diffusion coefficient is defined from the equation:

$$F = D\frac{dC}{dx} \, \text{kg/m}^2/\text{s} \tag{10.2}$$

where F is the flux in kg/m^2/s, D is the diffusion coefficient in m^2/s, C is the concentration in kg/m^3, x is the position.

Thus dC/dx is the concentration gradient.

Considering a small element of the system, the rate at which the concentration changes with time will be proportional to the difference between the flux into it and the flux out of it:

$$V\frac{dC}{dt} = A \, \Delta F \tag{10.3}$$

where V is the volume of the element, A is the cross-sectional area, ΔF is the change in flux from one side of the element to the other and,

$$\Delta F = L\frac{dF}{dx} \tag{10.4}$$

where $L = V/A$ is the length of the element. Thus, from equations (10.3) and (10.4):

$$\frac{dC}{dt} = \frac{dF}{dx} \tag{10.5}$$

and thus from equation (10.2):

$$\frac{dC}{dt} = D\frac{d^2 C}{dx^2} \tag{10.6}$$

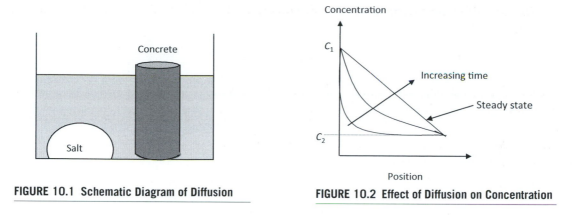

FIGURE 10.1 Schematic Diagram of Diffusion

FIGURE 10.2 Effect of Diffusion on Concentration

These equations are well suited to numerical modelling, and can be solved analytically in many situations (not easily).

Considering a system with concentrations C_1 and C_2 on each side. See Fig. 10.2.

Looking first at the long-term (steady state) solution, the system will eventually reach a point where the concentration stops changing. Thus $dC/dt = 0$ and therefore dC/dx is constant with distance (because $d^2C/dx^2 = 0$) and the graph is a straight line.

Before this happens, the rate of change of concentration with time (dC/dt), and thus the curvature of the concentration versus position curve (d^2C/dx^2), will progressively decrease. dC/dt will also increase with D, that is, the system will reach a steady state sooner if the diffusion coefficient is higher (the flux will also be greater).

10.5 ADSORPTION IN A POROUS SOLID

When considering transport of ions in a porous material, it is essential to consider adsorption at the same time, because in many situations the bulk of the ions that enter into a barrier will be adsorbed before they reach the other side.

The concentration of ions in a porous solid (in which the pores are filled with fluid) may be measured in two different ways:

- C_1 kg/m^3 is the concentration of ions per unit volume of *liquid* in the pores, and
- C_s kg/m^3 is the total concentration (including adsorbed ions) per unit volume of the *solid*.

When the chloride concentration in a concrete sample is measured, there are various different systems that can be used:

If the "acid soluble" concentration is measured by dissolving the sample in acid, this will extract all of the chlorides and C_s is obtained.

If the "water soluble" concentration is measured, only the ions in solution will come out (assuming the test is too short for adsorbed ions to dissolve), and C_1 is obtained. Alternatively "pore squeezing" can be used to squeeze the sample like an orange (using very high pressures) to obtain C_1.

It can be argued that the ions in solution are the only ones that will cause corrosion.

The ratio of the two concentrations is the capacity factor α:

$$\alpha = \frac{C_s}{C_1} \tag{10.7}$$

A simple approximation of the amount of material that is adsorbed onto the matrix may be obtained by assuming that at all concentrations, it is proportional to the concentration of ions in the pore fluid (note that this implies that the adsorption is reversible). Thus α is constant for all concentrations. This approximation works best for low solubility ions. Chlorides have a solubility of about 10%, so α is not exactly constant, but it is a reasonable assumption for modelling.

Equation (10.5) for the rate of change of concentration will be for the total concentration:

$$\frac{dC_s}{dt} = \frac{dF}{dx} \tag{10.8}$$

thus

$$\frac{dC_1}{dt} = \frac{1}{\alpha}\frac{dF}{dx} \tag{10.9}$$

From this it may be seen that a high value of α will make the concentration change much more slowly, that is, if chlorides are penetrating into a wall, it will delay the start of corrosion of the steel.

Adsorption will also affect pressure driven flow, which was discussed in Chapter 9.

It is necessary to make a clear distinction between absorption and adsorption. The term absorption is used to describe processes such as capillary suction, and osmosis that may draw water into concrete. Adsorption, which is discussed in this section, is the term used for all processes that may bind an ion (temporarily or permanently) in the concrete and prevent it from moving. These processes may be chemical reactions or a range of physical surface effects.

10.6 DIFFUSION WITH ADSORPTION

Because there are two different ways of measuring concentration in an adsorbing system, there are also two different ways of measuring diffusion:

The apparent diffusion coefficient D_a (that is what can be measured by testing the solid using measurements of total concentration C_s) is defined from:

$$F = D_a \frac{dC_s}{dx} \, \text{kg/m}^2/\text{s} \tag{10.10}$$

and the intrinsic diffusion coefficient (that is the diffusion coefficient for the pore solution and could be measured on the liquid extracted from the solid) is defined from:

$$F' = D_i \frac{dC_1}{dx} \, \text{kg/m}^2/\text{s} \tag{10.11}$$

where F' is the flux per unit cross-sectional area of the liquid in the pores. Thus:

$$F = \varepsilon D_i \frac{d C_i}{d x} \, \text{kg/m}^2/\text{s} \tag{10.12}$$

where ε is the porosity.

By integrating these (or by inspection) it may be seen that:

$$\frac{\alpha}{\varepsilon} = \frac{D_i}{D_a} \tag{10.13}$$

for a typical concrete pore solution $D_i = 5 \times 10^{-12}$ m²/s. If $\alpha = 0.3$ and $\varepsilon = 8\%$ this gives $D_a = 1.3 \times 10^{-12}$ m²/s.

10.7 ELECTROMIGRATION

It was noted in Section 10.2 that the ions in solution have a net charge. Thus, if an electric field is applied across the solid, the negative ions will move towards the positive electrode (Fig. 10.3).

If the electrical conductivity of the sample is measured (see Chapter 6), then the diffusion coefficient may be calculated from equation (10.14) that is known as the Nernst–Einstein equation. This means that test methods that use electrical voltages to drive the chlorides through concrete (and save a lot of time) can be used to estimate how fast they will go through by simple diffusion with no voltage. Unfortunately, this will only be an approximation because the other charged ions in the sample will affect the electric field, so a full computer model is needed for an accurate solution.

$$D = \frac{RT\sigma}{z^2 F_a^2 C} \tag{10.14}$$

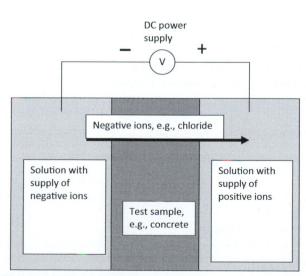

FIGURE 10.3 Schematic Diagram of Electromigration

where D is the diffusion coefficient in m^2/s, R is the gas constant (= 8.31 J/mol/°K) (see Section 7.4), T is the temperature (°K), σ is the measured electrical conductivity $(\Omega m)^{-1}$ (see Section 6.6.3), z is the valence (or valency) of the ion, that is, the number of electronic charges on each ion. This will be -1 for a Cl^- ion or 2 for a Fe^{++} ion (see Section 7.8), F_a is the Faraday constant $= 9.65 \times 10^4$ C/mol, C is the concentration (kg/m^3).

10.8 CONCLUSIONS

- Diffusion is driven by a concentration gradient.
- Electromigration is driven by a voltage gradient.
- Adsorption occurs when ions become bound to the matrix.
- The apparent diffusion coefficient is measured on solid samples with pore water in them.
- The intrinsic diffusion coefficient is measured on pore water extracted from a sample.
- Electromigration may be measured as a resistivity using alternating current.

TUTORIAL QUESTIONS

1. A 150 mm thick concrete wall is fully saturated with water with no pressure differential. On one side of it, the water contains salt at a concentration of 7% by mass, and on the other side, the water is kept at zero concentration. The capacity factor for the salt is 0.512, the apparent diffusion coefficient is 1.5×10^{-12} m^2/s and the porosity is 8%.
   a. Calculate the total mass of salt in each square metre of wall in the steady state.
   b. Calculate the mass of salt leaving each square metre of the wall per second in the steady state.
   c. Draw three graphs showing the changes in salt "flux" across the thickness of the wall:
      - Early in the transient phase
      - Late in the transient phase
      - In the steady state

   d. Describe the circumstances under which capillary suction would contribute significantly to the transport of chlorides.

   Solution:

   a. If the salt concentration is 7% by mass then each 1000 kg (i.e., each cubic metre) has 70 kg of salt in it.
   Thus C_1 on the high concentration side = 70 kg/m^3.
   This is a total per cubic metre of liquid. Within the solid the porosity is 8% thus at the surface near the salt solution the concentration in the pores per unit volume of solid is $70 \times 8\% = 5.6$ kg in solution.
   $\alpha = 0.512$, thus $C_s = \alpha C_1 = 36$ kg/m^3 (10.7)
   This will be made up of 5.6 kg in solution in the pores + $(36 - 5.6) = 30.4$ kg adsorbed.
   In the steady state, the flux must be the same at all points in the wall (otherwise the concentration would be changing because there would be more flux into an element than out

of it). Thus from equation (10.1) the concentration changes linearly with distance (as shown in Fig. 10.2).

- – The total concentration on the high side = 36 kg/m³.
- – Thus the average = 18 kg/m³.
- – Thus the total in the 0.15 m thick wall = 18 × 0.15 = 2.7 kg.

b. $F = 1.5 \times 10^{-12} \times 36/0.15 = 3.6 \times 10^{-10}$ kg/m²/s   (10.10)

c.

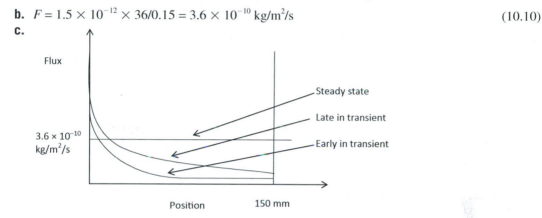

Note: The equations will give an infinite value for diffusion at position 0 at early times because the concentration gradient is technically infinite. The limiting factor is the supply of chlorides from solution.

d. Capillary suction works when water (with dissolved salt) enters dry pores so it would only be significant if there was wetting, and drying with the drying penetrating to a depth comparable to the cover. Capillary suction can often transport chlorides into the outer surface so that they can diffuse inwards, but in this situation (with the wall fully saturated) this will not happen.

2. a. A 100 mm concrete cube is placed in water with the water level just above the base of the cube, and the top surface is kept dry by evaporation. The sides of the cube are sealed. What is the theoretical rate of flow of water up the cube in mL/s?

b. If the water contains 5% salt what will be the flux of salt up the cube in the steady state?

c. If a second cube is tested in similar conditions, but with a constant flow of clean water across the top what will the steady state flux now be?

d. For each experiment describe the following:

(1) the effect of changing the capacity factor of the concrete,

(2) the salt concentration in the cube in the steady state,

(3) the ultimate effect on the cube if the experiment is continued for a long time.

For this question use the following data:

- Typical diameter of pores in concrete = 0.015 μm
- Surface tension of water = 0.073 N/m
- Coefficient of permeability of concrete = 10^{-12} m/s

- Porosity of concrete = 8%
- Intrinsic diffusion coefficient of salt in water = 10^{-12} m^2/s

Solution:

a. The flow in this cube is caused by capillary suction and is controlled by the permeability. The pressure head from capillary suction,

$$h = 2 \times 0.073/(0.75 \times 10^{-8} \times 1000 \times 9.81) = 1984 \text{ m} \qquad (9.7)$$

Thus the Darcy velocity,

$$v = 10^{-12} \times 1984/0.1 = 1.98 \times 10^{-8} \text{ m/s} \qquad (9.4)$$
$$\text{Thus the flow} = 1.98 \times 10^{-8} \times 0.1 \times 0.1 = 1.98 \times 10^{-10} \text{ m}^3\text{/s} = 1.98 \times 10^{-4} \text{ mL/s} \qquad (9.1)$$

b. If the salt concentration is 5% by mass then each 1000 kg (i.e., each cubic metre) has 50 kg of salt in it.

$$1.98 \times 10^{-10} \times 50 = 9.9 \times 10^{-9} \text{ kg/s}$$

c. For this second cube, the transport mechanism is diffusion because there is a concentration gradient but no pressure gradient.

$$C_1 = 50 \text{ kg/m}^3$$

$$F = 0.08 \times 10^{-12} \times 50/0.1 = 4 \times 10^{-11} \text{ kg/m}^2\text{/s} \qquad (10.12)$$
$$\text{Total mass flow} = \text{flux} \times \text{area} = 4 \times 10^{-11} \times 0.1 \times 0.1 = 4 \times 10^{-13} \text{ kg/s} \qquad (10.1)$$

d. (1) Increasing the capacity factor will increase the length of the transient in both cubes but will not affect the steady state.

(2) The first cube will be fully saturated to the concentration of the solution but the second will have a linear concentration drop to the top.

(3) In the first cube, the salt will build up near the top of the cube and, as the pressure from it increases, the corners will fall off (this is often observed on structures in dry climates with saline groundwater). Nothing will happen to the second.

3. Two 150 mm thick concrete walls (A and B) are fully saturated with water. The capacity factor for salt is 1.52, the apparent diffusion coefficient is 2×10^{-12} m^2/s and the coefficient of permeability is 5×10^{-11} m/s. On one side of the walls the water contains salt at a concentration of 7% by mass. Wall A has no pressure differential but on the opposite side to the salt it is kept washed with clean water. Wall B has no concentration gradient but has a pressure drop of 200 kPa across it.

a. Describe the processes, which transport the salt through the concrete in each wall.

b. Draw graphs showing the Flux and Concentration (C_1) of salt across the walls in the steady state. In each graph, distance should be on the x-axis, and concentration or flux on the y-axis. The value of all intercepts with the axes should be calculated and shown on the graphs.

c. What is the total mass of salt in each square metre of each wall in the steady state?

Solution:

a. Wall A is diffusion and wall B is permeability control.

b. If the salt concentration is 7% by mass then each 1000 kg (i.e., each cubic metre) has 70 kg of salt in it.

C_1 on the high concentration side = 70 kg/m³

$C_s = \alpha C_1 = 106.4$ kg/m³ (10.7)

Wall A

$F = 2 \times 10^{-12} \times 106.4/0.15 = 1.42 \times 10^{-9}$ kg/m²/s (10.10)

Wall B

200 kPa = 20 m head (5.2)

Darcy velocity $v = 5 \times 10^{-11} \times 20/0.15 = 6.66 \times 10^{-9}$ m/s (9.4)

Volumetric flux = 6.66×10^{-9} m³/s through each square metre of wall (9.1)

Flux = concentration × Darcy velocity = $6.66 \times 10^{-9} \times 70 = 4.66 \times 10^{-7}$ kg/m²/s (10.1)

	Wall A		**Wall B**	
Position	Flux	Concentration	Flux	Concentration
0	1.42×10^{-9}	70	4.66×10^{-7}	70
150	1.42×10^{-9}	0	4.66×10^{-7}	70

All the graphs are straight lines.

c. Wall A

- Total on high side $C_s = 106.4$ kg/m³
- Average = 106.4/2 = 53.2 kg/m³ (because there is a linear drop across the wall)
- Thus total in wall = 53.2 × 0.15 = 7.98 kg

Wall B: total in wall = 106.4 × 0.15 = 15.96 kg (because the concentration does not drop across the wall)

4. The mix design of concrete elements that are exposed to chlorides is changed to include pulverised fuel ash (PFA). This increases the capacity factor for chlorides from 1,200 to 24,000 and reduces the porosity from 10% to 8%. The intrinsic diffusion coefficient is 5×10^{-11} m²/s.

a. Calculate the apparent diffusion coefficient before and after the change is made.

b. Describe the changes to the flux of chloride ions both in the transient phase (shortly after exposure) and in the long-term steady state.

Solution:

a. Before:

$D_a = (5 \times 10^{-11} \times 0.1)/1200 = 4.16 \times 10^{-15}$ m²/s (10.13)

After:

$D_a = (5 \times 10^{-11} \times 0.08)/24,000 = 1.66 \times 10^{-16}$ m²/s (10.13)

b. During the transient phase the concentration of chlorides in the pore solution is increasing. Thus, if the distribution ratio is high, substantial adsorption of chlorides onto the cement matrix will occur. This will, in turn, control the concentrations in solution and restrict the rate of penetration.

Thus, in the transient (that will last for working life of structure) a major reduction in flow rates will occur because most of the chlorides entering the element are absorbed onto the matrix.

In the steady state, there is no change in the concentration in the pore solution so the distribution ratio is not relevant. The ions will diffuse through the concrete. Thus, the flux depends on porosity, i.e., it gets reduced by 20%.

5. A 50 mm thick concrete barrier has pure water on one side, and 80 kg/m^3 salt solution on the other side. D_i is 2×10^{-12} and the porosity is 15%. What is the flux in kg/m^2/s in the steady state? What is the velocity of the ions in solution 20 mm from the pure water side?

Solution:

$$\text{Flux } F = 0.15 \times 2 \times 10^{-12} \times 80/0.05 = 4.8\text{E} - 10 \text{ kg/m}^2/\text{s} \tag{10.12}$$

- C_1 will have linear gradient because F is constant (steady state)
- Thus $C_1 = 80 \times 20/50 = 32 \text{ kg/m}^2$

$$\text{Thus the Darcy velocity } v = 4.8\text{E} - 10/32 = 1.5 \times 10^{-11} \text{ m/s} \tag{10.1}$$
$$\text{Thus the seepage velocity } v_s = 1.5 \times 10^{-11}/0.15 = 10^{-10} \text{ m/s} \tag{10.3}$$

NOTATION

A Area (m^2)
C Concentration (kg/m^3)
C_s Total concentration per unit volume of solid (kg/m^3)
C_1 Concentration in solution per unit volume of liquid (kg/m^3)
D Diffusion coefficient (m^2/s)
D_a Apparent diffusion coefficient
D_i Intrinsic diffusion coefficient
F Flux (kg/m^2/s)
F_a Faraday constant $= 9.65 \times 10^4$ C/mol
L Length (m)
Q Volumetric flux (m^3/s)
R Gas constant $= 8.31$ (J/mol°K)
t Time (s)
T Temperature (°K)
V Volume (m^3)
v Velocity (m/s)
x Distance (m)
z Valence of an ion (i.e., the charge on it divided by the charge of an electron)
α Capacity factor
ε Porosity
σ Electrical conductivity $(\Omega\text{m})^{-1}$

IONISING RADIATION

11.1 INTRODUCTION

Most radiation, such as radiant heat or microwaves, will simply raise the temperature of an object (see Section 4.9). However, ionising radiation has enough energy to remove electrons from atoms or molecules, and thus create charged ions (see Section 7.8). At even higher energy, it may remove protons and neutrons, and thus break up the nucleus of an atom. Ionising radiation is dangerous at high doses, and civil engineering structures, normally made of dense concrete, are used to contain it, and protect people from it.

11.2 TYPES OF IONISING RADIATION

Some sorts of radiation are normally called "rays" and others are called "particles." In nuclear physics, it is shown that mass and energy are the same thing, so these conventions are not important.

α *Radiation* is ions of helium gas (He^{++}). Helium is a very light gas, but in terms of radiation α particles are relatively heavy. They can cause a lot of damage when they hit things, but are stopped by a few tens of mm of air.

β *Radiation* is high-energy electrons. These can penetrate a bit more than α particles, but are stopped by a thin barrier of steel or concrete.

γ *Radiation* is high-energy electromagnetic radiation. Light, radiant heat, and radio waves are electromagnetic waves, but γ rays have a far higher frequency and energy. They are highly penetrating and require a thick concrete or steel barrier to stop them. We get continual γ radiation from the sun.

X-rays are electromagnetic radiation at a slightly lower energy than γ radiation with consequently less penetration.

High energy (fast) *Neutrons* tend to go straight through things but lower energy (thermal) neutrons are more likely to be absorbed, and are thus more dangerous. Neutrons tend to be reflected from (bounce off) surfaces. In the design of nuclear containment, sharp bends in pipes and ducts are used to stop neutrons travelling along them.

11.3 SOURCES OF RADIATION

Typical sources of ionising radiation that may affect the design or construction of structures are:

- Nuclear reactors: If an atom has too many neutrons to form a stable isotope (see Section 7.1), it will spontaneously decay into two or more lighter atoms and emit radiation. Some isotopes decay so slowly that they are found in ore, such as uranium (a heavy atom with atomic number = 92), from mines. A process called enrichment is used to isolate these radioactive isotopes from the rest of the ore. They are then used as fuel for reactors. Reactors produce massive amounts of neutrons.
- Radioactive sources: These are small samples of unstable isotopes that spontaneously emit γ radiation. Sources are used for materials testing such as checking the integrity of welds in steel structures.
- Accelerators: These are electrical devices such as x-ray machines or larger devices that produce γ radiation.
- Natural radiation: There is natural radiation throughout the environment. This may be harmful if it builds up, for example, in the build-up of radon gas (an α emitter) under the floors of houses built on granite formations. Vents should be provided to permit air circulation to disperse this.
- Radioactive waste: This contains unstable material and comes in three classifications:
  - Low level waste: For example, contaminated protective clothing.
  - Intermediate level waste: This is more active than the low level waste, but is generally not significantly heat generating. It includes the containers from spent reactor fuel.
  - High level waste: This is the most active, and is produced in very small quantities.
- Other radiation: When radiation hits an object it will very often produce more radiation of the same or a different type. For example, a γ ray hitting an atom will often produce a β particle. Radiation from these processes is called "secondary radiation".

11.4 HALF-LIVES

This concept is used for unstable isotopes that spontaneously produce radiation. These sources do not last forever, their power decreases slowly. Some types will produce high power for a short time while others will produce lower power for longer. Their power at any time may be calculated from their initial power and their "half-life" $T_{0.5}$, that is the time taken for their power to drop by half.

Thus, the decay is exponential and can be expressed as follows:

$$X = Ae^{-kt} \tag{11.1}$$

where X is the intensity in Watts, t is time in seconds, and A is the intensity in Watts at $t = 0$, and:

$$k = \log_e(2)/T_{0.5} \tag{11.2}$$

where $T_{0.5}$ is the half-life in seconds.

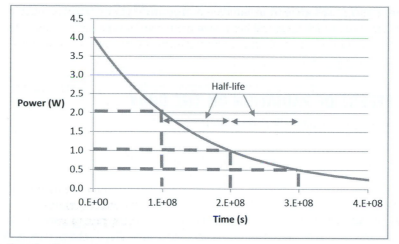

FIGURE 11.1 Radioactive Decay

Figure 11.1 illustrates the concept of the half-life. It is plotted using equations (11.1) and (11.2) for an initial power $A = 4W$ and a half-life of $T_{0.5} = 10^8$ s. It may be seen that during any interval of 10^8 s the power reduces by half.

11.5 THE EFFECT OF RADIATION ON MATERIALS

A radiation particle incident on a sample of material will produce one or several of the following effects:

- Transmission: A certain proportion of any radiation incident on a material will pass through it without any effect.
- Scattering: Some radiation particles will be scattered in materials. They may then come out of the material in any direction, even back in the direction they came from.
- Energy dissipation: This is the main effect of radiation, it heats things up. Part or all of the energy of the incident radiation is lost to the material. In this process, ionising radiation is acting in the same way as lower energy radiation.
- Chemical reactions: Radiation will stimulate a number of chemical reactions. For example, it will break up water molecules so hydrogen and oxygen gas are produced. Alpha particles may strip electrons from molecules and produce helium gas. The harmful effects of radiation on the human body are caused by chemical reactions.
- Nuclear reactions: High-energy radiation, especially neutrons, will stimulate nuclear reactions in which nuclei are broken down or added to, thus forming different elements.
- Production of secondary radiation (see Section 11.3).

The amount of energy absorbed from radiation is measured in Gray (J/kg). Concrete is not damaged significantly by doses up to 10^8 Gray, rubber is stable up to 10^5 Gray.

11.6 THE EFFECT OF RADIATION ON THE BODY

Radiation may enter the human body in two ways:

- Incident radiation (e.g., γ) may be able to penetrate the skin.
- Radioactive materials may be ingested in air or food.

The physical effect of radiation on the human body is similar to that on any other material, and includes all of the processes listed in Section 11.5. This may cause a number of effects, especially cancer. Different types of radiation are more harmful than others, and the rate of exposure has a significant effect. Human doses are measured as dose equivalent in Sievert (J/kg). The Sievert is defined from the equation:

$$E = QDN$$

where E is the dose in Sievert, D is the dose in Gray (J/kg), Q is a quality factor that varies between 1 and 20 depending on the type of radiation, N is a modification factor that takes account of the distribution of energy throughout the dose.

If you work near radiation, you should be issued with a dosimeter that will record your exposure in microsievert.

11.7 SHIELDING

The normal function of construction materials in relation to radiation is shielding. The radiation must be prevented from reaching the outside of the structure with sufficient intensity to cause damage or injury. Concrete is the most economic material to use for this, and its effectiveness depends on its density so a high-density aggregate (sometimes steel shot) is used. Calculations of shielding are based on a concept similar to the half-life. A given thickness of shielding is required to halve the transmission. The equations are very similar to those for the half-life: the intensity is given by:

$$X = Ae^{-kd} \tag{11.3}$$

where d is distance into the shielding in metres, and A is the intensity at $d = 0$ (i.e., at the surface) in Watts, and:

$$k = \log_e(2)/d_{0.5} \tag{11.4}$$

where $d_{0.5}$ is the distance in m required to produce a 50% drop in intensity.

11.8 CONCLUSIONS

- Most radiation is harmless (e.g., radio waves). Ionising radiation should be contained with shielding.
- Ionising radiation is of several types with quite different properties.
- There are many different sources of radiation that may be encountered in construction.
- Radioactive sources decay with a constant half-life.
- The effect of radiation on the body depends on the energy and factors for the type of radiation and the type of dose.
- Increasing the thickness of radiation shielding by a given distance will decrease the transmitted energy by a fixed proportion.

TUTORIAL QUESTIONS

Note that these calculations could be simplified by using other units for distance and time (e.g., years). This is, however, not recommended. Base units should be used for all calculations.

1. A concrete wall is being irradiated with 40 W of radiation. If the power is reduced to 20 W at a depth of 150 mm what is the power at a depth of 200 mm?

   Solution:

   In equation (11.3) $A = 40$ W (the intensity at the surface)
   The power reduces by half in 150 mm thus $d_{0.5} = 0.15$ m
   $$k = \log_e(2)/0.15 = 4.62 \tag{11.4}$$
   $$X = 40\, e^{(-4.62 \times 0.2)} = 15.88 \text{ W} \tag{11.3}$$

2. A concrete wall is being irradiated with 50 W of radiation. If the power is reduced to 20 W at a depth of 200 mm, what is the power at a depth of 300 mm?

   Solution:

   $$20 = 50 \times e^{-(k \times 0.2)} \tag{11.3}$$
   Thus, $\log_e(20/50) = -k \times 0.2$
   And $k = 4.58$
   $A = 50$ W
   $$X = 50 \times e^{-(k \times 0.3)} = 12.65 \text{ W} \tag{11.3}$$

3. A source is emitting 50 W of radiation. If the power is reduced to 35 W after 2 months, what is it after 6 months?

   Solution:

   2 months $= 5.26 \times 10^6$ s
   $$35 = 50 \times e^{(-k \times 5.26E6)} \tag{11.1}$$
   Thus $k = 6.78 \times 10^{-8}$
   6 months $= 1.58 \times 10^7$ s
   $$X = 50 \times e^{(-6.78E-8 \times 1.58E7)} = 17 \text{ W} \tag{11.1}$$

4. A source is emitting radiation with an initial power of 10 W and a half-life of 7 years. What is its power after 20 years?

Solution:

7 years $= 2.21 \times 10^8$ s

$k = \log_e(2)/2.21 \times 10^8 = 3.14 \times 10^{-9}$ (11.2)

20 years $= 6.31 \times 10^8$ s

$A = 10$ W

$X = 10 \times e^{(-3.14E-9 \times 6.31E8)} = 1.38$ W   (11.1)

NOTATION

A Initial radiation intensity (W)
D Radiation dose (Gray = J/kg)
d Depth (m)
$d_{0.5}$ Distance required to produce a 50% drop in radiation intensity (m)
E Radiation dose (Sievert)
e Mathematical constant = 2.718
k Radiation decay constant
N Modification factor for radiation dose
Q Quality factor for radiation dose
t Time (s)
$T_{0.5}$ Radiation half-life (s)
X Radiation intensity (W)

VARIABILITY AND STATISTICS

CHAPTER OUTLINE

12.1 INTRODUCTION

Whenever observations are made of material properties, they will form statistical distributions. This occurs because the readings are not absolutely precise. It is, therefore, apparent that in a sufficiently large sample, there will always be one or more results below any given acceptance level. The required properties must be specified for any construction project, but it may be seen that specifying zero failures at any acceptance level is not practical. The only practical solution is to specify a percentage defect rate. That is a percentage of failures that is considered to be acceptable.

The results that are obtained from most measurements will form a normal distribution. In this chapter, this distribution is discussed, and there are two questions that must be answered:

1. How far above the failure level does the average measurement have to be, in order to get a given percentage of defects? This is answered in Section 12.3.
2. How likely is it to observe a relatively large number of failures in a small sample taken from a parent population with a low percentage of defects? This is answered in Section 12.4.

The discussion is relevant to any material that is being tested in any units, and is illustrated by looking at the example of testing concrete on site using MKS units. The procedure for testing cubes is discussed in Chapter 22, and gives a result in MPa for each cube that is tested.

12.2 SAMPLING

Figure 12.1 shows a histogram for a set of results that might be obtained if 10 concrete cubes from the same mix were tested for strength (it was actually generated in a spreadsheet using the random number generator to simulate 10 people tossing a coin 100 times).

Figure 12.2 shows the effect of increasing the number of samples gives a more regular distribution that approximates to the curve in Figure 12.3.

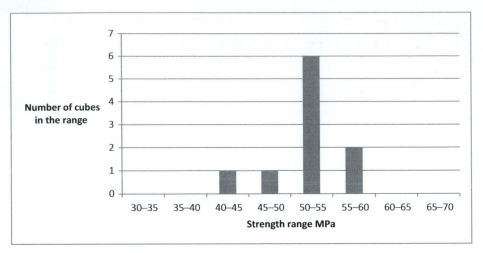

FIGURE 12.1 Strength Results for 10 Cubes

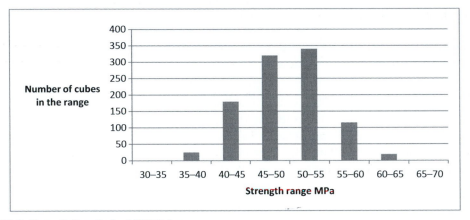

FIGURE 12.2 Strength Results for 1000 Cubes

This is called a normal distribution, and is observed almost every time a material property is measured. It is not absolutely accurate, but is close enough for analysis. The true distribution for cubes is normally skewed slightly to the right, because it is more likely to get a very strong outlying result than a very weak one.

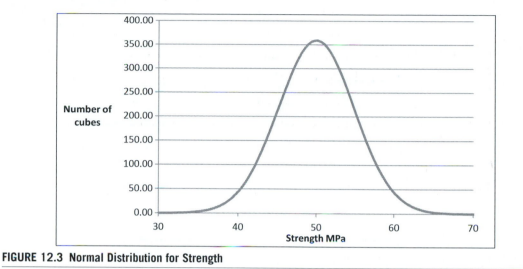

FIGURE 12.3 Normal Distribution for Strength

12.3 DISTRIBUTIONS

In order to calculate how far above the failure level the average measurement has to be in order to get a given percentage of defects, it is necessary to know the standard deviation σ of the measurements. The standard deviation is defined as:

$$\sigma = \sqrt{\left(\frac{\sum(x_i - m)^2}{n}\right)}$$ (12.1)

where x_i are the observations, n is the number of observations, $\sum x_i$ denotes the sum of all the values, and m is the mean (average) defined from:

$$m = \frac{\sum x_i}{n}$$ (12.2)

The standard deviation is a measure of the spread of the results. Two other statistics are derived from it:

- The standard error is the standard deviation divided by the mean
- The variance is σ^2

Figure 12.4 shows two distributions with different standard deviations. Both have a mean strength of 50 and the area under the graphs, that corresponds to the total number of samples tested, is the same. The high standard deviation would occur where there is poor quality control giving the wide variation between samples.

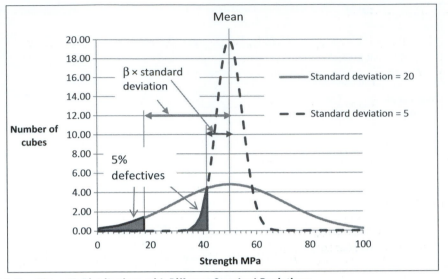

FIGURE 12.4 Two Strength Distributions with Different Standard Deviations

The two shaded areas in Figure 12.4 represent 5% of the area under each curve so 5% of a large number of samples would lie within this region. The permitted percentage failure level is given by equation (12.3).

$$\text{mean} = \text{permitted percentage failure level} + \beta \times \text{standard deviation.} \tag{12.3}$$

Where β is the "reliability index" and is obtained from Table 12.1 and the term "$\beta \times$ standard deviation" is referred to as the margin.

For 5% defectives β is 1.64. Thus, for the data set with a standard deviation of 5, the defective region is below a strength of $50 - 5 \times 1.64 = 42$ MPa. For the standard deviation of 20, it is below 17 MPa.

Table 12.1 Reliability Index Values for Various Values of Percentage Failures

Percentage Failure Permitted	β Value
16	1.00
10	1.28
5	1.64
2.5	1.96
2	2.05
1	2.33

12.4 PROBABILITY

Calculating how likely it is to observe a relatively large number of failures in a small sample taken from a parent population with a low percentage of defects is a question of probability.

Calculations of probability are based on equations (12.4) and (12.5):

The probability of multiple events all occurring
= the probabilities of the individual events multiplied together.   (12.4)

The probability of any one of several alternative events occurring
= the probabilities of the individual events added together.   (12.5)

If the probability of each outcome is the same, equation (12.5) means that the probability is multiplied by the number of ways the outcome can occur.

If just three cubes are tested with a probability of failure of 5% (= 0.05) and a probability of passing of 95% (= 0.95) for each cube the possible outcomes are given in Table 12.2.

The probability of an outcome in Table 12.2 is calculated as the probabilities for each sequence calculated from equation (12.4) and then the alternatives are added using equation (12.5). Note, that the probabilities in the table add up to one. It is certain that one of the outcomes will occur.

Now consider a site on which three cubes are tested every day. For one cube to fail, there is a probability of 0.1354 or 13% (from Table 12.2) so this is not considered remarkable.

However, the probability of this happening on each of five days in a row is:

$$0.1354^5 = 0.000045 \text{ so this is unlikely and a cause for concern.}$$

This is different from the probability of having one failure on just one of the five days. The probability of this is calculated as:

Probability of getting one failure on one day × probability of not getting this on four
days × number of days on which the failure could occur (i.e., the number of possible sequences).
= $0.1354 \times (1 - 0.1354)^4 \times 5 = 0.37$ or 37% which is probable
For more than one cube to fail on any given day, the probability may be calculated from values in Table 12.2 and is $0.0071 + 0.0001 = 0.0072$ or 0.7%, so this is unlikely.

Table 12.2 Outcomes for Three Cubes P = Pass, F = Fail

Outcome	Sequence	Probability
All three pass	P P P	$0.95^3 = 0.8574$
Two pass, one fail	P P F P F P F P P	$0.95^2 \times 0.05 \times 3 = 0.1354$
One pass, two fail	F F P F P F P F F	$0.05^2 \times 0.95 \times 3 = 0.0071$
All three fail	F F F	$0.05^3 = 0.0001$

However, if the site continues for 100 days of working, the probability of this happening once is:

$0.0072 \times (1 - 0.0072)^{99} \times 100 = 0.353$ or 35% that is also probable and should not necessarily be a cause for concern.

12.5 CORRELATIONS

Consider the example in Figure 12.5.

This shows a plot of exam mark against attendance at lectures for a construction materials class. What we want to know is whether it is necessary to attend the lectures to pass the exam. What statistics can tell us is what the probability is that this is a purely random distribution with no "correlation" at all. To calculate this, we work out a statistic called r^2 (the coefficient of determination). The computer package tells us that $r^2 = 0.52$ and therefore $r = 0.72$. Table 12.3 tells us that for 80 observations the 1% significance value of r is 0.283, that is if r was 0.283, there would be a 1% chance of a distribution showing this amount of correlation occurring by chance. r is well above this value, so the chance if this being a random distribution is even less, that is, the correlation is proven. Note that we have not answered our original question; it could just be that the "better" students attend the lectures and also do well in the exams, we have not proved cause and effect.

Exactly the same situation applies in the study of materials. Almost all of the properties of concrete (e.g., strength, permeability, frost resistance) correlate, but this does not prove that they affect each other. The strength is a good indicator of durability, and has been used for a long time as the only test that is routinely used to assess the quality of hardened concrete. However, the relationship between strength and durability is based on experimentally observed correlations. High strength does not cause high durability. A low water to cement ratio causes generally both high strength and

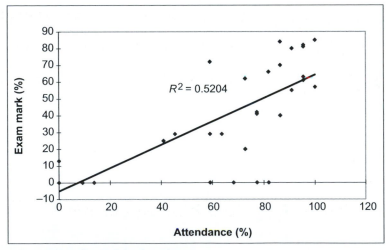

FIGURE 12.5 Exam Mark Versus Percentage Attendance

Table 12.3 Some Values of *r* for Two Variables

No. of Data Points	*r* for 5% Significance	*r* for 1% Significance
5	0.754	0.874
10	0.576	0.708
20	0.423	0.537
30	0.349	0.449
40	0.304	0.393
50	0.273	0.354
60	0.250	0.325
70	0.232	0.302
80	0.217	0.283
90	0.205	0.267
100	0.195	0.254

high durability, but there are many exceptions. Section 14.2 discusses the lack of effective tests to measure durability directly.

12.6 CONCLUSIONS

- It is not practical to specify zero failure for a material test, so it is necessary to use statistics for specifications.
- Most experimental results will form a normal distribution.
- The reliability index is measured in standard deviations from the mean and gives the number of failures that are to be accepted in a batch.
- The probability of multiple events all occurring = the probabilities of the individual events multiplied together.
- The probability of any one of several alternative events occurring = the probabilities of the individual events added together.
- Correlations do not prove cause and effect.

TUTORIAL QUESTIONS

1. Concrete cubes are being tested and the mean strength is 40 MPa and the standard deviation is 5 MPa. What is the strength below which 5% of the strengths will lie?

Solution:

$$40 - (1.64 \times 5) = 31.8 \text{ MPa} \tag{12.3}$$

2. Railway sleepers (railroad ties) are required to have a length of 3 m with 5% defects. If the standard deviation of the observed lengths is 5 mm, what average length is required?

   Solution:

   $3000 + 1.65 \times 5 = 3008.2$ mm   (12.3)

3. If 10 concrete cubes are tested from a concrete known to have a 5% failure level, what is the probability of one failure?

   Solution:

   $0.05 \times 0.95^9 \times 10 = 0.315$ (12.5)

4. Steel beams are manufactured with a failure level of 2%.
   a. What is the expected number of failures in a sample of 20 beams?
   b. What is the probability of one failure in the sample?

   Solution:

   Expected number $= 20 \times 0.02 = 0.4$
   Probability $= 0.02 \times 0.98^{19} \times 20 = 0.272$ (12.5)

5. Concrete is delivered to site with a characteristic strength of 25 MPa and a percentage defect rate of 5%. If 6 cubes are made, what is the probability of:
   a. No failures
   b. 1 failure
   c. If six cubes are made each day, what is the probability of getting one failure per day on each of three consecutive days?
   d. If the observed standard deviation of the test results is 5 MPa, what is the mean strength?

   Solution:

   a. $0.95^6 = 0.735$ (12.4)
   b. $0.95^5 \times 0.05 \times 6 = 0.232$ (12.5)
   c. $0.232^3 = 0.012$ (12.4)
   d. $25 + (5 \times 1.64) = 33.2$ MPa   (12.3)

6. **a.** Explain why the strength of materials such as bricks and concrete must be specified in terms of a statistical failure rate.
   b. Bricks are tested for strength. If the average is 15 MPa and the standard deviation is 3 MPa, what is the strength below which 5% would be expected to lie?
   c. If five of the bricks are tested, what is the probability of one strength being below the 5% level?
   d. If three further sets of five are tested, what is the probability of all three sets having one each below the 5% level?

Solution:

a. Because the strengths will form a normal distribution that is non-zero at all strengths.

b. $15 - (1.64 \times 3) = 10.08$ (12.3)

c. $0.95^4 \times 0.05 \times 5 = 0.203$ (12.5)

d. $0.203^3 = 0.008$ (12.4)

7. Bricks are tested for strength. If the average is 3 ksi and the standard deviation is 0.4 ksi, what is the strength below which 5% would be expected to lie?

Solution:

$3 - (1.64 \times 0.4) = 2.34$ ksi (12.3)

NOTATION

m Mean
n Number of samples
x_i Observations
β Reliability index
σ Standard deviation
σ^2 Variance

USE OF TEST RESULTS

CHAPTER OUTLINE

13.1 INTRODUCTION

The purpose of this chapter is to show how the theory from Chapter 12 can be put to practical use. Two examples are given. They both relate to concrete in MKS units but the methods would be fully applicable to other materials and units. The first relates to testing concrete samples on site (see Section 22.5 for the test methods). The second relates to designing structures to achieve durability. Both examples are about deciding when to take additional measures (that involve additional cost) to ensure that the structure performs adequately during its design life. When testing concrete on site, a decision must be made about how many cube failures can be accepted before deciding to improve the mix; probably by adding more cement. The second example is presented in Section 13.6. When designing a structure, the durability can be calculated from a number of parameters such as permeability, none of which are accurately known. A very uneconomic design will be produced if the worst case is assumed for all of them. Students may wish to refer to Chapter 17 that introduces the terms used for concrete in the examples.

13.2 SOURCES OF VARIATIONS IN CONCRETE STRENGTH TEST RESULTS

When concrete samples are tested for strength the variations come from a large number of different sources, for example,

Daily variations in concrete supplied to site:

- Changes in cement
- Changes in aggregate
- Changes in batching control
- Changes in temperature

Variations in concrete between successive batches (loads):

- Changes in water content
- Changes in haul times

Sampling variations within a load:

- Changes in aggregate content between two samples taken from the same load.
- Operator error, for example, effect of sampling from the end of a load.

Testing variations:

- Variations in the position of the aggregate in the cubes.
- Operator error, for example, changes in loading rate, or dirt on the platens.

Each of these variations will have a variance associated with it (see Section 12.3). These are additive, that is, the total variance is the sum of the variances from the different sources. Note that standard deviations are not additive.

13.3 MAKING DECISIONS ABOUT FAILING TEST RESULTS

Consider the set of 30 cube results given in column 2 of Table 13.1 that represent 10 sets of three cubes taken from 10 different loads (typically, they might be from 10 days of concrete placing). The target mean for these was 50 and the target standard deviation was 5. The simplest form of control chart is shown in Figure 13.1 that simply shows individual cube results. The dashed lines are at 1.64 standard deviations from the mean, and thus 95% of the samples would be expected to lie between them (see Table 12.1). Only two out of 30 samples lie outside so the control might be assumed to be satisfactory. The weakness of this chart is that it cannot show when a decision needs to be made.

The better method is CUSUM, in which the target mean strength is subtracted from each of observed strengths to give a positive or negative difference. These differences are then added to give a Cumulative SUM (see columns 3 and 4 in Table 13.1) that is plotted in Figure 13.2. This shows that, although the average of all 30 was close to target at 49.9, there was a series of poor results from cubes 15–25. The dashed line shows a "mask" that can be moved on screen to indicate when change is needed.

13.4 IDENTIFYING THE SOURCE OF THE PROBLEM

While the basic CUSUM plot can indicate when action is needed; it cannot be used to identify the source of the problem. The problem could come from two distinct areas. It could be from the concrete itself (the first two lists in Section 13.2) or from the testing (the second two lists). Problems with the testing are likely to increase the spread (range) of results between the three cubes in a set that are made from the same batch of concrete while problems with the concrete itself would not. This spread can be measured as a standard deviation of each set of three results. In Table 13.1, column 5 shows the standard deviation of each set, and columns 6 and 7 show the CUSUM analysis of them that is plotted in Figure 13.3. This does not show any particular trends so it is likely that the concrete itself was at fault.

Table 13.1 Analysis of a Set of Cube Results in MPa						
Column	**2**	**3**	**4**	**5**	**6**	**7**
Cube Number	**Strength**	**Strength - Target Mean**	**CUSUM**	**Range (SD of Set)**	**Range - Mean**	**CUSUM Range**
1	49	−1	−1	4.73	1.07	1.07
2	51	1	0			
3	42	−8	−8			
4	47	−3	−11	5.57	1.91	2.98
5	58	8	−3			
6	54	4	1			
7	46	−4	−3	3.06	−0.60	2.38
8	48	−2	−5			
9	52	2	−3			
10	49	−1	−4	1.15	−2.50	−0.12
11	47	−3	−7			
12	47	−3	−10			
13	55	5	−5	5.13	1.48	1.36
14	58	8	3			
15	48	−2	1			
16	51	1	2	2.52	−1.14	0.22
17	48	−2	0			
18	46	−4	−4			
19	46	−4	−8	1.15	−2.50	−2.29
20	48	−2	−10			
21	46	−4	−14			
22	49	−1	−15	4.73	1.07	−1.22
23	47	−3	−18			
24	40	−10	−28			
25	47	−3	−31	7.00	3.34	2.13
26	60	10	−21			
27	58	8	−13			
28	53	3	−10	1.53	−2.13	0.00
29	55	5	−5			
30	52	2	−3			
Mean	49.90			3.66		
Standard deviation	4.80			2.05		

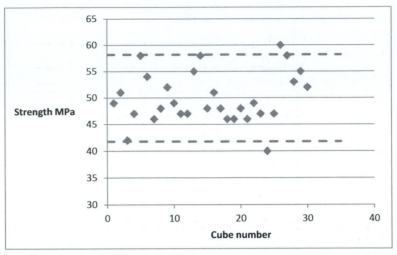

FIGURE 13.1 **A Basic Control Chart**

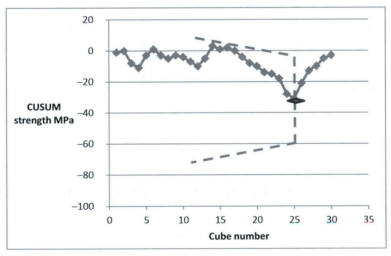

FIGURE 13.2 **A CUSUM Chart**

13.5 MULTIVARIATE ANALYSIS

CUSUM methods only work well when a large number of tests on a single type of concrete are being analysed. If there are many different mixes being made, there will be insufficient data to develop a different cusum plot for each individual mix. In order to overcome this, the strengths can be adjusted up or down using historic data on the different mix types to fit them all to a single model. Adjusting results from mixes with cement replacements or admixtures to match those of plain cement mixes is, however, very unreliable.

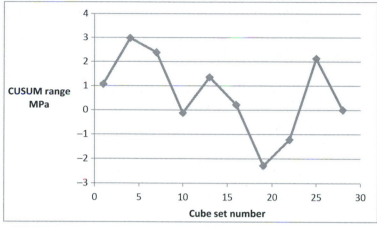

FIGURE 13.3 Plot of CUSUM Range

More powerful systems require a lot more data collection to understand all the factors that could affect the strength. These are called predictor variables and could be as follows:

- Water to cement ratio
- Cement content
- Cement type
- Cement replacement percentage
- Admixture percentage
- Aggregate content
- Aggregate grading
- Slump
- Temperature
- Haul time

A model can be created based on past data to predict the strength. A simple form of this could take the form of:

$$\text{Strength} = (x_1 \times \text{predictor 1}) + (x_2 \times \text{predictor 2}) + (x_3 \times \text{predictor 3})......\text{etc}........ \qquad (13.1)$$

Where x_1, x_2, x_3, are constants obtained from a process called multiple regression on past test results. Further terms such as $(\text{predictor 1})^2$ are included to make the model accurate.

All test results are entered into the model and action is needed if the strength deviates from the predicted value. This will indicate that an unexpected additional factor is affecting the strength, and is a cause for concern on site.

13.6 DESIGNING FOR DURABILITY

When designing major structures, clients increasingly ask designers to carry out a durability assessment rather than just relying on the provisions in local codes and standards. For a concrete structure, this could be a calculation of the depth to which salt (chloride) will penetrate during the design life;

because if it reaches the cover depth for the steel reinforcement, it will cause corrosion (this is discussed in Section 25.3). The durability may be increased by using a higher grade of concrete or increasing the depth of cover to the steel, but both of these will add to the cost.

This calculation will require values for the following variables (see Section 25.9 for the method of calculation):

1. The cover depth
2. The permeability
3. The rate at which the permeability reduces with time
4. The coefficient of diffusion
5. The rate at which the coefficient of diffusion reduces with time
6. The salt concentration in the local environment
7. The critical chloride concentration necessary for corrosion.

None of these will be accurately known. However, it will be possible to estimate a mean values and a standard deviation for each of them and from these a 5% defectives level. What the designer needs to know is the values of the permeability and diffusion coefficient that must be specified in order to achieve the necessary durability with 95% certainty. The penetration depth could be calculated with all six variables at their 5% defectives level, but this would only have a probability of happening of 0.05^6 ($= 1.5 \times 10^{-8}$) and would give highly uneconomic requirements.

The problem is solved by calculating a number (typically, three) different possible values for each variable.

Figure 13.4 shows the normal distribution divided into three equal areas that thus have equal probability. The boundaries lie 0.43 standard deviations either side of the mean. The arrows show the average values for each area. These lie at the mean and 1.1 standard deviations to either side and are representative of each of the three different outcomes.

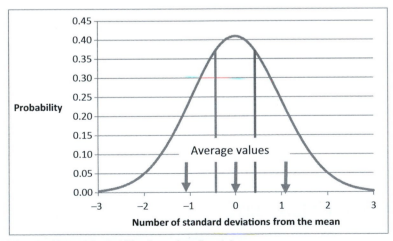

FIGURE 13.4 Dividing the Normal Probability Curve into Equal Areas

If three values for all six variables are obtained there are 3^6 (= 729) different combinations of them that could occur with equal probability. If the chloride penetration depth is calculated for all of these and the mean and standard deviation calculated for them a value of β can be obtained as follows from equation (12.3):

$$\text{Design value} = \text{mean} + \beta \times \text{standard deviation} \tag{13.2}$$

Thus, if the mean chloride penetration depth is 40 mm and the standard deviation is 10 mm; a cover depth of 50 mm will give a β value of 1.

If a 5% probability of failure is considered appropriate then Table 12.1 shows that β should be 1.64. For critical elements β must be 3 or more to reflect the consequences of failure.

13.7 CONCLUSIONS

- CUSUM is a method for analysing trends in test results, and deciding when to take action.
- CUSUM range can be used to identify the cause of data trends.
- Multivariate analysis is a more powerful method, but requires more data collection.
- When designing structures for long design lives, a statistical approach may be used to calculate a reliability index.

TUTORIAL QUESTION

1. Prepare a CUSUM chart from the following data:

Cube Number	Strength
1	55
2	43
3	59
4	51
5	42
6	44
7	51
8	46
9	48
10	54
11	48
12	42
13	44
14	58
15	63

Solution:

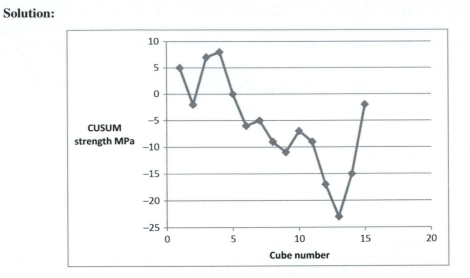

NOTATION

β Reliability index

SPECIFICATIONS AND STANDARDS

CHAPTER OUTLINE

14.1 INTRODUCTION

All materials and services that are used in a construction project will be specified by the client, in order to achieve the quality they require. This may be done either by describing them in detail in the specification, or requiring compliance with either standards or building codes. This chapter considers the different types of specifications that are used for materials, and the quality assurance (QA) processes that may be used to achieve compliance with them.

All standards and building codes are available for download from the internet, but a fee is normally charged that goes towards the cost of preparing them.

Standards also include design codes that are central to the study of structures, but are not considered here.

The example of ready-mixed concrete is used in this chapter to illustrate the use of specifications and standards. The use of concrete on site is introduced in Chapter 17.

14.2 SPECIFICATIONS

14.2.1 THE DESIGNER, SUPPLIER, AND CONTRACTOR

When a project is designed, the designer must decide what properties are required for the concrete. The strength, and possibly other properties that influence the durability, need to be specified. The designer will then write a specification and send it out to potential contractors with the rest of the design for tendering (alternatively, the client or a management contractor may do this). The contractors will then generally send out the concrete specification to local ready-mix producers to get them to price it. When the contract is awarded to a contractor, they will, in turn, award a supply contract to a ready-mix company.

For the purposes of the supply of concrete, there are, therefore, three parties involved:

- the *designer* (i.e., the *specifier* acting for the client, i.e., the *purchaser*),
- the *supplier* (i.e., the *producer* - the ready-mix company),
- the *contractor*.

Acceptance of concrete is normally based on compressive strength results for cube or cylinder samples taken before placing, so, provided the cubes are made correctly, and the concrete is placed and cured as required by the specification, the contractor can pass all of his responsibility on to the supplier or the specifier. If the cubes or cylinders fail, the contractor can pass responsibility to the supplier for not meeting the specification. If they pass, but the concrete has some other problem, such as cracking or poor durability, the contractor can pass responsibility to the specifier for an inadequate specification.

This may change, if effective tests to measure the potential durability of *in situ* concrete are developed. Some existing tests for this are described in Section 27.4, but none of these are regarded as sufficiently accurate for widespread use. If *in situ* durability tests are adopted, the contractor will be given far greater responsibility for the quality of the concrete.

14.2.2 TYPES OF SPECIFICATIONS

Specifications may be *performance specifications* or *method specifications*. A performance specification requires a particular performance. In the case of concrete, this is usually a 28 day strength because, as noted previously, there are not yet any reliable *in situ* tests for durability. A method specification stipulates a particular method for production, for example, the mix design. The ultimate performance is thus the responsibility of the specifier, provided the producer has adequately adhered to the method. In the United Kingdom, most concrete is specified by performance (strength). In many other countries, concrete is often specified by method.

Performance specifications are:

- simpler to write,
- more difficult to price,
- simpler to supervise, and
- more difficult to comply with than method specifications.

14.3 STANDARDS

14.3.1 THE NEED FOR STANDARDS

Standards are standard specifications that have been developed so that every client does not have to write their own specification for each component of a structure. They cover the test methods used to measure the quality, and also specify the tests needed for each component, as well as the test results required. They are written by committees with representatives of the relevant producers and clients on them, as well as, academics in the field, and are often the subject of lengthy negotiation. A producer can gain significant commercial advantage if a standard is set, that they can comply with, but a competitor cannot.

14.3.2 STANDARDS ORGANISATIONS

Most countries issue their own standards. Some major ones are:

- British Standards (BS)
- American Society for Testing and Materials (ASTM)
- Deutsches Institut für Normung (DIN) standards are German.

European Standards (EuroNorms or EN's) which are authorised by the European Community to replace national standards from European countries. Before an EN is published, a draft European standard (ENV) is usually published. Some European standards have national annexes that are appendices to the standard that specify acceptable regional variations that allow for different climate, and construction practice.

Agrément Certificates are given by the British Board of Agrément, or European equivalents, for products for which there is no appropriate national standard. If a national/international standard is produced, the agrément certificate is not renewed.

ISO standards are issued by the International Standards Organisation, and are intended for use worldwide. An example is the slump test for concrete that is covered by an ISO, so the national standards will generally refer to it, rather than providing a detailed description.

14.3.3 THE BENEFITS OF INTERNATIONAL STANDARDS

There are clear benefits to the development of international or regional (such as European) standards, because they help suppliers reduce cost by being able to market the same product in different places, and help designers specify projects in other countries. However, the process of developing these standards is slowed down, as countries seek to defend their economic or cultural interests.

14.4 BUILDING CODES

A building code, or building control, is a set of rules that specify the minimum acceptable standards for buildings and other structures. These are established by countries or regions, and take account of regional variations, such as climate and seismic risk. They will cover a number of issues, such as safety (including fire hazards), and environmental concerns, particularly, the insulation in cold climates that will determine the running cost, and carbon footprint of the building.

Building codes will frequently refer to standards, and require particular performance based on standard tests. Unlike standards, they are generally legal requirements in the areas where they apply.

14.5 REPEATABILITY AND REPRODUCIBILITY

When a particular compliance value is set for a test results, it is necessary to know how variable the results are likely to be. Also, when several tests are carried out, it is important to know whether differing values mean that the tests are being carried out incorrectly. Standards will, therefore, give two indicators of accuracy for tests:

- *Repeatability* is a measure of the expected variation, if a test is carried out several times in succession by the same person, in the same laboratory.
- *Reproducibility* is the measure of the expected variation, if the same material is tested in a number of different laboratories.

14.6 QUALITY ASSURANCE
14.6.1 DEFINITION OF QUALITY ASSURANCE

QA is a management system that aims to ensure compliance with standards, codes, specifications, and general good practice. It is defined as a "management process designed to inspire confidence in the product or process."

The chief features of a QA system are:

1. The existence of a quality manager not involved with production or marketing.
2. A comprehensive system of procedures for every process involved (quality manual).
3. Detailed records of all inspections.
4. Effective training arrangements for all personnel.
5. A formal procedure for identifying and rectifying substandard goods or operations.

Some definitions of QA include the whole quality process; however it is generally considered that it is the management process, while quality control is the actual testing process that it manages.

14.6.2 CERTIFYING ORGANISATIONS

Manufacturers may choose to have their processes inspected by a certifying organisation that is licensed by the standards authority. This inspection will include both the quality of the product being produced at the time of the visit, and the QA system in place to maintain it. If the process passes the inspection, the products are licensed to display a certification mark to confirm the quality. Once a license is issued, licensees are regularly audited, and are subject to surveillance visits to ensure continuing compliance. If a product does not have a certification mark, the manufacturers may simply claim that they have tested it and it meets the standard. The purchaser will then have less confidence in the quality, but it will probably still comply.

14.6.3 **THE EUROPEAN CE MARK**

In Europe, the Conformité Européenne (CE) mark is slightly different, in that construction products are required by law to carry the CE mark, but, in its simplest form, it can be a self-declaration of compliance by a manufacturer with no third party certification. The CE marking is focused on safety-critical aspects.

14.6.4 **QUALITY CULTURE**

It must be observed that, while specifications and QA, are a key part of the quality process, there are many other factors involved. In order to achieve quality, a strong quality culture and management is needed on a construction site, and over-reliance on procedures and forms may be counter-productive. No system can give an absolute assurance of quality.

14.7 **CONCLUSIONS**

- The designer, the supplier, and the contractor all have different responsibilities in construction.
- Standards are required to avoid the need for a full specification for each component of a structure.
- Standards are issued by most countries and several international organisations.
- International standards are beneficial for international trade.
- QA is a management system to ensure compliance with quality standards.
- A strong quality culture is the key to quality construction.

REPORTING RESULTS

15.1 INTRODUCTION

This chapter describes how to prepare a report. This may be a report of a laboratory exercise or it may be a paper for publication or even a book, the principles remain the same. Chapter 16 discusses the structure of a research programme, and the report proceeding from it, and also the ways in which results are published. This chapter focuses on the detail of how work should be presented.

The key point about a report is that it should communicate with the reader. It is particularly important to check diagrams and graphs, to make sure that when somebody looks at them, they can understand what you are trying to tell them.

15.2 GRAPHS

15.2.1 THE PURPOSE OF GRAPHS

Graphs are generally the most important part of a materials report.

The graphs in this section have been drawn with Microsoft Excel. However, there are many other packages with similar capabilities. Computer spreadsheet packages are essential tools for this type of work, and should be used to plot most of your graphs. It is important, however, that you do not consider the computer responsible for the graph, and just let it plot with all of the default settings. If the package cannot do what you require, you must find another or plot by hand.

The reasons for drawing graphs are:

- To convey a clear and immediate impression (note, however, that this impression will be incorrect if the scales, are incorrect).
- To smooth data.
- To fit a curve so an empirical relationship can be found.
- To observe changes in data over a range (e.g., time).
- To compare experiment with theory.
- To compare results with reference sources.
- To reject data points. This must always be carefully justified. Points may only be rejected due to clear anomalies.

Two examples of graphs that communicate results (and are taken from papers) are given in Figs 15.1 and 15.2.

Figure 15.1 shows an example of how a graph may be used to compare experimental values with results calculated from an equation (note that log scales have been used – see below). This graph clearly shows that the permeability data from the two sources (gas and water permeability) does not give the same result, but the equation from the reference over-estimates the difference.

Figure 15.2 shows a simpler graph. The analysis for this graph was carried out using the histogram function in Excel. This graph clearly shows that there are two outlier results in the data, but, with that exception, it approximates to a normal distribution (see Chapter 12).

15.2.2 THE GRADIENT OF A GRAPH

Obtaining the gradient of points that lie in an approximate straight line

Figure 15.3 shows a basic analysis to get the gradient from two data sets. They both have the same gradient (= 2), but the upper line does not go through the origin, so it has a constant term (= 1). This gives the Y intercept of a graph, and is the value of Y at the point where the straight line through the points crosses the Y-axis. A nonzero intercept is often not relevant, but in some instances (e.g., viscosity measurements – see Section 8.2) it is significant.

Figure 15.4 shows a set of real data for the compression of a concrete sample. The modulus may be read off as 14.3 E10 or 14.3 GPa. Since this is a materials calculation (rather than structural analysis), the minus sign is ignored because the modulus is known to be positive. The negative slope has been created by the setting of the displacement transducer in the test. The constant term (that gives the intercept) is also an artefact of the transducer setting, and is ignored. In this case, the transducer was held in a clamp, and the constant simply depended on exactly how it was positioned in the clamp (see Fig. 3.9).

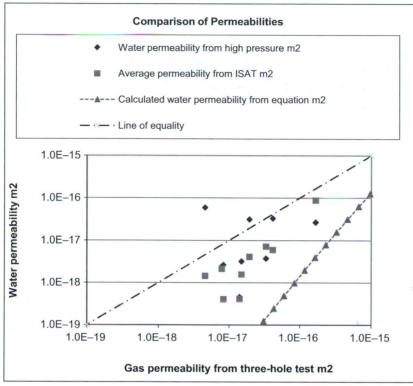

FIGURE 15.1 Comparison of Permeability Results

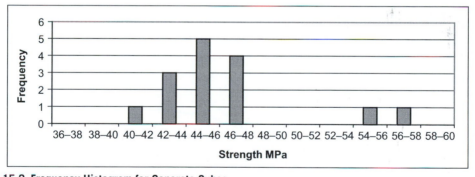

FIGURE 15.2 Frequency Histogram for Concrete Cubes

15.2.3 OBTAINING THE GRADIENT OF A CURVE

The gradient between each successive pair of points may be obtained as shown in Fig. 15.5 (students with a knowledge of calculus may recognise this as numerical differentiation). This obviously changes at different points on the curve.

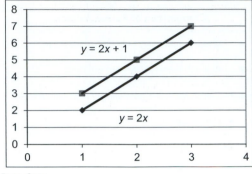

FIGURE 15.3 Line Fits to Two Data Sets

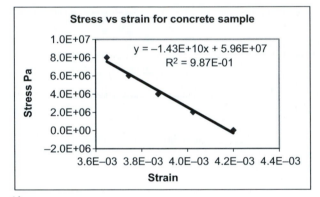

FIGURE 15.4 Regression Line

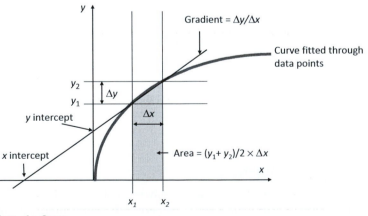

FIGURE 15.5 Gradient of a Curve

Figure 15.6 shows experimental results in which a gradient has been obtained for just part of the data. This can be done by placing a second series of selected data points on the same graph.

15.2.4 ERROR BARS

Typical error bars are shown in Fig. 15.7. These show the statistical confidence limits on the data. The figure shows the effect that this has on the confidence in the gradient. Error bars on a graph are a good sign that the author has considered the reliability of the data, but they are notorious for being calculated incorrectly. Normally, they should extend either side of the points by one standard deviation.

Note that the uncertainty in the gradient and intercept will depend on the grouping of the data, as well as, the size of the error bars. *Interpolations* within the range of the data are more accurate than *extrapolations* that go beyond the range.

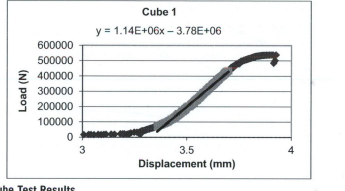

FIGURE 15.6 Typical Cube Test Results

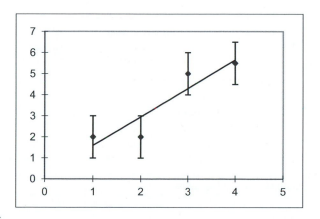

FIGURE 15.7 Error Bars

15.2.5 **LOG SCALES**

The effect of plotting to log scales may be seen in Fig. 15.8. Plotting to log scales has the effect that the points 1, 10, 100, 1000, etc. are equally spaced along the axis. Use a log scale if the points are spread over several orders of magnitude, for example, several between 0 and 10, and some between 1,000 and 10,000. Graphs may be log-log (with both axes on log scales) or log-lin (with just one axis on a log scale). The log function is available on all spreadsheets, and log scales may be

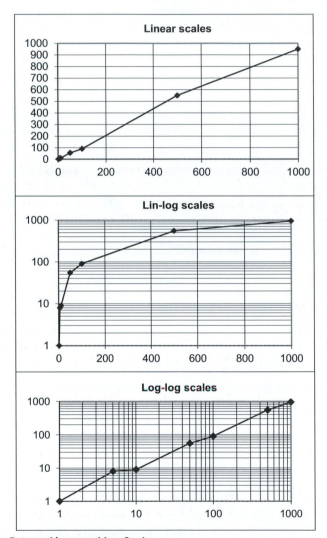

FIGURE 15.8 The Same Data on Linear and Log Scales

selected on any graph in Excel. Watch out for the interpretation of log-log graphs, almost any data looks fine on one.

Remember that the gradient from a log scale graph cannot be used without further calculation. If $y = ax^b$, then $\log(y) = \log(a) + b.\log(x)$, thus, if $\log(y)$ is plotted against $\log(x)$, the gradient is b and the intercept is $\log(a)$.(see Section 1.5)

The use of log scales is one example of *rectification* that is the use of an algebraic relationship to produce a straight line. For example, if something is known to depend on the square root of time, then it should be plotted against the square root of time. The advantages of rectification are:

- The slope and intercept may be obtained.
- Changes in behaviour are easier to see on lines, than curves.

It may be seen from Fig. 15.8 that, when the data is plotted on linear scales, most of the lower values cannot be seen.

15.3 REFERENCES

All reports should have cited references. Any fact that is stated, and does not come from your own experimental observations or calculations, should be referenced to the source from which it came. The "citation" in the main text refers the reader to the reference list at the end of the report. The reference list should contain sufficient information about each item for any reader to locate it, for example, if it is a book, it should state the publisher and the year of publication.

There are two different ways of citing references:

- The numerical method.
- The Harvard Method.

In the numerical method, each citation in the text is numbered, and the references are listed in number order at the end. For longer reports and many papers, the "Harvard" system is used. In this system, a name and date is given in the text, for example, "Neville (1981) p.250," and at the end the references are listed in alphabetical order, for example, "Neville, (1981), "Properties of Concrete," Pitman, London."

These two methods are used internationally in all publications, throughout the world, and new variations on them may look good to you, but they are simply incorrect.

When using references do not:

- List references "second hand," that is, uplifted from elsewhere, without being read to see what is actually in them.
- List any reference (e.g., a book) more than once in the same list.
- Quote anything directly from any source such as a book, paper, or website, without making very clear exactly what you have quoted, and stating the source (this is called plagiarism).
- Include a website address in your text as part of the citation. The website address should only be in the reference list.

15.4 HOW TO GET GOOD MARKS FOR YOUR MATERIALS LAB REPORTS

Or write a good paper

15.4.1 THE SUMMARY (MAY BE CALLED AN ABSTRACT OR SYNOPSIS)

- The summary should stand alone, and make sense when read without the rest of the document.
- Summarise main points from your report, including the objectives and conclusions.
- Write this after writing the rest of the report.

15.4.2 THE INTRODUCTION

- Briefly outline the purpose of the lab. This should be the scientific purpose, not the education.
- Include some background to the topic, and at least one reference.
- Introduce the other sections of your report.

15.4.3 THE EXPERIMENTAL METHOD

- State what was done (not who did it, or when they did it).
- If you have confirmed that the experiment was carried out using a procedure from a standard, you may reference the standard, rather than providing a full description or a diagram.
- Include diagrams of apparatus and samples, where necessary. These are far more informative than text.
- Do not present a list of apparatus, including items such as buckets.
- Do not write this as a set of instructions. It should be a record of what was done.
- Separate "method" that describes your detailed experiments, from "methodology" that describes the overall strategy for your experiments, and is probably not needed in a lab report.

15.4.4 THE RESULTS AND ANALYSIS

- Present your results in tables or graphs, but not both.
- Give a concise description of what analysis has been done, and how it has been done. Do not discuss the results, or draw conclusions from them, until the next section.
- Present the method, the equations, and the results. Do not write out all your calculations.
- Tables and graphs should all be inserted into the text.
- All tables and graphs must be numbered and have a title, and must be referred to (and explained) in the text.
- In graphs, generally put what you set (the *dependant variable*) on the *X*-axis (the *abscissa*), and what you measure (the *independent variable*) on the *Y*-axis (the *ordinate*) (except for stress-strain curves).
- Always put titles and scales on both axes, and check that the font size is adequate.
- Choose your scales carefully. Remember these can be set manually, do not rely on the default setting.
- Always check at least one point by hand, to make sure the data is plotted correctly.
- When you want an *X-Y* graph, make sure you select the correct type and don't end up plotting *Y* against reading number (if this common error occurs, it will be revealed by checking one data point).

- When plotting bar charts, all of the bars must be the same width.
- All graphs must show all the data points. If you plan to dismiss an "outlier," it must be shown and discussed.
- You may put up to eight graphs per page, but take care to ensure that the font size on them is still at least 10 point, and the spread of the data can be clearly seen.
- Gradients should be obtained from graphs by showing the equations for trend lines. Format these equations to display sufficient decimal places.
- Consider carefully (and discuss) whether your trend lines should go through the origin. This can be set manually.
- If the gradient is required from a selected part of the data (as in Fig. 15.6), when using Excel, the "paste special" button may be used to super-impose a second data set on the graph.

15.4.5 THE DISCUSSION

- A key part of your discussion will be to compare your results with those from references.
- Support your discussion with cited references.
- All quotes must be clearly marked, and the source given as a reference.
- Do not describe what can be seen on a graph or in a table. Discussion means making relevant points about it.

15.4.6 THE CONCLUSIONS

- Summarise the main points from the discussion.
- Do not introduce new material or ideas.
- This is often best presented as a list of about five key points.

15.4.7 THE REFERENCE LIST

- All references in your list must be cited in the text. Other general background material goes in the bibliography (but a bibliography now serves little purpose, because search engines can produce similar lists).
- Any standards that are cited in the text must be listed in the reference list, complete with their full titles.

15.4.8 WHAT NOT TO INCLUDE

- Many school exercises require an "evaluation" that usually starts "I think that the experiment went very well." This is not required at University level. If anything went wrong, it should be noted when the method is described, and the consequences should be evaluated in the discussion.
- Try to avoid using appendices. Nobody will read them.

15.4.9 THE PRESENTATION OF THE TEXT

- This is the part of the report that the reader will follow from beginning to end, so it must be written in full sentences, for example, "The results are shown in Table 3," or "the gradient from Figure 6 was used in equation (3) to calculate the modulus in Table 3," not "Results in Table 3."

- A report consists of four basic elements:
  - text
  - figures (these may be diagrams, graphs, photographs)
  - tables
  - equations

  There should be nothing else. Any "fragments," such as bits of text that do not form sentences, should be modified to be included in one of these four, or removed.
- Short, sharp sentences that only contain relevant facts should be used, for example, "The length and width were measured," not "When we had finished measuring the length, we started to measure the width."
- All figures, tables, and equations should be numbered and referred to in the text, in order to direct the reader to them, or there is no point in including them. Always refer to figures and tables by number. Say "the results are in Figure 1," not "the results are in the figure below." When printed out, it may be on the next page.
- All text must be in third person, that is, "the concrete was mixed," not "we mixed the concrete," or "I mixed the concrete." Say "it may be seen," rather than "you may see." Also, do not name people; say "the length was measured," not "the technician measured the length."
- All text must be in past tense, that is, "the concrete was mixed," not "the concrete is mixed," "the concrete will be mixed," or "mix the concrete."
- Reports should have a sub-heading approximately every 10–20 lines, to keep them structured. Never present a whole page with no headings. Numbering the sections (as in this book) is often a good idea.
- Use spelling and grammar checkers. If you are in the United Kingdom, the language must be set to English (UK). The default setting is normally English (US).
- Use plain English, for example, *use* not *utilise*, *utilisation* or *usage*. Long, complicated words do not impress when simple ones could be used.

15.5 HOW TO PUBLISH A PAPER ON MATERIALS
15.5.1 SELECTING YOUR JOURNAL

The selection of a journal or conference is discussed in Section 16.3, where the quality of different references is considered. Excellent data is often wasted at second-rate conferences. If you believe you have good results, send them to a good journal. You should also research the time to publication for your chosen journal. It is not unusual for journals to have a queue of over a year, for papers that have been fully reviewed, and accepted to appear in print. However, many journals are releasing papers online while they are in the queue, and this is counted as publication, as they can be searched and cited.

15.5.2 THE LITERATURE REVIEW

This will be required for a publication, and generally follows the introduction. It is important to clarify exactly what it is for, and give it focus. It has become very easy to obtain copies of considerable numbers of publications on most topics, but simply picking a random selection of them (e.g., those available for free download), and describing their contents is pointless. The review should only provide background information that is necessary for the paper, and introduce ideas, and sources that will be used in the discussion.

15.5.3 SUBMISSION AND RE-SUBMISSION

Papers that are sent to a journal will be assigned to an editor, who will appoint reviewers. The reviewers will then submit their comments back to the editor, who will check them through for any contradictions, and send them back to the author with a decision. The following points should be noted, if you get a paper back from a journal.

- If your paper is rejected, read the comments and correct it, but don't give up. All reviewers have different ideas, so just send it to another journal and try again.
- If the decision is "major corrections," this means that the paper will go back to the reviewers, but if it is "minor corrections," your corrections will generally just be checked by the editor.
- The reviewers are generally not paid for their review, so always start your response by thanking them. You may feel inclined to criticize their comments, but you should never do this.
- You must provide a detailed response in which you must list each comment that has been made, and say exactly what you have done to the paper in response. All comments require some changes to the text. Many authors provide detailed responses to the reviewer, and then do nothing to the paper. Your detailed responses must be added to the paper. Even if the reviewer's comment is completely incorrect, an extra line or two should be added in the text, to ensure that other readers don't make the same mistake.
- It is helpful to reviewers to highlight your changes to the paper in some way, to make them easy to find.

15.6 VERBAL PRESENTATION

Most research will be presented at progress meetings or conferences. Following a few simple guidelines can make this far more effective.

- All verbal presentations should be supported by a "PowerPoint" or similar presentation. Use this to remind you of what to say. Do not use notes or "prompt cards." Never read anything out.
- When presenting text, a single slide should not contain more than six lines, with six words in each line. If you stick to this rule, it will generally help avoid the common error of reading out the text on the slides.
- Graphs should follow the guidelines given in Section 15.2. Pay particular attention to font sizes for the scales and axis titles.
- Use plenty of photographs, if you have them. Unlike graphs, these are effective even when made very small to fit with text.
- Avoid all special effects, such as words "feeding in" from the side of the screen.
- Be aware that colours vary considerably on some projectors. Use strongly contrasting colours for text.

15.7 CONCLUSIONS

- The purpose of a report is to communicate with the reader.
- Graphs are often the most important part of a materials report. They should communicate very clearly.

- References must be presented using the numerical or Harvard methods.
- Each part of a report must contain specific elements. It is important to ensure that everything is presented in the correct place.
- In a good published paper, the literature review should be very focused to make it contribute to the rest of the paper.
- Never read out text in a verbal presentation.

TUTORIAL QUESTIONS

1. What is the velocity in the following graph?

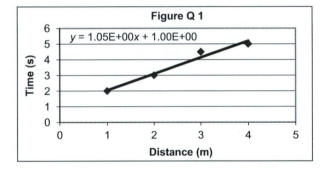

Solution:

Velocity = distance/time = 1/1.05 = 0.953 m/s

2. In the following graph, what is the (a) velocity, (b) the distance at time = 0 and (c) the time at distance = 0?

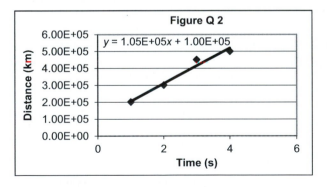

Solution:

a. Velocity = 1.05×10^5 km/s = 1.05×10^8 m/s
b. When $x = 0$, $y = 10^5$ km = 10^8 m,
c. $y = 1.05 \times 10^5 x + 10^5 = 0$.
　　 Thus $x = -0.95$ s

3. What is the modulus of elasticity in the following graph?

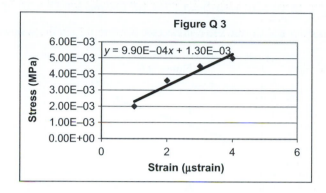

Figure Q 3

$y = 9.90E{-}04x + 1.30E{-}03$

Solution:

Modulus $= 9.9 \times 10^{-4}$ MPa/μstrain $= 9.9 \times 10^{8}$ Pa

4. What is the modulus of elasticity in the following graph?

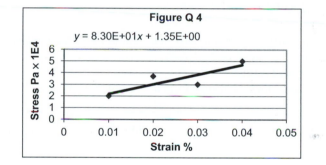

Figure Q 4

$y = 8.30E{+}01x + 1.35E{+}00$

Solution:

Modulus $= 8.3 \times 10 \times 10^{4}$ Pa/(strain/100) $= 8.3 \times 10^{7}$ Pa

5. What is the modulus of elasticity in the following graph?

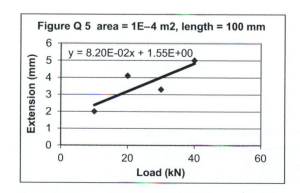

Figure Q 5 area = 1E−4 m2, length = 100 mm

$y = 8.20E{-}02x + 1.55E{+}00$

Solution:

Modulus $= [1/(8.2 \times 10^{-2})] \times (0.1/10^{-4}) = 1.22 \times 10^4$ kN/mm $= 1.22 \times 10^{10}$ Pa   (2.8)

6. What is (a) the viscosity and (b) the yield stress (see Section 8.2) in the following graph?

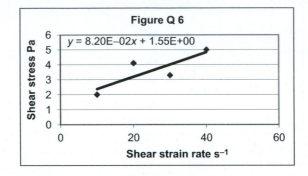

Solution:

a. Viscosity 8.2×10^{-2} Pa s   (8.1)

b. When $x = 0$, $y = 1.55$ Pa

7. When the ISAT test is carried out (for details, see Section 27.4.2), the results fit approximately to the relationship:

Flow $= A \times t^n$, where t is time and A and n are constants.

The following graph shows the results from an experiment. What are A and n?

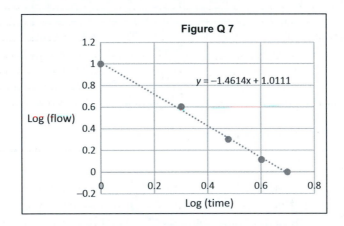

Solution:

Taking logs:

$\log(\text{flow}) = \log(A) - n \log(t)$

Thus, $n = 1.4614$

and

$\log(A) = 1.0111$, thus, $A = 10.256$

TESTING CONSTRUCTION MATERIALS

16

CHAPTER OUTLINE

16.1 INTRODUCTION

It is essential that all construction professionals should have a clear understanding of how to initiate research into material properties, and locate, and use reports on them. The mistakes made with high alumina cement and calcium chloride admixtures (see Chapters 18 and 24) were basically caused by a lack of understanding in this area. The research had been reported, but reference was not made to it, and this resulted in the failure of many structures. The details of how to write reports are considered in Chapter 15. This chapter discusses the components of an experimental programme and, in particular, the literature search which should be done at the start of it.

In order to illustrate the ideas, a typical programme investigating a new secondary (waste) material, for use as a replacement for some of the cement in concrete, is considered. A large number of papers have been published on this topic. There are obvious economic and environmental benefits in replacing cement with a secondary material (see Chapter 18), and avoiding the need to dispose of it elsewhere.

A key concept for both the experimental programme, and the publication, is reproducibility. Research is of little value if nobody is able to get the same results in other labs. Thus, the experiments must include sufficient analysis of the materials, and the publication must include a full description of the experiments.

16.2 HOW TO FIND REFERENCES

A review of previous work is the starting point for any research programme.

Publishing research is an enormous industry that is essential for technical progress. The conventions for publication are similar throughout the world. Publications may be found from a *literature search*. This may be done with:

- Search engines. Some general search engines have specialist services for academic use, such as Google Scholar. This is very quick and easy to use, and quite powerful.
- There are several academic databases, such as Scopus, that can be searched for a fee.
- Searches also include citation searches. If you find an interesting paper, this will list other papers that have cited it, that is, used it as a reference. These other papers will clearly be more recent.
- The references in a paper can be found, but these will clearly be older then the paper itself.

When searching using search engines or databases, it is often necessary to search for complete phrases, and these should be put in quote marks. A search for "strength of air entrained concrete" (in quotation marks, as shown) will be far more useful than entering the words without quotation marks, because this will find results for strength, air, entrained, and concrete, as individual words.

16.3 TYPES OF REFERENCES
16.3.1 PAPERS FROM REFEREED JOURNALS

These are the best, because they are peer reviewed by reviewers who are experts in the subject.

There are two basic types of journal:

- Traditional journals in which readers must pay to see the contents. These are printed as paper copy for libraries, but in reality, most of their income (and readers) come from the downloads of individual papers.
- Open source (free to view) online journals, in which their income comes from authors who pay to have their work published. These may be good quality and use rigorous review, but there are some that conduct almost no review, and will publish virtually anything, provided the fee is paid.

Some funding agencies insist that work that they have paid for should be available free to readers. Thus, some traditional journals permit authors, whose papers have been reviewed and accepted, to pay a fee to make them free to download.

Journals are ranked by a system called "impact factor" that counts the average number of times the papers in it have been cited (i.e., used as a reference) by other papers. A paper may have a large number of citations for various reasons:

- It may be an important work, give details of useful methods, or a good review of a topic (methods and review papers normally get more citations)
- It may be of poor quality, but easy to find and download, or it may have been well publicised by authors who have persuaded others to cite it. Reviewers rarely check the references in a paper.
- It may even be incorrect, and authors may be referencing it to point this out.

This system of counting citations is thus far from perfect, but it is the best there is to judge quality. In general, journals run by the professional institutions, such as the Institution of Civil Engineers, the American Society of Civil Engineers, the American Concrete Institute, etc. and the traditional journals, such as Cement and Concrete Research, and Construction and Building Materials, have good papers. However, even the best journals sometimes publish very poor papers.

16.3.2 PAPERS FROM CONFERENCES

The review process for these is often minimal. Authors pay to attend conferences and present their papers, so the organisers aim to accept as many as possible. Conference proceedings are often published as a printed set of abstracts, with the full papers on a disc. However, it is becoming more common to make the papers open source on the internet.

16.3.3 COMPANY WEBSITES AND LITERATURE

This is basically advertising, and it is not externally reviewed, but it can still be valuable.

16.3.4 BOOKS

Text books are normally based on published research. They often have a few errors in them. Some books are research monographs, and the contents may be new and unchecked.

16.3.5 WIKIPEDIA

This is a source that is often discussed. Many people consider it unreliable, because it can be changed by any user; however, it is not subject to the same commercial pressure as some other sources. It contains a wealth of good information and, provided it is used as just one source among many, it can be very valuable.

16.3.6 NEVER USE A SINGLE SOURCE

While the vast majority of published work is of good quality, the results in some papers may be poor, because most researchers are under considerable commercial pressure to publish papers and, in particular, to publish results that will help them obtain further funding. Universities and many other organisations may not carry out any internal review. Academics are free to send off anything they want for publication.

The essential point is never rely on references from a single source, always look for independent confirmation of results. A consequence of this requirement for confirmation is that there is a demand for data from experiments that repeat what has been done before. This work is of value, and should be accepted for publication.

16.4 DEFINING THE OBJECTIVES OF A RESEARCH PROGRAMME

The research objective for any programme must clearly state the chosen application. Considering the example of a cement replacement material, few new products for use in concrete will be suitable for the full range of uses of Portland cement. Many may only be suitable for low strength concrete, or possibly masonry blocks. This will reduce their economic value, but will probably still be worth pursuing because low-strength products have numerous applications, including house foundations and road subbases. Indeed, it can be argued that very large amounts of "structural grade" cement are wasted on these uses.

The basic objective of the research must be to investigate the use of the chosen material, with a view to either dismissing it as unsuitable, or promoting its application. Both of these options must remain open, and researchers should not be penalised for negative results. This is a risk for industry; spending money on a negative result is hard to justify, but is an inevitable risk in genuine research. Such results are of use and should be published.

16.5 CARRYING OUT A RESEARCH PROGRAMME

16.5.1 CHOOSING THE TESTS

The choice of tests to carry out, and numbers of samples to use, will depend on time and resources. The key consideration must be the requirements of potential users for the specific application. For the example of a new cementitious material, these will include:

1. Materials characterisation. This is needed to make the work reproducible.
2. Strength. This is the standard test for concrete, and correlates well with many durability-related properties.
3. Leaching. This is a measure of the likely pollution that may be caused if the concrete gets wet. If harmful chemicals are going to leach out, the concrete cannot be used in most applications.
4. Durability related tests. If the concrete is to have steel reinforcement in the chosen application, this will probably focus on preventing corrosion.

Where possible, tests from published standards should be used. This will make the work easier for other labs to repeat. However, nonstandard tests may be used, provided they are adequately described in the report.

16.5.2 ANALYSING THE MATERIALS

This is a key step toward making the work repeatable. If another person is to be able to repeat the work, they must know exactly what materials were used. This is why all good papers provide a full physical, and chemical analysis of the cement, and the proposed cement replacements.

16.5.3 PREPARING THE SAMPLES

In a real construction environment, materials are never in optimum condition. Materials prepared on site (e.g., concrete) suffer most, but even prefabricated items may suffer surface damage. The difficulties with simulating site conditions are that all sites are different, and that all scientific experiments must be designed to be repeatable. One solution is to try to simulate both best, and worst conditions. It is important to ensure that preparation methods do not bias the experiment toward one type of sample.

The geometry of the sample may affect durability – for example, a high surface area/volume ratio will promote sulphate attack on concrete samples. The effect of this may be easily calculated, so it may be used to accelerate the experiment.

16.5.4 EXPOSURE TO THE ENVIRONMENT

In almost any materials, research durability is a key concern. For the example of cement replacements it is paramount. In general, real exposure experiments are not very useful, because they are too slow to be completed, even during a three-year programme. Deterioration must therefore be accelerated with, for example, heat, pressure, applied voltages, or precontaminating the samples. Each of these methods must be used with care. Sometimes materials give the reverse effect from that which would be expected; for example, heat slows down sulphate attack, and mixing chlorides into wet concrete makes it less permeable and thus more durable in some environments. Even greater care is required to ensure that the environment is realistic. How realistic would it be to use a sulphate solution five times more concentrated than any found on site?

16.6 THE STATISTICAL BASIS
16.6.1 THE NEED FOR CONTROL SAMPLES

It is rare to carry out an experiment simply to see if one material is adequate for an application (radioactive waste disposal research is an exception to this). In general, experiments are aimed at improving a material or method. Thus, the main objective will be to determine whether the new product will be as good as current products for the chosen application. Thus, is will be necessary to make *control* samples using current methods. For the example of a cement replacement, the control samples will only have cement in them. However, as discussed in Section 18.3.1, cement replacements are often used in larger quantities than the cement they replace, if the same strength is required. Thus, a decision must be made as to whether the control will have the same strength, or the same total mass of cementitious material.

16.6.2 THE NULL HYPOTHESIS

This is a statement such as "Cement replacement X does not reduce the durability of this product in this application." The experimental data is then used to show that there is a probability of 95% of this being true.

16.6.3 THE DETAILED METHOD

Experiments may be multivariate or bivariate. Variables in an experiment on sulphate attack might be: water/cement (w/c) ratio, exposure time, sulphate concentration, temperature, cement type, curing.

Bivariate means change just one variable and measure another – for example, vary the w/c ratio, and measure the strength.

Multivariate experiments involve changing several variables, and testing whether they interact, that is, whether changing one makes the result more sensitive to changes in another. They are difficult to analyse, but can be very powerful. Bivariate relationships can be investigated by plotting one variable against another, and looking for correlations (see Section 12.5), but multivariate analysis needs more complex statistics (see Section 13.5).

An example of a null hypothesis for a multivariate method is: "this change to the concrete mix will not make its susceptibility to sulphate attack more sensitive to increases in the water/cementitious ratio."

16.6.4 PRESENTATION OF THE RESULTS

Using a spreadsheet package, it is possible to produce large numbers of different graphs, showing the data in different ways. Some of these, particularly if log scales are used, will lead to quite different interpretations (see Section 15.2.5). For this reason, it is technically correct to decide on the method of presentation before the data is obtained. This will mean that it will be presented without bias.

16.7 THE PUBLICATION

Section 15.5 gives some general guidance on publication. When writing a paper, it is important to remember that the purpose of the publication is to provide a source that will be useful to others. Recent guidelines for papers on cement replacements in concrete suggested that they should include:

1. An informed discussion of the source of the material, including the availability. *Any potential uses would need this information.*
2. A physical and chemical analysis of the material, including estimates of the range of values that might occur in the supply.
3. Test results for strength and leaching of the product.
4. A report on a site trial. *A large-scale trial will often reveal problems, such as excessive heat of hydration that are not apparent at lab scale.*
5. An unbiased discussion of the problems that may be expected before the product is brought to market. *The researcher may not be aware of all problems that may arise, but they should discuss any that are apparent.*
6. An analysis of the long-term consequences of introducing the proposed technology. *Durability may often be estimated by studying similar materials from the literature.*

16.8 CONCLUSIONS

- Journal papers are the most reliable source of research information.
- Whenever data from a publication is used, it is essential to check it against another source.
- The objective of a research programme must include the possibility of finding a negative result.

- All experimental work should be carried out and reported in a way that will enable the results to be reproduced in another lab.
- A null hypothesis should be defined for all research.
- Multivariate analysis may be needed in materials research.

TUTORIAL QUESTIONS

Students should study Chapters 17–29 of this book before attempting these questions.

1. A number of large sewage sludge incineration plants have been constructed, and you have been awarded a three-year contract to determine whether the ash is suitable for use in the concrete for highway bridges. Describe the major parts of your programme.

   Solution:

   Literature survey: consider relevance and integrity of sources.
   Initial trials: does it act as a pozzolanic material or is it just a filler? Does it have a substantial water demand?
   Establish candidate mix designs to take forward for further testing.
   Null hypothesis: the durability is the same as the control.
   Cast samples and test for:

   a. Chloride ingress: Test in cell with chlorides on one side, and measure penetration.
   b. Sulphate attack: Expose samples to sulphates, and look for loss of integrity.
   c. Reinforcement corrosion: Cast steel into samples, and look for corrosion. Might be accelerated with anodic polarization. Could be detected with linear polarization, rest potential, or breaking open samples.
   d. ASR: Cast mortar bars with reactive sand, and expose to warm moist environment. Look for expansion.
   e. Carbonation: Expose to carbon dioxide (dehumidified), and either measure shrinkage, or break open and use phenolphthalein.

2. Polymer-modified gypsum-cladding panels for buildings are made by mixing gypsum with a thermosetting polymer resin, and glass fibre reinforcement. The panels are cast 15 mm thick.
   You are responsible for the design of a major commercial development in a city centre location, and your client has expressed an interest in using the panels. Outline a 6 month research programme to determine whether they are suitable. The manufacturer is prepared to supply a panel for testing, and to give you access to a building that was built with them 5 years ago.

   Solution:

   Define objectives: what is the design life of the building? What are the alternatives?
   Objective: to show that panels will perform as well as precast concrete.
   Review previous work: much of this may be in manufacturer's literature. Look for refereed journals, and for confirmation of results from different labs. Check to see if tests are relevant to this application.
   Decide on method: look at the existing building for obvious signs of distress. The range of tests will depend on budget. A method of accelerated aging will be required; it could be heating (under

pressure?), exposure to concentrated chemicals (e.g., sulphur dioxide), or UV light. It is probably best to set up a sample panel on the site as soon as possible, and cut bits off it for lab tests. Possible lab tests:

a. Mechanical: look for loss of strength in the glass.
b. Chemical attack: test environment on site to decide which chemicals to use.
c. Loss of colour: this may be crucial.
d. Thermal expansion and cracking.
e. Thermal conductivity for insulation.
f. Frost attack: check permeability and absorption.

Survey of existing structure: the range of tests depends on what the owner will accept. Look for signs of differences between panels, indicating poor control of manufacture.

3. When designing a large concrete structure to be built overseas, you find a substantial source of pozzolanic material close to the site, and are considering it as a cement replacement in the concrete.
   a. Describe how you would carry out a literature search to find information on the performance of concrete made with the pozzolanic material. Describe the type of literature you would expect to find, indicating the relative merits of each type.
   b. Describe two experiments that you would propose to determine the durability of the concrete, and give reasons why the two you have chosen are the most suitable tests for this purpose. These may be either laboratory or *in situ* tests.

   Solution:

   a. See Section 16.2
   b. Possible tests for durability of pozzolanic mix (the student should describe two tests):
      Permeability to liquid/gas: lab test – this is good indicator of durability.
      Surface absorption (ISAT): probably used *in situ* – concrete must be dry.
      Gas migration (Figg): good *in situ* test.
      Potential survey: used *in situ* – results can be misleading.
      Resistivity measurements.
      Phenolphthalein test for carbonation: good, simple *in situ* test
      Linear polarisation: requires cast in probe for *in situ* use.

4. You are responsible for the materials specification for a major new concrete road in a hot country. Cement is very expensive in the area, but a supplier has offered to supply you with a sufficient quantity of a far cheaper alternative material to make your concrete. Initial trials with test cubes have shown that this new material hydrates in a similar way to normal cement, and develops sufficient strength. It is a by-product slag from a metal refining process that has just started, and contains boron and zinc, rather than the calcium and silicon in normal cement. A trial load of 5 m^3 of concrete made with the new material has been placed, and looks adequate.

   You have 1 year to decide whether to specify the new material. You have access to full laboratory facilities, and to the concrete placed in the trial pour. Describe the main elements of your research programme, and explain why each part is necessary.

Solution:

Literature Review: this is a new process, so there is unlikely to be much published material, but it is always worth checking. It could be in literature about cement and concrete, or it could be in papers about metal refining.

Test programme: (all tests run with control samples using ordinary cement).

Strength tests: should be used to check strength gain, any loss of strength under wet, dry, hot, cold conditions. Also, check variability between samples.

Workability tests: check for workability retention in hot conditions. Check for compatibility with admixtures (e.g., do they work, is there strength loss?).

Chemical resistance: test for loss of strength when exposed to fuel spills, acids, alkalis.

Reinforcement durability: it may be that this material does not protect steel, so a different (e.g., polypropylene) reinforcement will have to be used. If steel is to be used, check the climate… if there is no rain, corrosion won't be a problem. If needed, check for corrosion using LPRM on samples in salt solution.

Analysis: show whether the new material is as strong, as durable, as normal cement, and also show whether the sensitivity to hot, cold, changes in w/c, changes in aggregate, changes in mix procedure, is increased.

Conclusions: conclusions must show a degree of certainty from statistics, for example, the probability of strength being lower than control.

5. A new coating system is being proposed for concrete that is intended to protect structures from exposure to marine environments. You have been awarded a 6 month contract to evaluate the material for use on a concrete bridge. Describe the major parts of your programme. You may assume that you have access to a structure on which the coating has recently been used.

Solution:

a. Define the general objectives: for example, which properties are important (relate to use as bridge – mostly reinforcement corrosion).
b. Review previous work: should describe types and sources of literature, and precautions to be taken (e.g., try to find two sources for important items).
c. Identify available resources and time limits.
d. Choose a general method: for example, linear polarisation, sulphate exposure.
e. Establish a null hypothesis: "This coating is as good as siloxane."
f. Prove the null hypothesis: must be established statistically to 95% certainty.

INTRODUCTION TO CEMENT AND CONCRETE

CHAPTER OUTLINE

17.1 INTRODUCTION

The purpose of this chapter is to introduce the terms that are used for concrete, and to give an outline of what it is, what it is used for, and what problems need to be overcome when using it. All of the different materials concepts are discussed in more detail in later chapters, but it is necessary to understand what all the terms mean, in order to understand the context.

17.2 CEMENT AND CONCRETE

The word "*cement*" is used to cover a wide variety of materials that, when mixed, may be formed into shapes, and then set to become solid. This includes adhesives and many other materials. In civil engineering terms, cement generally means "*hydraulic cement,*" a cement that sets due to a reaction with water, called *hydration*. It is important to appreciate that cement hydration is a chemical reaction, and not just a drying out process. Cement will hydrate and set under water. When cement reacts with water, the main products of the reaction are Calcium-Silicate Hydrate gel (CSH gel), and lime.

When cement is mixed with water, the mixture is known as *cement paste* (or *unsanded grout*). When cement is mixed with sand and water, the mixture is known as *mortar* (or *sanded grout*). Mortar

155

is used to lay bricks. When cement is mixed with larger stones, sand, and water, the mixture is known as *concrete*. The term concrete is also sometimes used to describe a mixture of other binders and aggregate, for example, asphaltic concrete.

The sand is known as *fine aggregate*. This is particles of stone, or other material, generally less than 5 mm (0.2 inch) across, if it is to be used in concrete, but smaller if it is to be used to lay bricks. The larger stones (particles generally between 10 mm and 20 mm (0.4 inch and 0.8 inch)) are known as *coarse aggregate*. For most concrete works, these two types of aggregates are added separately to the mix, but for small projects *all-in aggregate* may be used that is a combination of both. The aggregate normally makes up about 75% of the concrete by weight. When the concrete is placed in *formwork*, also known as *shutters*, it is *vibrated* with a mechanical device to expel the air, and the aggregate moves away from them, leaving cement paste exposed in order to give a smooth surface. When cement is mixed with coarse aggregate and water (without the sand), the mixture is known as *no-fines concrete* or *pervious concrete*, because water can flow through it.

17.3 USES OF CEMENT

The terms in Table 17.1 are used to describe the major applications for cement.

17.4 STRENGTH OF CONCRETE

The strength of concrete is normally measured by compressing 100 mm or 150 mm (4 inch or 6 inch) cubes or cylinders in a hydraulic testing machine. The maximum stress recorded before crushing is the strength "*grade*," typically 15–40 MPa (2–6 ksi). The cubes are cast in metal or plastic moulds that are *struck* (removed) the day after casting. Once struck from the moulds, the cubes are stored in water until

Table 17.1 Applications for Cement

Terms Used	Applications
In situ concrete	Concrete that is placed in its final location.
Precast concrete	Concrete items moved into position after casting. Precasting offers considerable advantages because it is carried out in a controlled environment, protected from the elements (Fig. 17.1).
Concrete masonry	Building blocks, or a type of brick made with concrete, rather than clay.
Building mortars	Used for laying masonry.
Soil stabilisation	Cement is mixed with wet or soft soil to enable it to take higher loads, such as construction plant (Fig. 17.2).
Grout	This is a mixture of cement and water, and sometimes sand, used to fill small gaps, such as those beneath machine bases, or cracks in rocks.
Waste containment	Cement is mixed with harmful wastes to prevent them from escaping into the environment.

FIGURE 17.1 Lifting Precast Pretensioned Concrete Flooring Units onto a Steel Frame

testing (normally at 28 days). If a series of cubes from the same batch of concrete are stored at different temperatures, strength gain curves, like those in Fig. 17.3, may be obtained. High temperatures will give high early strengths; low temperatures will give the best long-term strength.

In order to achieve the best strength from concrete, it must be *cured* well. This involves keeping it at the right temperature and protected from frost, and, very importantly, preventing it from drying out too quickly before the hydration reaction has completed. Concrete should be kept as wet as possible after casting.

The strength of concrete can be increased by reducing the amount of water in the mix, and thus reducing the *water to cement (w/c) ratio*. Unfortunately, reducing the water content will reduce the *workability*, which means it will prevent the mix from flowing, and make it hard to place properly. Some concrete is not intended to flow. Figure 17.4 shows *semidry* concrete that only has just enough water in it for the cementitious material to hydrate, and must be compacted with a roller.

FIGURE 17.2 Soil Stabilisation

The cementitious material is being spread and mixed into the soil.

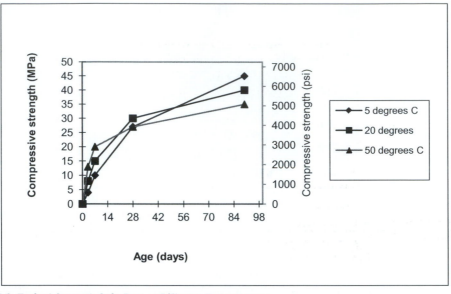

FIGURE 17.3 Typical Strength Gain Data at Different Temperatures

17.5 REINFORCED CONCRETE

Reinforced concrete is a *composite* material. This means that it is made up of different constituent materials with very different properties that complement each other. In the case of reinforced concrete, the component materials are almost always concrete and steel. The steel is the reinforcement. Other reinforcement, such as glass fibre or polypropylene, is used for specialised applications.

Concrete is strong in compression. Steel is strong in tension and compression, but in compression a steel bar that is thin enough to be economic will buckle. A simple reinforced concrete structure therefore uses steel in tension, and concrete in compression. See Fig. 17.5.

FIGURE 17.4 Semidry Concrete for a Road Foundation

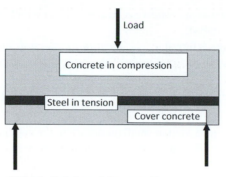

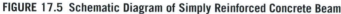

FIGURE 17.5 Schematic Diagram of Simply Reinforced Concrete Beam

It may be seen from Fig. 17.5 that as the beam is bent, the steel is loaded in tension. The closer the steel is to the bottom of the beam, the more efficiently it will take the load. However, it is necessary to maintain a specified *cover depth* of concrete between the steel and the outer surface of the concrete, in order to protect the steel from corrosion.

17.6 PRESTRESSED CONCRETE

In order to create more efficient structures, the steel in concrete is sometimes stressed so that it is under considerable tension at all times, and the surrounding concrete is thus in compression. The structural analysis of this is complex, but it has the effect of making many types of structure carry more load while using less concrete.

Two different systems are used:

* *Pretensioned* concrete is almost always precast, and contains steel wires (*tendons*) that are held in tension while the concrete is placed and sets around them (see Fig. 17.1).
* *Post-tensioned* concrete has ducts through it. After the concrete has gained strength, tendons are pulled through the ducts and stressed. The ducts are then filled with grout. Bridges built in this way have experienced considerable corrosion of the tendons, so *external post-tensioning* may now be used in which the tendons run along the outer surface of the concrete.

17.7 CEMENT REPLACEMENTS

Cement is expensive, so it is frequently replaced in concrete with materials that are by-products of other industrial processes, and are thus a lot cheaper, and have a low carbon footprint. These materials can only normally be used to replace part of the cement, because they react with the lime produced when the cement reacts with water. The common ones are:

Pulverised fuel ash (PFA). This is the ash from burning pulverised coal for power generation. It is typically used for up to 40% replacement.

Condensed Silica Fume (CSF) also known as *Microsilica*. This is a by-product of silicon steel manufacture. It is only typically used for up to 10% replacement, but can give very high strength concrete.

Ground Granulated Blastfurnace Slag (GGBS). This is a slag from steel production. It is typically used for up to 50% replacement.

The PFA and CSF are known as *pozzolans* because they react with lime and set. The GGBS reaction needs the lime, but only as a catalyst.

The cement replacements are referred to as *cementitious* materials and the water/cementitious ratio is the ratio of the mass of water to the total mass of the cement and cement replacements.

17.8 ADMIXTURES

Admixtures are chemicals that are added in small quantities to the concrete to achieve specific benefits. They are generally expensive, so they are only used when the benefits are significant. Two examples are as follows:

1. *Air entrainers* are used to entrain microscopic air bubbles into the concrete, and protect it from frost
2. It was noted in Section 17.4 that adding water to concrete, in order to improve the workability, will reduce the strength. This problem may be solved by using a *plasticiser* or *superplasticiser,* also known as a *water reducer* that can increase the workability without adding water. If high doses of superplasticisers are used, the mix will become *self-compacting* and can be placed without the need for vibration.

17.9 ENVIRONMENTAL IMPACT

Cement production is one of the largest industries in the world. Annual world production in 2013 was approximately 4 GT (of which, about half was in China). It is produced in kilns at around 1400°C (2500°F), and approximately 750 kg (1650 lb) of CO_2 are released for each tonne (2205 lb.) that is made. This is 400 m^3 (524 yd^3) of gas; and CO_2 normally forms only 0.04% of air. Cement production is responsible for around 7% of worldwide greenhouse gas emissions. The only larger contributors are electricity generation and transport.

Most of the CO_2 released when cement is made comes from the chemical reaction. The limestone from which it is made is calcium carbonate, and when this is heated it gives off carbon dioxide, and forms calcium oxide. Thus, while some benefit has been obtained by making the process more energy efficient, there is little scope for any further reductions.

Numerous different solutions to this problem have been proposed, including carbon capture and inventing completely new cements; but the only effective method currently used to achieve significant reductions in greenhouse gas emissions is the use of cement replacements.

One process that has not been extensively investigated is carbon sequestration. In this process, the hydrated cement reacts with CO_2 in the air, slowly reversing some of the processes that took place in the kiln when the cement was made (this is the carbonation process; it also causes reinforcement corrosion, see Section 25.3.2). It is estimated that this may reduce the carbon footprint of the cement industry by 3–5%.

17.10 DURABILITY
17.10.1 TYPES OF DURABILITY PROBLEM

Durability is a major concern for concrete structures. Many of them do not last for their *design life.* A main objective of this book is to show how this situation could be improved. Some major causes of

Table 17.2 Major Durability Problems with Concrete

Name of Problems		Solutions
Sulphate attack	Sulphates from the ground or other sources react with the cement paste.	Use sulphate resisting cement or blastfurnace slag.
Freeze-Thaw	In cold climates, repeated cycles of freezing and thawing take the surface off the concrete.	Use an air-entraining admixture.
Alkali-silica reaction	A chemical reaction between alkalis in the cement and silica in the aggregate.	Use a low alkali cement or a nonreactive aggregate.
Reinforcement corrosion	The steel reinforcement corrodes.	No easy solution.

durability problems are listed in Table 17.2. By far, the most serious of these is reinforcement corrosion. Unlike the others, it has no simple solution. It is discussed in the next section.

17.10.2 REINFORCEMENT CORROSION

Corrosion is the process in which iron is lost from the steel; it normally then combines with water and oxygen to form rust. In most circumstances, the factors that control corrosion are therefore:

* The availability of oxygen.
* The availability of water.

Both water and oxygen move rapidly through concrete, so it is evident that steel in concrete is not normally protected by excluding oxygen and water. The steel in concrete is actually protected by the alkalinity of the lime, and CSH gel formed during cement hydration.

There are some circumstances, however, when the protection to the steel is inadequate:

* The alkalinity of concrete can be destroyed by carbon dioxide. This process is known as *carbonation* (this is the carbon sequestration reaction mentioned in Section 17.9).
* If there are chlorides present (e.g., from road salt or sea water), they will break down the protection to the steel, even in an alkaline environment.

The history of the chloride mechanisms is:

1824	Portland Cement patented.
1837	Published paper notes that poor workmanship may cause premature deterioration.
1890s	Reinforced concrete in increasing use with steel reinforcement.
1907	Corrosion studies from New York reported in journal papers.
1912	Published paper identifies chlorides as the main cause of the corrosion.

A century later, construction with poor durability continues to date throughout the world.

Corrosion of reinforcement in concrete costs taxpayers very large sums in repairs, and disruption to roads and other infrastructure. In 2004, the cost to the US economy was estimated at $150 billion.

17.10.3 PROTECTION FROM CORROSION

In order to protect the steel, both chlorides and carbon dioxide must be kept away from it. The *transport properties* of the cover concrete are the permeability, diffusion coefficient, and other properties discussed in Chapters 9 and 10. The steel must be protected with an adequate thickness of cover concrete, with transport properties low enough to ensure that the carbonation and chlorides do not reach it during the design life. This is normally achieved with a low water/cement ratio in the mix design. This requires constant effective supervision of the work to check the concrete quality, the cover depth, and the curing and is not easily achieved.

17.11 CONCLUSIONS

- Cement is a powder that hydrates when mixed with water.
- Mortar is a mix of cement, water and fine aggregate.
- Concrete is a mix of cement, water and fine and coarse aggregate.
- Reinforced concrete normally contains steel reinforcing bars.
- Cement production has a major environmental impact, which may be reduced by using cement replacements.
- The main concern with the durability of concrete structures is the corrosion of the steel reinforcement.

CEMENTS AND CEMENT REPLACEMENT MATERIALS

CHAPTER OUTLINE

18.1 INTRODUCTION

In this chapter, cements and the different compounds that occur in them are discussed. These compounds are important because they are responsible for the different properties of cement, and, by varying their proportions, cements can be made for specific applications. The most commonly used cement replacements are also discussed. The term Ordinary Portland Cement (OPC) is commonly used to describe the standard cement that is used for normal applications. The different cements and the cement replacements are used in order to achieve lower cost, better durability, different colours or, increasingly, a lower environmental impact than OPC.

18.2 CEMENTS
18.2.1 THE STAGES OF CEMENT PRODUCTION

The production and hydration of cement may be considered in five stages that take it from a raw material in the quarry to a hydrated mix in a structure as follows:

1. Component elements – these are in the raw materials.
2. Component oxides – these are produced in the kiln.
3. Cement compounds – these are formed from the oxides.
4. Portland cements – the properties of these are determined by the compounds.
5. Hydration products – produced when the cements react with water.

These different stages are discussed below except for the hydration that is in Chapter 20.

18.2.2 COMPONENT ELEMENTS

Traditionally, the raw materials for cement manufacture were clay and chalk but, with the industry working at a vast scale of production for over 100 years, a very wide range of different source materials have been used. These materials need to have the correct proportions of the different chemical elements in them to make the required type of cement. The proportions of the elements will not change, so a chemical analysis of concrete in a structure, that may be many years old, will reveal the type of cement that was used.

18.2.3 COMPONENT OXIDES

The raw materials are mixed together, and ground and fed into a kiln where they combine with oxygen. The component oxides of cement each have single letter abbreviations in cement chemistry.

Names	Abbreviations	Chemicals
Lime	C	CaO
Silica	S	SiO_2
Alumina	A	Al_2O_3
Iron	F	Fe_2O_3
Water	H	H_2O
Sulphuric anhydride	$\bar{S}$	SO_3
Magnesia	M	MgO
Soda	N	Na_2O
Potassa	K	K_2O

The proportions of these oxides in the cement are determined by the raw materials. It is of note that the symbols of cement chemistry are different from normal chemistry, for example, in normal chemistry C is carbon, but in cement chemistry, it is lime. This difference has come about because the cement industry has been working for so long, that the symbols for cement chemistry were in use before the standard symbols for normal chemistry were established in the first half of the twentieth century.

18.2.4 CEMENT COMPOUNDS

As the cement progresses along the kiln, it is heated to higher temperatures; and, before reaching the maximum of around 1400°C, the oxides combine to form four different compounds.

The four major cement compounds are:

Tricalcium silicate	C_3S
Dicalcium silicate	C_2S
Tricalcium aluminate	C_3A
Tetracalcium aluminoferrite	C_4AF

The approximate proportions of these in the cement may be calculated from the Bogue equations:

$$C_3S = 4.07C - 7.6S - 6.72A - 1.43F - 2.85\bar{S}$$
$$C_2S = 2.87S - 0.754(C_3S)$$
$$C_3A = 2.65A - 1.69F$$
$$C_4AF = 3.04F$$

These equations were established experimentally, and are used when a cement sample has been analysed to give the proportions of the different oxides. All values in them are percentages. They tend to overestimate C_2S and underestimate C_3S. Not all of the lime will be used to form compounds, and some will remain as free calcium oxide (see Section 7.9) that will then form calcium hydroxide in the hydrated cement. Similarly, free soda and potassa will be present. These three compounds make hydrated cement strongly alkaline.

The properties that the compounds give cements are as follows:

- C_3S

    Rapid strength gain gives early strength (e.g., 3–7 days).
- C_2S

    Slow strength gain. This gives long-term strength.
- C_3A

    Quick setting. C_3A reacts with sulphate, resulting in sulphate attack on the concrete that causes it to crumble. It also reacts with chlorides to form chloroaluminates and "bind" them, and thereby protect the reinforcing steel.
- C_4AF

    Responsible for the grey colour of cement.

The speed of reaction of the different compounds may be seen from Table 18.1.

Table 18.1 Typical Data for Reaction of Cement Compounds

Compounds	Time to 80% Hydration (Days)	Heat of Hydration (J/g)
C_3S	10	502
C_2S	100	251
C_3A	6	837
C_4AF	50	419

18.2.5 PORTLAND CEMENTS

Cement consists of the four compounds in different proportions, together with some lime. When it comes out of the kiln, it is called clinker, and must be ground to make the final product. The fineness of grinding affects the rate at which the cement hydrates. It is measured as the total surface area of the particles, in m^2/kg, and is calculated from a measurement of the permeability to air of a compacted bed of cement, of standard thickness (the air will flow more easily through a bed of coarser particles because the gaps between them will be larger).

Gypsum is used to slow down the setting of the C_3A, and is added during grinding. The gypsum may be in the form of anhydrite or dihydrate (see Section 7.10). A mixture of equal parts of the two may be used.

By varying the proportions of different compounds, cements for different applications may be produced:

- Ordinary Portland cement (OPC).

    This is the cheapest and most commonly used cement. The name "Portland" was given to it by its inventor (John Aspdin of Leeds, UK, in 1824) because he wished to imply that the cast material had the properties of Portland stone.
- Sulphate-resisting cement.

    Sulphate attack (from sulphates in the ground, the sea etc.) is due primarily to the effect of sulphates on C_3A. Sulphate-resisting Portland cement thus has low C_3A (maximum 3.5%). Sulphate-resisting cement offers poor protection against chlorides, because the C_3A binds them. Note that GGBS may be used for sulphate resistance (see Section 18.3.2), and this gives good resistance to chlorides.
- Rapid hardening cement.

    This has a fineness of not less than 350 m^2/kg (compared with 275 m^2/kg for OPC), and a high C_3S content. It has a strength approximately 50% higher than OPC at 3 days, but similar long-term strength. Heat evolution rates are high.
- Low heat cement.

    This contains relatively small percentages of C_3S and C_3A, and is intended for mass concrete pours where heat is a problem. The cement replacements GGBS or PFA can delay heat production at lower cost, and have therefore largely replaced it.
- White cement.

    This has a low C_4AF content to give white cement for architectural purposes. GGBS gives a reasonably white concrete.
- Hydrophobic Portland cement.

    This is OPC mixed with a water repellent for long-term storage.

Table 18.2 Typical Proportions of Compounds in Cement

Compounds	OPC	Sulphate-Resisting Cement	Rapid Hardening Cement	Low Heat Cement	White Cement
$C_3S\%$	55	43	60	30	51
$C_2S\%$	16	36	11	46	26
$C_3A\%$	11	3	12	5	11
$C_4AF\%$	9	12	8	13	1
Free CaO%	0.8	0.4	1.3	0.3	0.2

- Masonry cement.

  This is OPC with fine mineral filler and an air-entraining admixture to give frost resistance to masonry mortar (see Chapter 24 for more information about admixtures).

  Typical compositions are in Table 18.2.

18.2.6 OTHER CEMENTS

There are some other cements that are different from Portland cement.

- High alumina cement (HAC).

  This is also known as Fondu cement or Calcium Aluminate Cement, and is highly sulphate resisting, and has a high early strength. Unfortunately, after some years in a warm moist environment, the strength is partially lost. This "conversion" process is due to a crystalline transformation of one of the hydrates, and is irreversible. A number of structures with HAC are still in use, so engineers will often test for it when inspecting a building. It is currently used for applications such as fast repairs to airport runways, where high early strength or chemical resistance is needed, and it will not convert.
- Super-sulphated cement.

  This is made from GGBS, gypsum, and a small percentage of OPC. It is highly resistant to sulphates, but uneconomical – partly because it has a short shelf life. It is mentioned here because it forms the basis of a number of new cements that are being developed using waste materials to give a low carbon footprint (see Section 41.5)
- Expansive cements.

  These cements promote a reaction similar to sulphate attack, and expand with considerable force after setting. They may be used for demolition. For expansion with less pressure (for void filling) use aluminum or zinc additives (see Section 24.7.2).

18.2.7 CHANGES IN CEMENT

Over a period of many years, the manufacturers increased the C_3S in cement in response to economic demands for high 28 days strength with low cement contents. Typical figures are in Table 18.3.

Table 18.3 Historic Data for Cement Compound Percentages

Year	1910	1960	1980
$C_3S\%$	25	47	53
$C_2S\%$	45		25

This has had the effect of reducing long term strength gain and durability, because the required 28 days strength may be achieved with a higher water to cement ratio. Many structures from the 1980s are now suffering durability problems. However, this may be balanced in new structures by including a cement replacement that will make the blend less reactive, and thus increase strength gain after 28 days.

18.3 CEMENT REPLACEMENT MATERIALS (ALSO KNOWN AS MINERAL ADMIXTURES)

18.3.1 INTRODUCTION

These are materials used as a replacement for part of the cement, in order to reduce cost or improve properties. Not all of the cement in concrete is replaced. for example, a mix containing 300 kg/m³ (500 lb/yd³) of cement could be modified to contain 200 kg/m³ (340 lb/yd³) of cement, and 130 kg/m³ (220 lb/yd³) of GGBS.

When looking at data on their effect of cement replacement on concrete properties, it is important to find out what replacement method was used. If an OPC mix has 300 kg/m³ (500 lb/yd³) of cement, we might add 300 kg/m³ (500 lb/yd³) of PFA (*addition*) to give a very high total cementitious content of 600 kg/m³ (1000 lb/yd³), and a strong and expensive mix. Alternatively, we might replace 150 kg/m³ (250 lb/yd³) of the OPC with PFA (*direct replacement*) to give a weak, cheap mix. As a third option, we might make a mix with, say, 200 kg/m³ (340 lb/yd³) of OPC and 200 kg/m³ (340 lb/yd³) of PFA to give the same 28 days strength at 20°C, as the original OPC mix (*equal strength replacement*). Each of these could be used with the OPC mix to produce graphs of the effect of, for example, 50% PFA mixes with very different conclusions.

These cement replacement materials are used very widely throughout the world. There are two ways in which this may be done: either the cement replacement is premixed into the cement when it is made, or it may be supplied separately and mixed at the concrete batching plant. Mixing the materials into the cement avoids the need for extra silos at the batch plant, and thus reduces costs, but it also reduces the flexibility available to mix designers who may wish to use different proportions for different applications. Different practices for this are common in different countries.

18.3.2 GROUND GRANULATED BLASTFURNACE SLAG (GGBS)

Slag is derived from the production of iron in blastfurnaces. The slag contains all of the compounds that would affect the purity of the iron. The slag is a hot liquid, and may be cooled in air, by mixing with water (foaming), or with high-pressure water jets at high water/slag ratios (granulation). Only granulation produces noncrystalline slag, and only this slag exhibits hydraulic properties, and is therefore suitable for use with cement. The other types of slag are used as aggregate. The granulated slag is ground before use.

GGBS is a latent hydraulic material that will hydrate in contact with water, but very slowly. It is activated when combined with Portland cement by the lime and other alkalis released by the hydrating cement. The lime and other alkalis act as a catalyst – they are not depleted.

Direct replacement may result in lower early strength, but higher later age strengths, depending on replacement level. When 50% is replaced, the same 28 days strength may be achieved as OPC, but the 7 days strength will be about 60% of the 28 days strength, compared to 80–90% with OPC. GGBS concrete must be cured well, but can give excellent durability. It is recommended for use as an alternative to sulphate-resisting cement for all but the most aggressive sulphate conditions. It also tends to reduce chloride transport by decreasing the diffusion and increasing the chloride binding.

18.3.3 PULVERISED FUEL ASH (PFA)

PFA is the ash from the burning of pulverised coal in power stations. The pulverised coal is blown into the furnace in a jet of air, and burns in seconds. About 20% of the ash fuses into large particles, and drops out of the flue gases to form furnace bottom ash. The remaining 80% (fly ash) is extracted with electrostatic precipitators, and the material for use with cement is selected from this on grounds of fineness, microstructure, and chemical content. Some of it is processed ("classified") using a cyclone to remove the largest particles that have a high carbon content from un-burnt coal.

PFA is a **pozzolanic** material. These materials react with lime to form hydrates, and are named after the town in Italy where natural pozzolans (produced by volcanic action) are found. This material was mixed with lime by the Romans to make concrete that has survived for 2000 years.

Some coal contains limestone that calcines in the furnace to form lime. If this "high lime" PFA is mixed with water and aggregate, and no other materials, the lime will react with the pozzolan and a low strength concrete can be made.

The effects of using up the lime in the chemical reaction are:

- The maximum percentage of PFA is limited to about 40%. After this, there is no lime left (unless a high-lime PFA is used).
- Lime is a weak component of hydrated cement, and also eventually dissolves in water. Removing it therefore theoretically increases strength and durability.
- Lime provides the hydroxyl ions that conduct electricity in concrete. Removing the lime therefore increases the resistivity and reduces corrosion.
- Lime is responsible for a lot of the alkalinity of concrete. Removing it therefore theoretically makes the concrete more susceptible to loss of alkalinity (e.g., carbonation) causing corrosion. However, the net effect of using pozzolans is generally to reduce reinforcement corrosion.

PFA will generally delay strength gain and, for equal strength replacement, it will improve durability.

18.3.4 CONDENSED SILICA FUME (CSF)

This is a highly reactive pozzolan, also known as microsilica, and is derived from the production of silicon steel. The production process is highly energy intensive, and is carried out in countries like Sweden where hydropower is available. The high reactivity can be used to obtain very high strengths, but means that great care must be taken with curing, etc. Various problems have been reported with this material, particularly at replacement levels above 10%. The hydration reaction is so strong that self-desiccation (i.e., depletion of all the available water) may occur at construction joints.

18.3.5 NATURAL POZZOLANS

These are mined in many parts of the world, and may be ground and used without further processing. In general, they are less reactive than PFA but make an excellent cheap cement replacement material.

18.3.6 LIMESTONE FLOUR

This is just finely ground limestone (calcium carbonate). It is used in composite cements. While limestone itself is assumed to be inert when used as an aggregate, the finely ground product does make some contribution to strength, so the cements achieve standard strengths.

18.3.7 COMPARING CEMENT REPLACEMENTS

Typical compositions of the main cement replacements are given in Table 18.4.

These are shown diagrammatically in Figure 18.1

The advantages and disadvantages of cement replacement materials are:

- Substantial reduction in carbon footprint, due to reduction in cement content.
- Low early strength (except CSF).
- Reduce cost of raw materials (assuming CSF is used to save cement).
- Increase cost of production and possibility of errors in mix proportions.
- May improve durability.
- Require better curing, and therefore increase cost of placing. GGBS can cause bleeding, but PFA generally improves cohesion.

Table 18.4 Typical Compositions				
	OPC	**PFA**	**GGBS**	**CSF**
C%	64	2.5	40	–
S%	20	50	35	98
A%	6	27	10	–
F%	3	10	0.5	0.2
$\bar{S}$%	2	0.8	–	0.45
M%	2	1.5	8.7	0.26
N%	0.2	1.2	0.4	–
K%	0.5	3.5	0.5	0.45

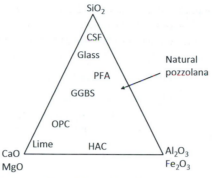

FIGURE 18.1 Schematic Representation of Cementitious Materials

- PFA and CSF produce a darker colour mix. GGBS gives an almost white colour (it may be a bit blue or green initially, but this soon fades).
- GGBS, PFA, and CSF are all industrial by-products that could be environmentally damaging if not mixed into concrete (substantial amounts of PFA have to be disposed of each year, see the discussion in Section 40.3.1).

18.4 CEMENT STANDARDS
18.4.1 US STANDARDS
American standard cement types are given in Table 18.5.

Types Ia, IIa, and IIIa have the same composition as types I, II, and III. The only difference is that in Ia, IIa, and IIIa an air-entraining agent is ground into the mix.

18.4.2 EUROPEAN STANDARDS
The European standard defines five types of cement (Table 18.6). These classes differ from the ASTM types.

A typical example of a cement to the European standard is:

Table 18.5 American Standard Cement Types	
Type I	General purpose
Type II	Moderate sulphate resistance
Type III	High early strength
Type IV	Low heat of hydration (slow reacting)
Type V	High sulphate resistance
White	White colour

Table 18.6 European Cement Types		
CEMI	Portland cement	Comprising Portland cement and up to 5% of minor additional constituents
CEMII	Portland-composite cement	Portland cement and up to 35% of other single constituents
CEMIII	Blastfurnace cement	Portland cement and higher percentages of blastfurnace slag
CEMIV	Pozzolanic cement	Portland cement and up to 55% of pozzolanic constituents
CEMV	Composite cement	Portland cement, blastfurnace slag or fly ash, and pozzolana

Portland-slag cement CEM II/A-S 42.5N

This is a type II cement. *A* indicates a high clinker content, *C* would be low. The *S* indicates a slag (GGBS) content, *V* would indicate PFA, *L* would indicate limestone flour, and *D* silica fume. 42.5 is the strength in MPa of a standard sample made with the cement. *N* indicates normal setting, *R* would indicate rapid hardening.

18.5 CONCLUSIONS

- The properties of cement are determined by the proportions of key elements in the raw materials, and these may be found by analysis of concrete structures.
- The elements in the raw feed form oxides, and these combine to form cement compounds.
- By varying the proportions of the compounds, cement may be made sulphate-resisting, rapid hardening, or white.
- Cement replacements are used to replace part of the cement in concrete.
- Pozzolanic materials, such as PFA and CSF, deplete the lime in the hydrated cement.
- US cement types are not the same as European cement types.

TUTORIAL QUESTIONS

1. a. The following oxide analyses have been obtained from three cements designated A, B, and C.

	A	B	C
C%	65	66.5	59.5
S%	23	21	21
A%	6	6	3
F%	2	2	4.5
$\bar{S}$%	2	2	2

Calculate the proportions of cement compounds in them

b. Identify which cement is (1) OPC, (2) sulphate resisting, and (3) rapid hardening.

c. Using the data in Table 18.1, calculate the total heat generated by 1 m³ of concrete made with a cement content of 300 kg/m³.

Solution:

a. Cement A

$C_3S = 4.07 \times 65 - 7.6 \times 23 - 6.72 \times 6 - 1.43 \times 2 - 2.85 \times 2 = 40.87$
$C_2S = 2.87 \times 23 - 0.754 \times 41.0 = 35.19$
$C_3A = 2.65 \times 6 - 1.69 \times 2 = 12.52$
$C_4AF = 3.04 \times 2 = 6.08$

The others are calculated in the same way to give:

	A	B	C
C_3S	40.87	62.18	50.27
C_2S	35.19	13.39	22.37
C_3A	12.52	12.52	0.35
C_4AF	6.08	6.08	13.68

b. A is OPC

B is rapid hardening (highest C_3S)

C is sulphate resisting (lowest C_3A)

c.

Heat in Joules for 300 kg/m³			
	A	B	C
C3S	6.16 E + 07	9.36 E + 07	7.57 E + 07
C2S	2.65 E + 07	1.01 E + 07	1.68 E + 07
C3A	3.14 E + 07	3.14 E + 07	8.66 E + 05
C4AF	7.64 E + 06	7.64 E + 06	1.72 E + 07
Total	1.27 E + 08	1.43 E + 08	1.11 E + 08

Note that the total heat from the rapid hardening cement is not what causes the problems in structures. The problem is that it is generated quickly, so it does not have time to dissipate.

2. An office block is to be constructed for a commercial client, and the start of construction is to take place in six months' time. The client has been very impressed by advertising for the use of High Alumina Cement with Blastfurnace Slag in the concrete for the frame.

a. Why would there be a commercial benefit from using this concrete mix?
b. Why should the client be concerned about the long-term performance of the concrete?

Solution:

a. The reason this blend was proposed was the idea that the long-term strength gain from the GGBS would compensate for the loss of strength due to conversion of the HAC. The benefits would have been: low cost, high early strength, and chemical resistance.
b. The concern is the conversion of the HAC in warm and moist environments. The ability of the GGBS to compensate for this would need very careful checking – but there would still be a risk.

3. Cementitious samples are obtained for oxide analysis from four different structures. Structures A and B are known to have been constructed without cement replacements. Structures C and D are known to have been constructed with the same cement as structure A but with 40% cement replacement.
Oxide percentages

Sample	A	B	C	D
C	64.15	62.00	39.49	54.49
S	21.87	22.50	33.12	27.12
A	5.35	3.00	14.01	7.21
F	3.62	4.20	6.17	2.37
S	2.53	4.00	1.84	1.52
M	2.00	2.00	1.80	4.68
N	0.20	0.20	0.60	0.27
K	0.50	0.50	1.70	0.50

a. Identify the cements and cement replacements used.
b. Describe typical applications for each mix.

Solution:

a. Use the Bogue equations on the two without cement replacements (A and B) and get:

Cement	A	B
C_3S	46.54	43.77
C_2S	27.68	31.57
C_3A	8.06	0.85
C_4AF	11.00	12.77

B has a very low C_3A, so it is sulphate resisting. A is OPC
Then, subtract the 60% of cement A from the percentages for C and D, and divide by 0.4 to get the following percentages for the replacement materials in them.

Sample	C	D
C	2.5	40
S	50	35
A	27	10
F	10	0.5
$\bar{S}$	0.8	0
M	1.5	8.7
N	1.2	0.37
K	3.5	0.5

Considering the three main cement replacement materials (PFA/GGBS/CSF): CSF is almost pure silicon. C may be identified as PFA form the low C/S ratio, D is GGBS.

b. Applications

OPC is used for general concrete work.

Sulphate resisting cement is used for sulphate resistance, but is unsuitable when chlorides are present. Typical uses: footings or ground slabs.

OPC/PFA: 40% is a typical replacement level. Use for economy, better cohesion, better durability, but lower early strengths. Uses: most general and heavy concrete work.

OPC/GGBS: 40% is lower than average replacement levels. Use for economy, durability, sulphate (and possibly frost) resistance. Uses: marine work and general concrete work.

4. In oxide analysis, the following proportions are found:

	Cement A	Cement B
Alumina	3.6%	4.4%
Iron	4.1%	0.3%

What may be concluded about these cements?

Solution:

Calculate the following from the Bogue equations:

	Cement A	Cement B
C_3A	2.6	11.15
C_4AF	12.46	0.91

A is low on C_3A, so it is sulphate resisting.
B is low on C_4AF, so it is white cement

5. For each of the following materials, describe (i) applications for which they are particularly suitable for use in concrete, and (ii) applications for which they are unsuitable for use in concrete.
   a. Sulphate resisting cement
   b. Rapid hardening cement
   c. PFA
   d. GGBS
   e. High alumina cement
   f. White cement.

   Solution:

   a. (i) Exposure to sulphates.
   (ii) Exposure to chlorides when reinforcement present.
   b. (i) Precast or rapid frame construction.
   (ii) Thick sections where heat may build up.
   c. (i) Low-cost or low-heat applications.
   (ii) Precast or rapid frame where high early strength required (unless used with RHPC).
   d. (i) Low heat or architectural applications.
   (ii) High early strength.
   e. (i) High early strength patch repairs (e.g., runways).
   (ii) All structural applications.
   f. (i) Architectural applications (and use with pigments).
   (ii) Most (it has a high cost).

AGGREGATES FOR CONCRETE AND MORTAR

19

CHAPTER OUTLINE

19.1 INTRODUCTION

The effect of aggregates in concrete may be summarised as follows:

Aggregates:

- Make up most of the mass and volume of concrete
- Cost substantially less than cement paste
- Shrink less than cement paste
- Do not generate heat of hydration
- Do not bleed (see Section 22.3)

- Control the density of concrete
- Help mix the water and cement paste (high shear mixing is required if there is no aggregate in the mix)

  In hydrated concrete, they:

- Have a lower permeability than cement paste
- Are stronger than most cement pastes
- Control surface hardness
- Control thermal properties
- May control appearance, especially if exposed (see Section 29.9.2)

It may be seen that increasing the aggregate content of a mix will improve the properties of concrete. Many durability specifications, however, set a minimum cement content that automatically limits the aggregate content. The reason for this is that a set water content may be required for workability, and increasing the cement content will reduce the water/cement ratio. A better way of reducing the w/c ratio is to reduce the water content with a water reducing admixture (Section 24.2).

19.2 ENVIRONMENTAL IMPACT OF AGGREGATE EXTRACTION

The extraction of aggregate from mines and quarries has a major impact on the environment. Worldwide, aggregates are quarried in far greater quantities than anything else that is mined, such as coal or iron ore. In densely populated areas, it is therefore rarely possible to open up new quarries. Sometimes, it is possible to extract aggregates while creating a useful lake (the Eton College boating lake, used for the 2012 London Olympics, being a classic example), but these opportunities are rare, and very few communities will accept new quarries. There are also some countries, such as Bangladesh, where there is no suitable material to be quarried. This leaves concrete producers with the high cost of transporting aggregate from more remote areas, unless they use artificial aggregates.

19.3 AGGREGATE SIZES

Aggregates range in size from 75 mm (3 in.) down to 0.07 mm (3×10^{-3} in.). Material larger than 5 mm (0.2 in.) is generally classified as coarse, and below 5 mm as fine. Use of aggregate in concrete larger than 25 mm (1 in.) is now quite rare. All-in aggregate is a mixture of coarse and fine, and is used on small projects.

In order to control the properties of concrete, all aggregate is sized ("graded"). Grading involves "sieving" to sort it into fractions of different sizes. Fine sizing is carried out by exploiting differences in particle velocities suspended in water.

19.4 MINED AGGREGATES
19.4.1 CRUSHED ROCK

This is obtained from sedimentary (e.g., limestone) or igneous (e.g., basalt) rocks. The rock is exposed by removal of overburden and then excavated, generally by drill and blast methods. The rock is then mechanically crushed and graded. The resulting aggregate is angular, and generally gives high strength, and low workability.

19.4.2 UNCRUSHED GRAVEL

This is obtained from the same igneous or sedimentary rocks, but they are broken up naturally by the action of water or ice (glaciers). The properties of sand and gravel are largely dependent on the parent rock from which they are derived, but weaker particles tend to be removed by attrition so the aggregate is slightly stronger than the parent rock. Uncrushed aggregate contains more rounded particles that give higher workability and lower strength.

There are many different types of deposit:

1. Alluvial deposits arise from running water.
2. Glacial deposits that are less uniform than alluvial deposits.
3. Marine deposits that may be either coastal (beach), or seabed deposits.

Seabed deposits are worked extensively around the United Kingdom, United States, and Japan. Dredging activities have led to a deterioration of the deposits, as a result of the aggregate being returned to the sea after the removal of the preferred sizes. There are also concerns about loss of material from beaches that has been blamed on aggregate extraction even at considerable distances away.

19.5 ARTIFICIAL AGGREGATES
19.5.1 LIGHTWEIGHT AGGREGATE

These are generally produced from industrial by-products such as PFA, slag, or colliery shale. They are normally intended to reduce the weight of concrete, and have appropriate trade names such as "Lytag" that is sintered PFA. The greatest use is in precast concrete building blocks, but they are also used in concrete – such as bridge decks where weight reduction is a major benefit. Concrete made with light-weight artificial aggregates generally has: low density, reduced strength, low thermal conductivity, and good fire resistance. Weaker materials, such as vermiculite, will produce a very light-weight mix that is only suitable for non-structural applications.

19.5.2 HEAVYWEIGHT AGGREGATE

This is used for radiation shielding, and also for kentledge – such as counterweights for cranes – and may be steel shot, or any other heavy material. For general shielding purposes, a concrete density of 2600 kg/m^3 (4400 lb/yd^3) is adequate, and a heavy granite or basalt may suffice. Steel shot aggregate will give very high densities and shielding, but is expensive, and precautions must be taken to prevent segregation.

19.5.3 RECYCLED WASTE

Many types of waste material are now being tested and used as aggregates, even sintered domestic refuse. Sometimes, it is acceptable to use concrete production as a means of encapsulation of toxic wastes that would otherwise incur high disposal costs. More commonly, crushed demolition hardcore may be used for lower cost concrete. With all of these aggregates problems may arise because they are not physically sound or chemically inert.

19.6 MAJOR HAZARDS WITH AGGREGATES

19.6.1 UNSOUND OR CONTAMINATED AGGREGATE

Some aggregates turn out not to be as inert as they are usually assumed to be. In general, aggregates, which fail are classified as "unsound." The failure may be caused by moisture or frost but usually the aggregate in concrete is well protected, so failure in a test on plain aggregate may not indicate failure in concrete. Contamination of aggregate with clay or salt (common in the Middle East) is also obviously harmful to concrete.

19.6.2 ALKALI AGGREGATE REACTIONS

Some aggregates react in the alkaline environment in concrete. This reaction is expansive, and may be catastrophic for structures. There are many different reactions:

1. Alkali silica reaction (ASR) with siliceous rocks.
2. Alkali silicate reaction.
3. Alkali carbonate reaction.

The alkalis present in Portland cement are released during normal hydration. The pore fluid although saturated with calcium hydroxide, is largely a mixture of sodium and potassium hydroxides that are primarily responsible for the high alkalinity (Section 18.2.4). Cement producers limit alkalis to levels set in the standards, but environmental controls on cement works have had the effect of raising alkali contents, as processes have been changed in order to reduce emissions.

Many rock types used as aggregates contain forms of silica that can react with these alkalis in pore water. This reaction forms an alkali–silica gel that is hygroscopic, and the imbibition of the pore water causes it to swell. Where conditions are favorable for the production of sufficient gel and its subsequent expansion, the expansive forces created can be high enough to crack the concrete, or cause other structural distress.

The expansion will not always damage a structure, and a small amount of expansion in a prestressed structure may even be beneficial. The reaction stops if the concrete is dry; so, sometimes it is possible to add weather protection, and keep the structure in service. Damage caused by ASR is rare, when compared with other forms of damage, such as that from reinforcement corrosion.

The expansion is slow, so tests take a long time, but aggregates may be checked for reactivity with a microscope in thin sections; but the presence of the white reaction product on the surface of aggregate particles does not always correlate with failure. Mortar bars made with reactive aggregate and alkaline paste may also be tested for long-term expansion (Fig. 19.1). The bars are typically 280 mm (11 in.) long by 20 mm (3/4 in.) square, and are cast with metal studs in the ends in special moulds, so the studs locate into a measurement frame, which accurately measures the length using a digital dial gauge. If

FIGURE 19.1 Mortar Bar Expansion Test

Before testing, the gauge should be set to zero with an invar steel calibration bar in place of the mortar sample.

concrete samples must be tested with larger aggregate, shorter 40 mm (1.6 in.) square test bars may be made, but they will normally expand more slowly.

19.6.3 AGGREGATE SHRINKAGE

If the aggregate shrinks in concrete, the concrete itself shrinks. In, for example, a cantilever, this can cause major deflections. The only way to test aggregate for it is to test a mortar bar with the aggregate in it, and dry it and test for long-term expansion, as for the ASR measurement (the shrinkage occurs on drying). The difficulty occurs because the results are very variable and, if there is one bad result, it takes a long time to repeat the test, and a decision must be made about continuing construction.

19.6.4 POP-OUTS

Aggregates (often artificial) containing substances such as pyrites, that expand in moisture, cause "pop-outs," where a particle of coarse aggregate falls out from the concrete surface. The term "Mundic" is used to describe the mine wastes used in South-West England that cause this problem.

19.7 PROPERTIES OF AGGREGATES

19.7.1 GRADING

The grading of an aggregate is the proportion of particles of different sizes in it, and is determined with sieves (Figure 19.2). To a first approximation, the water demand attributable to the aggregate is proportional to the surface area, so attempts are made to minimise this. At the same time, it is desirable to maximise aggregate packing. Figure 19.3 shows a set of grading limits. When an aggregate is tested, its grading curve is checked to see if it lies between the lines.

Gap grading is grading with some size ranges missing. This gives better theoretical packing. Figure 19.4 shows a grading curve with gap grading, and Fig. 19.5 shows schematically how gap grading helps packing.

19.7.2 FINENESS MODULUS

This is defined as 1/100 of the sum of the cumulative percentages retained on the 150, 300, and 600 μm, and 1.18, 2.36, and 4.75 mm (0.006, 0.012, 0.023, 0.047, 3/32, and 3/16 in.) sieves and up to the largest size used. Note that if all the aggregate is retained on, say, the 300 μm sieve, the retained percentage

FIGURE 19.2 Sieves Stacked in Sieve Shaker

The largest sieve size is placed at the top and the sample is placed in it. After the sieve shaker has been run, the mass remaining in each sieve is measured.

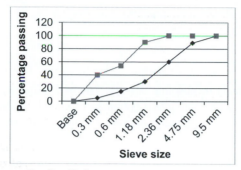

FIGURE 19.3 Fine Aggregate Coarse Grading Limits

(0.3 mm = 0.012 in. 0.6 mm = 0.023 in. 1.18 mm = 0.047 in. 2.36 mm = 3/32 in. 4.75 mm = 3/16 in. 9.5 mm = 3/8 in.)

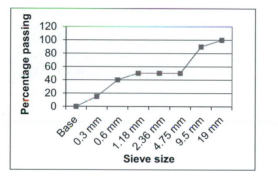

FIGURE 19.4 Gap Graded Aggregate

(0.3 mm = 0.012 in. 0.6 mm = 0.023 in. 1.18 mm = 0.047 in. 2.36 mm = 3/32 in. 4.75 mm = 3/16 in. 9.5 mm = 3/8 in. 19 mm = 3/4 in.)

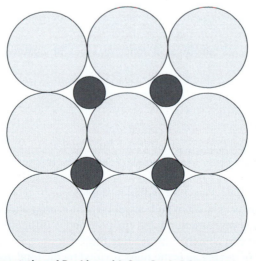

FIGURE 19.5 Schematic Representation of Packing with Gap-Graded Aggregate

The gaps between the large particles are only big enough for significantly smaller ones.

in the 150 μm sieve should still be entered as 100. The smaller the size of the fineness modulus, the finer the sand.

19.7.3 DENSITY

The density may be bulk density or absolute density. The bulk (dry-rodded) density is measured by compacting a sample of loose aggregate into a container of known volume, and measuring the mass. For the absolute density, the mass is normally measured saturated surface dry (Section 8.4).

19.7.4 MOISTURE CONTENT

This must be known in order to calculate the water requirement. It is obtained by measuring weight loss on drying.

19.7.5 STRENGTH

In general, the aggregate is far stronger than the paste, and the exact strength is not of great interest. The exceptions are very high strength paste and low strength (normally light-weight) aggregate. The strength may be measured either as a solid (e.g., a core of a parent rock), or as particles. If particles are crushed, the effect may be measured as the quantity of fines produced.

19.7.6 SHAPE/TEXTURE

This is obtained by inspection. Categories are:

1. Rounded
2. Irregular
3. Flaky
4. Angular
5. Elongated

Equidimensional particles are beneficial in concrete. Flaky/elongated particles are detrimental, because they reduce workability.

19.8 CONCLUSIONS

- Environmental concerns are limiting possibilities for aggregate extraction in many countries.
- Artificial aggregate may be made from secondary materials. It is often light-weight, and must be used with care.
- Heavyweight aggregate may be used for radiation shielding.
- Alkali aggregate reaction occurs when high alkali cement is used with a reactive aggregate.
- Aggregate shrinkage can cause structural deformation, and is difficult to test for.
- Aggregates normally have to comply with grading limits.
- The dry rodded density of aggregate is the mass divided by the total volume, including the gaps between the particles.

TUTORIAL QUESTIONS

1. A sample of aggregate is weighed on a balance that is arranged so the sample is placed in a cradle below it. A tank of water, positioned below the balance, may be raised up to submerge the sample in the cradle (Figure 8.3). The following observations are made:

Reading on balance in grammes	
0	Empty cradle suspended in air below the balance
−50	Empty cradle submerged in water below the balance
2000	Aggregate sample in cradle suspended in air
1200	Aggregate sample in cradle submerged in water

Calculate the absolute density.

Solution:

This calculation depends on the principle of Archimedes that states that the uplift, when the sample is submerged, is equal to the mass of water displaced.
First, the submerged mass is corrected for the 50 g uplift from the displacement of the cradle:
Corrected wet mass = 1200 + 50 = 1250 g
The uplift from the displacement of the aggregate
= dry mass–submerged mass
= 2000–1250 = 750 g = 0.75 kg
Thus, the volume of water displaced = uplift/density of water = 0.75/1000 = 7.5 × 10^{-4} m^3
The mass of the sample when weighed in air = 2000 g = 2 kg.
Thus, the density = mass/volume = 2/7.5 × 10^{-4} = 2667 kg/m^3

2. Three samples of aggregate are sieved to obtain the grading curves. A set of sieves of different sizes is stacked with the coarsest mesh at the top. The aggregate is then placed in the top sieve and the stack is placed in a sieve shaker. The amount remaining in each sieve after they have been shaken is shown in the table.

Sieve	Mass retained in sieve g		
	Sample A	Sample B	Sample C
19 mm	0	0	0
9.5 mm	150	1200	0
4.75 mm	200	0	0
2.36 mm	170	0	100
1.18 mm	150	0	200
600 μm	250	150	200
300 μm	200	200	100
150 μm	50	100	100
Base	35	54	16

a. Plot the grading curves.
b. Identify which samples are (i) sand, (ii) gap graded aggregate, and (iii) all in aggregate.
c. For the sand, calculate the percentage passing a 600 μm sieve, and the fineness modulus.

Solution:

a. First, the masses in the sieves are expressed as percentages of the total mass:

	Percentages		
	Sample A	**Sample B**	**Sample C**
19 mm	0.00	0.00	0.00
9.5 mm	12.45	70.42	0.00
4.75 mm	16.60	0.00	0.00
2.36 mm	14.11	0.00	13.97
1.18 mm	12.45	0.00	27.93
600 μm	20.75	8.80	27.93
300 μm	16.60	11.74	13.97
150 μm	4.15	5.87	13.97
Base	2.90	3.17	2.23

The percentages passing through each sieve are calculated by adding each row starting from the base. Thus, for sample A, the 7.05% passing through the 300 μm sieve is the sum of the 2.9% in the base and 4.15% in the 150 μm sieve. Similarly, the 23.65% passing through the 600 μm sieve is the sum of the 7.05% passing through the 300 μm sieve, and the 16.6% retained in it.

	Percentage passing		
Sieve	**Sample A**	**Sample B**	**Sample C**
19 mm	100.00	100.00	100.00
9.5 mm	87.55	29.58	100.00
4.75 mm	70.95	29.58	100.00
2.36 mm	56.85	29.58	86.03
1.18 mm	44.40	29.58	58.10
600 μm	23.65	20.77	30.17
300 μm	7.05	9.04	16.20
150 μm	2.90	3.17	2.23
Base	0	0	0

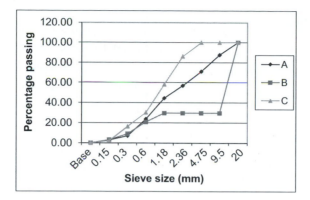

The key features of the grading curves are:
- The different sizes are shown equally spaced (or may be plotted on a logarithmic scale).
- All curves start at zero and end at 100%.

b. Sample A is all in aggregate, because the curve rises continuously.
Sample B is gap graded, because the curve has a horizontal section in the mid-range.
Sample C is sand, because the curve rises to 100% for the larger sizes.

c. The percentage passing the 600 μm sieve for sample C is 30.17 from the table above.

Sieve	Cumulative % retained
19 mm	0.00
9.5 mm	0.00
4.75 mm	0.00
2.36 mm	13.97
1.18 mm	41.90
600 μm	69.83
300 μm	83.80
150 μm	97.77
Total	307.26

The cumulative percentage retained in each sieve is calculated by adding each row starting from the top. Thus, the cumulative 41.90% in the 1.18 mm sieve is the sum of the 13.97% in the 2.36 mm sieve, and 27.93% in the 1.18 mm sieve. The fineness modulus is the total of 307.26 divided by 100 = 3.07.

3. A sample of aggregate is weighed on a balance that is arranged so the sample is placed in a cradle below it. A tank of water positioned below the balance may be raised up to submerge the sample in the cradle (Figure 8.3). The following observations are made:

Reading on balance in pounds	
0	Empty cradle suspended in air below the balance
−0.1	Empty cradle submerged in water below the balance
4	Aggregate sample in cradle suspended in air
2.4	Aggregate sample in cradle submerged in water

Calculate the absolute density.

Solution:

This calculation depends on the principle of Archimedes that states that the uplift, when the sample is submerged, is equal to the mass of water displaced.

First, the submerged mass is corrected for the 0.1 lb. uplift from the displacement of the cradle:

Corrected wet mass = 2.4 + 0.1 = 2.41 lb

The uplift from the displacement of the aggregate

= dry mass–submerged mass

= 4–2.41 = 1.59 lb.

Thus, the volume of water displaced = uplift/density of water = 1.59/1686 = 9.43×10^{-4} yd^3 (Table 1.6)

The mass of the sample when weighed in air = 4 lb.

Thus, the density = mass/volume = 4/9.43 $\times 10^{-4}$ = 4241 lb/yd^3

HYDRATION OF CEMENT

20.1 INTRODUCTION

When cement is mixed with water, it hydrates. This is a chemical reaction and can take place in the absence of air, and the environment in which it does take place has a considerable influence on the strength, and durability of the concrete. This chapter discusses the heat generated by the reaction, the effect of surplus water in the mix, and the effect of curing, in particular drying after casting. The key conclusion is that concrete should have a low water content when cast, but, as soon as the rigid hydrate is formed, it should be kept as wet as possible during curing. Adding extra water to the mix creates voids (pores) that provide pathways for the transport of ions such as chlorides into the concrete that will then corrode the reinforcement. However, adding water during curing will enable the hydration reaction to complete and reduce the porosity.

20.2 HEAT OF HYDRATION

Cement gives off about 0.5 MJ/kg of heat when it hydrates (Section 4.8). An idealised curve of heat output is shown in Fig. 20.1. GGBS and PFA delay heat generation (Fig. 20.2).

The maximum temperature of the concrete will depend more on the rate of heat generation, rather than the total heat generated. If the heat is produced more slowly, it will have time to dissipate into the surrounding environment. Thus, if PFA or GGBS is used, the peak temperature will be lower despite the fact that the total heat produced (i.e., the area under the curves in Figs 20.1 and 20.2) would be similar to OPC if the curves were drawn for a period of months.

The heat of hydration can be useful on sites in cool climates because it will prevent freezing, and will also accelerate the hydration itself that will accelerate strength gain, and permit the removal of false work. However, if the temperature rises significantly, the differential expansion can cause cracking. A

189

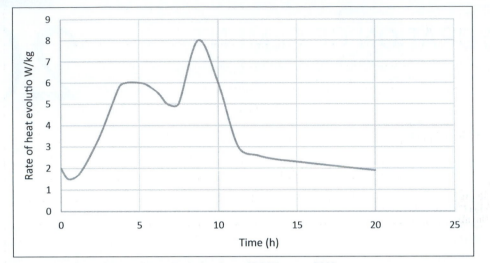

FIGURE 20.1 Typical Plot of Rate of Heat Evolution From OPC/Water at 20°C

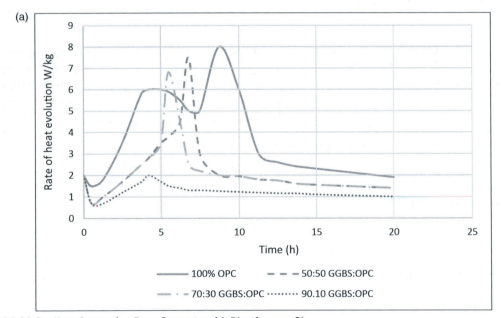

FIGURE 20.2a Heat Generation From Concrete with Blastfurnace Slag

(b)

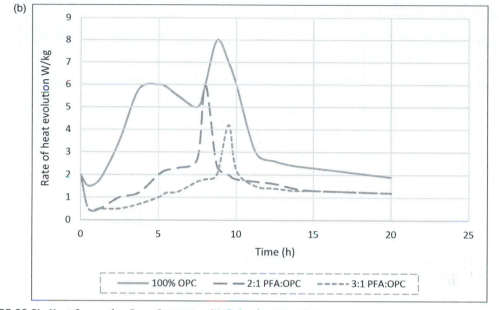

FIGURE 20.2b Heat Generation From Concrete with Pulverised Fuel Ash

general guide is that if there is a temperature difference of more than 20°C (36°F) between any two parts of a pour, then cracking may occur. In hot climates, ice may be added as part of the mix water.

In precasting works, it is common to apply additional heating, in order to accelerate strength gain. However, the temperature is generally kept below 80°C (175°F) because delayed ettringite formation (DEF) can occur, that causes long-term deterioration of the concrete (Section 25.8). Temperatures above 100°C (212°F) would normally cause the mix water to boil, but this can be avoided by applying pressure (Section 8.3), and a process called "autoclaving" is used to produce small precast items, such as kerbs, using temperatures around 150°C (300°F) to give very rapid rates of production. The hydrates produced in autoclaved concrete are quite different from those in normal concrete.

The heat of hydration may be measured on site by casting thermocouples into the concrete. These are inexpensive electrical probes that work by measuring the voltage across the junction of two different metals. The temperature in the pour may be checked with a hand-held reader, or a data logger. For quick checks of the concrete temperature, hand-held infrared thermometers that do not require contact with the material can be used. The output from thermocouples may be used to run a temperature-matched curing tank, in which tests cubes are held at the same temperature as the pour, so that when they are tested they give an indication of the strength of the concrete in the pour.

20.3 TYPES OF POROSITY

Concrete has a porosity of around 10% (Section 8.4 for a definition of porosity). The types of porosity are shown in Table 20.1.

Table 20.1 Size Ranges for Porosity

Size Range (m)	
10^{-10}–10^{-9}	The diameter of the ions such as, S^{2-}, Cl^-, O^{2-}
10^{-9}–10^{-8}	Gel pores: these are part of the hydrated cement structure, they are not interconnected, and do not affect durability.
10^{-8}–10^{-6}	Capillary pores: these are connected, and considerably affect durability. Their volume may be calculated (see below). They are formed by excess water that does not react with the cement, either due to a high w/c ratio, or insufficient hydration due to poor curing.
10^{-4}–10^{-3}	Entrained air bubbles: not interconnected. They are produced with air entraining admixtures to give frost resistance (Chapter 24).

The capillary pores are of great interest, because they influence the transport properties and, thus, the durability of concrete.

20.4 CALCULATION OF POROSITY

The following is a method for calculating the porosity of a hydrated cement sample, from knowledge of the mix proportions, and the dry weight of a hydrated sample. By far the easiest way to carry out these calculations is to complete a table as in Table 20.2. This calculation is presented for a sample with no aggregate. Some of the tutorial examples show how it is easily extended to include aggregate.

Table 20.2 Calculation of Porosity. Equation Numbers are in Brackets

	Before Hydration		After Hydration and Drying	
	Mass	**Volume**	**Mass**	**Volume**
Cement	M_C	$M_C/3.15$ (20.1)	$M_{UHC} = M_C - (4 \times M_{CW})$ (20.9)	$M_{UHC}/3.15$
Water	$M_W = M_C \times W$ (20.2)	M_W (20.3)	0	0
Hydrated cement	0	0	$M_{HC} = 5 \times M_{CW}$ (20.8)	$M_{HC}/2.15$
Pores	0	0	0	$V_T - M_{HC}/2.15 - M_{UHC}/3.15$
Aggregate				
Total	$M_C + M_W$ (20.4)	V_T	M_D	V_T

Table 20.3 Assumptions Used in the Calculation of Porosity

Specific gravity of unhydrated cement	3.15 kg/L
Specific gravity of hydrated cement	2.15 kg/L
Ratio of combination of water and cement in hydration reaction	4 parts cement to 1 part water (by mass)

The specific gravity is quoted in kilogrammes per litre because the density of water is 1 in these units.

The calculation uses the assumptions in Table 20.3.

Consider the hydration of a mass M_C kg of cement at a w/c ratio of W (with no aggregate).

The specific gravity of unhydrated cement is 3.15 kg/L

$$\text{Thus the volume of the cement is } M_C / 3.15 \, L \tag{20.1}$$

$$\text{The mass of water is } M_W = M_C \times W \text{ kg} \tag{20.2}$$

$$\text{The volume of water is also } M_C \times W \text{ L} \tag{20.3}$$

because the density of water is 1 kg/L

$$\text{Thus, the total wet volume is } V_T = M_C / 3.15 + M_W \text{ L} \tag{20.4}$$

Not all of the cement will hydrate. The extent of the hydration may be measured by weighing the hydrated sample after drying it.

If the dry mass is M_D kg

$$\text{The mass of free water that has dried off is } M_C + M_W - M_D \text{ kg} \tag{20.5}$$

$$\text{Thus, the mass of combined water is } M_{CW} = M_W - (M_C + M_W - M_D) \text{kg} \tag{20.6}$$

We now assume that the water combines with the cement in a ratio of 1:4. This assumption has been found to be quite good for a variety of OPCs

$$\text{The mass of cement which has hydrated is thus } 4 \times M_{CW} \text{ kg} \tag{20.7}$$

$$\text{The total mass of hydrated cement is } M_{HC} = 5 \times M_{CW} \text{ kg} \tag{20.8}$$

and the mass of unhydrated cement is the original mass – the mass that has hydrated

$$M_{UHC} = M_C - (4 \times M_{CW}) \text{kg} \tag{20.9}$$

The specific gravity of hydrated cement is approximately 2.15 kg/L.

The bulk volume of the hydrated sample may be assumed to be V_T, the same as the volume of the wet mix (i.e., zero shrinkage is assumed).

The volume of pores will be the total volume V_T – the volumes of the hydrated and unhydrated cement

$$\text{Pore volume} = V_T - M_{HC}/2.15 - M_{UHC}/3.15\,\text{L} \qquad (20.10)$$

and

$$\text{The porosity will be:} \frac{\text{Pore volume}}{V_T} \qquad (20.11)$$

Figure 20.3 shows the calculated volumes for a sample with a water/cement (w/c) ratio of 0.4 (see tutorial question 1). It is clear from this that it is essential to cure concrete well, so the hydration can continue, and the porosity will reduce.

Figure 20.4 shows the porosity for samples in which all of the cement has hydrated. This is almost impossible to achieve in practice, but it illustrates the substantial increase in porosity that will occur in all mixes, if the w/c ratio is increased.

In Figure 20.4, it may be seen that the porosity goes negative at w/c = 0.265. This means that there is no space remaining, and hydration will stop. This is actually observed in lab experiments at about this w/c ratio. Clearly, if the assumption is made that the cement and water combine at a ratio of 4:1, the w/c ratio cannot go below 0.25. In current projects using high range superplasticisers (Chapter 24), w/c ratios in the range 0.3–0.35 are being used to give very high durability.

20.5 INFLUENCE OF POROSITY

Figure 20.5 shows the classic relationship between w/c ratio and permeability that was developed over 50 years ago. The influence of porosity and permeability on the durability of concrete is discussed in Section 25.2.5. If extra water is added to a truck mixer on site, then it is clear that, as well as the strength, the durability of the concrete will be substantially reduced.

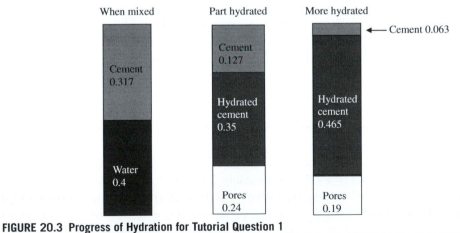

FIGURE 20.3 Progress of Hydration for Tutorial Question 1

Volumes are in litres. Note that the total volume is constant.

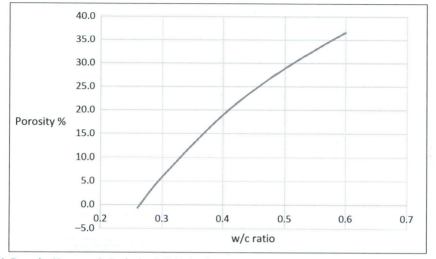

FIGURE 20.4 Porosity Versus w/c Ratio for Full Hydration

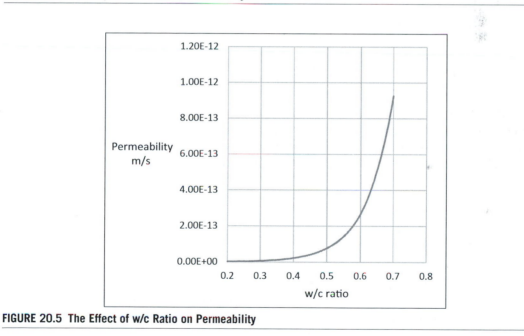

FIGURE 20.5 The Effect of w/c Ratio on Permeability

Figure 20.6 shows details of the reduction of porosity with continuing hydration, shown in Figure 20.3. These curves show that, as hydration proceeds, almost all the larger pores are filled, and there is a minor increase in the volume of the smallest pores at long ages. These results were obtained by mercury intrusion, in which mercury is forced into the sample at high pressures, and the size of pore it will enter is proportional to the pressure applied. Thus, it is conventional to draw the graph with the smallest pores on the right hand side (as shown), because these correspond to the highest pressures.

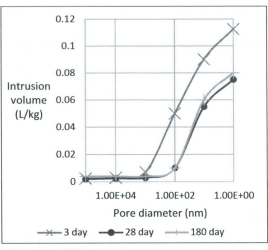

FIGURE 20.6 Cumulative Pore Volume at Different Ages

20.6 CURING

Even if the concrete has a good low w/c when it is placed, it will still not make a durable structure unless it is cured well. Curing has two distinct functions:

1. To stop the concrete from drying out during hydration. If this occurs, there will be a significant loss of durability. The damage is normally irreversible. If concrete is dried, the cement grains will form an impermeable layer of hydration product around them, and this will prevent any further hydration, however many times it is wetted again.
2. To retain heat at the surface. This may be done for the following reasons
    a. To prevent frost damage (below 5°C, 40°F)
    b. To increase early strength
    c. To reduce temperature gradients

While low temperatures will reduce early strength, the effect is not permanent if the concrete has not frozen, and it is subsequently kept at higher temperatures. Thus, a sample that is kept at 5°C (40°F) will not fully hydrate (particularly if it contains a pozzolan), but even after several months it will hydrate further when the temperature is raised.

Because they hydrate more slowly, cements containing pozzolans and GGBS generally require longer curing. It is therefore essential that these concretes are identified on site, and cured adequately. The pozzolanic reaction will then create additional hydration products that can block some of the pores between the cement grains, and achieve good durability.

Methods of curing are discussed in Chapter 26. Typical specified minimum curing times are 3 days for OPC, and 5 days for pozzolanic mixes.

20.7 CONCLUSIONS

- Heat is evolved during the hydration of cement in concrete, and this may cause cracking and other problems.
- Cement replacements may be used to reduce the effect of the heat of hydration.
- Capillary porosity in hydrated cement is caused by the water needed to give workability.
- The porosity may be calculated from the weight loss, when a hydrated sample is dried.
- The porosity is a key indicator of durability.
- Curing is essential for durability, and is carried out to retain heat and prevent drying.

TUTORIAL QUESTIONS

1. a. A sample of cement paste is made with a w/c ratio of 0.4. The sample is cured and dried, and the dry weight is 15% greater than the original mass of cement. What is the porosity?

   b. If an identical sample is allowed to hydrate until the dry weight is 20% greater than the original mass of the cement, what does the porosity become?

What percentage of the cement has hydrated in each case?

Solution:

a. In these questions, it is always necessary to consider a fixed quantity of cement. This is necessary to do the calculations, but the amount considered will not affect the calculated porosity. Thus, we consider 1 kg of cement ($M_C = 1$)

The solution is shown in the table below
Considering the materials before hydration

The volume of the cement $= 1/3.15 = 0.32$ L   (20.1)

The mass of water $= 1 \times 0.4 = 0.4$ kg   (20.2)

The volume of the water $= 0.4$ L   (20.3)

Thus the total wet volume $= 0.32 + 0.4 = 0.72$ L   (20.4)

The final mass M_D is 15% greater than the original mass of cement.
Thus $M_D = M_C \times 1.15 = 1.15$ kg
The mass of free water that had dried off, i.e., the water lost, is the total initial
mass – the dry mass $= 1 + 0.4 - 1.15 = 0.25$ kg   (20.5)

Thus, the mass of combined water in the hydrated cement (HC) (i.e., the rest of the water)
$M_{CW} = 0.4 - 0.25 = 0.15$ kg   (20.6)

It is assumed that the cement and water combine in the ratio 4:1.
Thus the mass of cement in the hydrated cement $= 4 \times 0.15 = 0.6$ kg   (20.7)
and the total mass of the hydrated cement $M_{HC} = 5 \times 0.15 = 0.75$ kg   (20.8)
The mass of unhydrated cement will be $M_{UHC} = 1 - 0.6 = 0.4$ kg   (20.9)

The volume of unhydrated cement is calculated from the specific gravity:

volume = 0.4/3.15 = 0.13 L

and similarly for the hydrated cement volume = 0.75/2.15 = 0.35

The pore volume is obtained by subtraction = 0.72–0.13–0.35 = 0.24 L (20.10)

The porosity is the pore volume/the total volume = 0.24/0.72 = 0.337 = 33.7% (20.11)

The fraction hydrated is 0.6/1 = 60%

	Before Hydration		After Hydration	
	Mass (kg)	Vol (L)	Mass (kg)	Vol (L)
Cement	1.0	0.32	0.4	0.13
Water	0.4	0.40		
Hyd Cem	0.0	0.00	0.75	0.35
Pores	0.0	0.00		0.24
Total	1.4	0.72	1.15	0.72
Water lost	0.25			
Water in HC	0.15		porosity %	33.7
Cem in HC	0.60		% hydrated	60
Total HC	0.75			

b.

	Before Hydration		After Hydration	
	Mass (kg)	Vol (L)	Mass (kg)	Vol (L)
Cement	1.0	0.32	0.2	0.06
Water	0.4	0.40		
Hyd Cem	0.0	0.00	1.0	0.47
Pores	0.0	0.00		0.19
Total	1.4	0.72	1.2	0.72
Water lost	0.20			
Water in HC	0.20		porosity %	26.3
Cem in HC	0.80		% hydrated	80
Total HC	1.00			

The final volumes are shown in Figure 20.3

2. Assuming that the specific gravity of unhydrated cement is 3.15, and for hydrated cement it is 2.15, draw a graph of porosity against w/c ratio for fully hydrated cement paste mixes. From this graph, obtain a minimum w/c ratio for full hydration.

Solution:

To solve this question, the porosity must be calculated at various w/c ratios and plotted.
Considering w/c = 0.3
The calculation is completed as in question 1, but for full hydration. Thus, the mass of cement in the hydrated cement is the total mass of cement, and there is no cement remaining after hydration. Thus, the dry mass is not needed for the calculation.

	Before Hydration		After Hydration	
	Mass (kg)	**Vol (L)**	**Mass (kg)**	**Vol (L)**
Cement	1.0	0.32	0.0	0.00
Water	0.3	0.30		
Hyd Cem	0.0	0.00	1.25	0.58
Pores	0.0	0.00		0.04
Total	1.3	0.617		0.62
Water in HC	0.3		porosity %	5.8
Cem in HC	1.0		% hydrated	100
Total HC	1.3			

This calculation is repeated for different w/c ratios

w/c	porosity %
0.26	−0.7
0.3	5.8
0.4	19.0
0.5	28.9
0.6	36.6

These are plotted in Figure 20.4. The porosity goes negative (i.e., no space for further hydration) at approximately w/c = 0.265.

3. A concrete sample is made with a w/c ratio of 0.3, and a cement content of 300 kg/m^3, and an aggregate content of 2010 kg/m^3. A sample is dried, and has a density of 2355 kg/m^3. What percentage of the cement has hydrated, and what is the porosity?

Solution:

This calculation is similar to question 1, but it is easiest to consider 1 m^3 (1000 L) of concrete rather, than 1 kg of cement.
The water content is obtained from the w/c ratio: water = 300 × 0.3 = 90 kg

This permits calculation of the total initial mass.
The aggregate volume is obtained by subtraction:

volume = 1000–95–90 = 815 L.

The aggregate mass and volume remain constant during hydration and drying. The calculation is then completed as in question 1.

	Before Hydration		After Hydration	
	Mass (kg)	**Vol (L)**	**Mass (kg)**	**Vol (L)**
Cement	300	95	120	38
Water	90	90		
Hyd Cem	0	0	225	105
Pores	0	0		42
Aggregate	2010	815	2010	815
Total	2400	1000	2355	1000
Water lost	45			
Water in HC	45		porosity %	4.2
Cem in HC	180		% hydrated	60
Total HC	225			

NOTATION

M_C Mass of cement (kg)
M_w Mass of water (kg)
M_D Dry mass (kg)
M_{CW} Mass of combined water (kg)
M_{HC} Mass of hydrated cement (kg)
M_{UHC} Mass of unhydrated cement (kg)
V_T Total volume (L)
W Water/cement ratio

CONCRETE MIX DESIGN

21

21.1 INTRODUCTION

21.1.1 THE BASIC PROCESS

"Mix design" or "mixture proportioning" is used to calculate the quantities of the different constituents required to achieve the properties that are specified for a batch of concrete. The design properties are, normally, the workability and the compressive strength. The slump test that measures workability, and the compressive test for strength, are described in detail in Chapter 22. In this chapter, two methods of mix design are considered; the first is common in the United Kingdom (Design of Normal Concrete Mixes, Building Research Establishment report BR331), and the second in the United States (Standard Practice for Selecting Proportions for Normal, Heavyweight, and Mass Concrete, American Concrete Institute Standard 211.1).

The basic steps of mix design are as follows:

- Establish the water/cement (w/c) ratio for the required strength. The approximate relationship between strength and w/c ratio is:

$$\frac{\text{Water}}{\text{Cement}} \propto \frac{1}{\text{Strength}}$$

- Establish the total water content for the required workability.
- Divide the water content by the w/c ratio to calculate the cement content.
- The only remaining component is the aggregate. This must fill the remaining volume.

21.1.2 SPECIFYING FOR DURABILITY

Referring to the definitions in Chapter 14, basic mix design works with a performance specification. This means that the performance (i.e., the strength and workability) is specified, and it is up to the concrete supplier to design a mix to achieve it. The main problem with this is that, while stronger concretes are often more durable, simply depending on strength specifications to achieve durability has resulted in numerous structures failing to perform adequately during their design life. This problem is addressed by bringing in an element of method specification, by requiring a maximum w/c ratio, or minimum cement content for a given type of exposure. These values would be substituted for the calculated values during the mix design process. Some clients also specify performance in tests for durability, such as the "rapid chloride" test (see Section 22.8.3). However, these requirements are met by testing trial mixes, rather than any specific method of calculating a mix design.

21.1.3 USING WATER REDUCERS AND CEMENT REPLACEMENTS

The key problem that the concrete mix design must address is that adding water will increase the workability, but decrease the strength. The only simple way to increase both the workability and strength is to add more cement, and this increases the cost. The solution to these conflicting requirements is to use a plasticising admixture (see Section 24.2).

Most concrete also contains cement replacements of the type described in Chapter 18. These bring benefits of reduced cost, reduced environmental impact, and often improved durability.

The initial methods presented in this chapter do not, however, include the use of cement replacements or admixtures. Modifications to these basic mix designs to include these materials are discussed in Sections 21.4, 21.5, and 21.6.

21.1.4 MIX DESIGN IN INDUSTRY

In practical applications, mix designs are almost always developed by modifying existing designs to overcome specific problems. It is, however, important to understand the principles of the design process, in order to see the effect of any adjustments to the material proportions. For industrial use, it may be seen that the detailed methods are well suited to calculation on spreadsheets or similar programmes. A number of packages are available for this.

21.1.5 MEASUREMENT OF STRENGTH

The strength results of concrete obtained by testing will form a statistical distribution, as described in Chapter 12. To obtain the design strength, it is necessary to calculate the margin, which is then added to the *characteristic strength* (that comes from the structural design) to form the *target mean strength* (that is used in the mix design), as in equation (12.3). This procedure is the same, no matter what method of mix design is being used.

The US method is based on cylinder strengths, and the UK method is based on cube strengths. The difference between these is discussed in Section 22.5.5.

21.1.6 WATER IN THE AGGREGATE

These methods assume that the aggregate is in a saturated surface dry condition. If the aggregates are wetter or drier than this, the mix quantities, especially the water content, must be adjusted to allow for the differences.

21.1.7 TRADITIONAL MIXES

Historically, many concrete structures were built with 1:2:4 mixes (1 part cement to 2 parts sand to 4 parts coarse aggregate), and many of them survive today in good condition. If the data is not available for a full mix design, a 1:2:4 mix may work well for minor works.

21.2 UK MIX DESIGN
21.2.1 STAGE 1. CALCULATING THE TARGET MEAN STRENGTH

From the specified proportion of defectives, and the known or assumed standard deviation (see Fig. 21.1), calculate the margin and add this to the characteristic strength to obtain the target mean strength using equation (12.3).

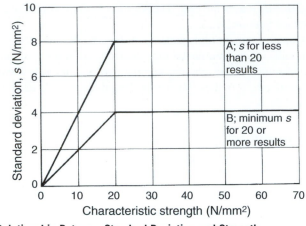

FIGURE 21.1 Assumed Relationship Between Standard Deviation and Strength

© IHS, reproduced from BRE BR 331 with permission.

21.2.2 STAGE 2. INITIAL STRENGTH VALUE

For the specified materials and test age, find the strength that would be expected with a w/c ratio of 0.5 from Table 21.1.

21.2.3 STAGE 3. DRAW THE CURVE FOR THE W/C RATIO

On Fig. 21.2 plot the point found at stage 2 above. Draw a curve through this point parallel to the existing curves on the figure. These curves represent the relationship between strength and w/c ratio for the particular materials, and test age, being used.

21.2.4 STAGE 4. OBTAIN THE W/C RATIO

On the curve drawn in Fig. 21.2, locate the target mean strength, and read off from the bottom axis the required water/cement ratio.

Table 21.1 Approximate Compressive Strengths (MPa) of Concrete Mixes Made With a Free w/c Ratio of 0.5

Type of Cement	Type of Coarse Aggregate	Age (Days)			
		3	**7**	**28**	**90**
Ordinary Portland (OPC) or Sulphate Resisting (SRPC) EN Class 42.5	Uncrushed	22	30	42	49
	Crushed	27	36	49	56
Rapid Hardening (RHPC) EN Class 52.5	Uncrushed	29	37	48	54
	Crushed	34	43	55	61

© IHS, reproduced from BRE BR 331 with permission.

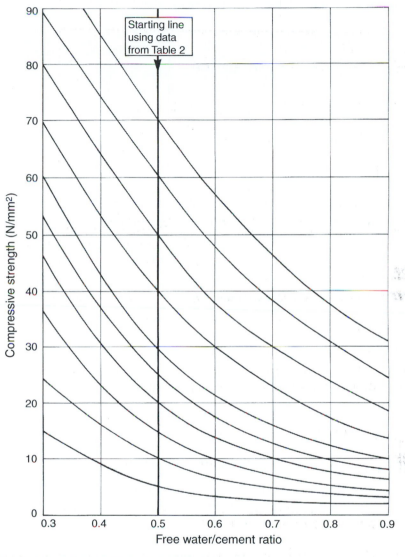

FIGURE 21.2 Relationship Between Strength and w/c Ratio

21.2.5 STAGE 5. CHOICE OF WATER CONTENT

From Table 21.2, obtain the free water content required to give the specified workability.

21.2.6 STAGE 6. CALCULATE CEMENT CONTENT

Calculate the cement content as the water content divided by the w/c ratio.

Table 21.2 Approximate Free Water Contents (kg/m³) Required to Give Various Levels of Workability

Slump (mm)		0–10	10–30	30–60	60–180
Maximum Aggregate Size (mm)	**Type of Aggregate**				
10	Uncrushed	150	180	205	225
	Crushed	180	205	230	250
20	Uncrushed	135	160	180	195
	Crushed	170	190	210	225
40	Uncrushed	115	140	160	175
	Crushed	155	175	190	205

© IHS, reproduced from BRE BR 331 with permission.

21.2.7 STAGE 7. ESTIMATE WET DENSITY

Using the known or assumed value for relative density (specific gravity) of aggregate, from Fig. 21.3, obtain an estimate of the wet density of the concrete. Note that a given aggregate density of 2500 kg/m³ would be used in Fig. 21.3 as a relative density of 2.5.

21.2.8 STAGE 8. CALCULATE AGGREGATE CONTENT

Calculate the total aggregate content by subtracting cement and water content from the concrete density.

21.2.9 STAGE 9. OBTAIN PROPORTION OF FINE AGGREGATE

For the specified fine aggregate grading (given as the percentage passing a 600 μm sieve), obtain an estimate for the proportion of fine aggregate in the total aggregate, depending on the maximum aggregate size, workability, and w/c ratio from Fig. 21.4.

21.2.10 STAGE 10. CALCULATE FINE AND COARSE AGGREGATE QUANTITIES

Calculate the fine aggregate content by multiplying the total aggregate content by the proportion of fine aggregate. Calculate the coarse aggregate content by subtracting fine aggregate content from total aggregate content.

21.2.11 STAGE 11. CALCULATE QUANTITIES FOR A TRIAL MIX

The quantities for each material are multiplied by the volume required for the trial mix.

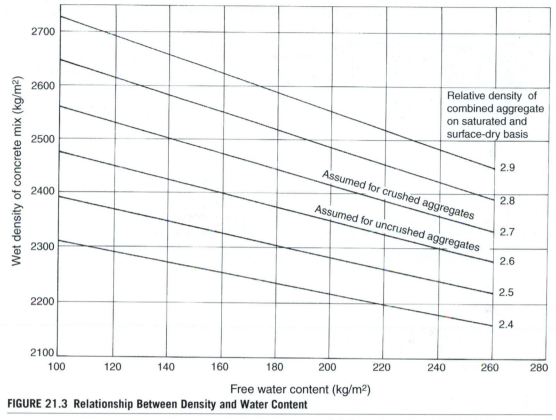

FIGURE 21.3 Relationship Between Density and Water Content

© IHS, reproduced from BRE BR 331 with permission.

21.3 US MIX DESIGN
21.3.1 STEP 1. CHOICE OF SLUMP
The workability is selected to suit the application. Table 21.3 gives guidance for suitable values.

21.3.2 STEP 2. CHOICE OF NOMINAL MAXIMUM SIZE OF AGGREGATE
The maximum aggregate size will also depend on the application. Most ready-mixed concrete has a maximum aggregate size of 20 mm (3/4 in.), and sizes above 40 mm (1.5 in.) are rarely used, except in large pours for very specialised applications, so they are not considered in this discussion.

21.3.3 STEP 3. ESTIMATION OF WATER CONTENT
Table 21.4 gives the water contents in kilogrammes per cubic metre required to achieve the necessary slump. It also gives the amount of entrapped air expected in the mix. This is not entrained air for

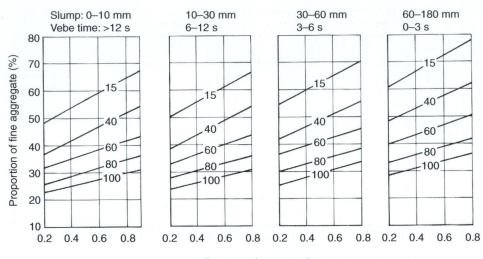

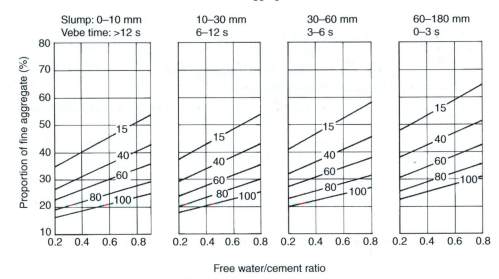

FIGURE 21.4 Relationship Between Proportion of Fine Aggregate and w/c Ratio

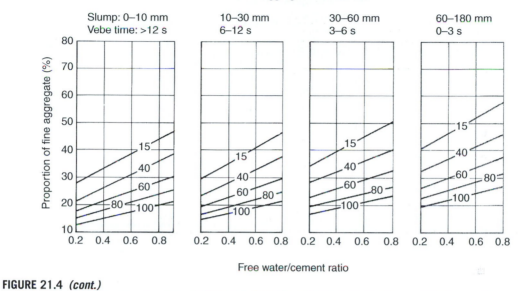

FIGURE 21.4 *(cont.)*

Table 21.3 Recommended Values for Slump		
	Slump mm (in.)	
Type of Construction	**Maximum**	**Minimum**
Reinforced foundation walls and footings	75 (3)	25 (1)
Plain footings, caissons, and substructure walls	75 (3)	25 (1)
Beams and reinforced walls	100 (4)	25 (1)
Building columns	100 (4)	25 (1)
Pavements and slabs	75 (3)	25 (1)
Mass concrete	75 (3)	25 (1)

Data taken from American Concrete Institute Standard 211.1 and reproduced with the permission of the ACI.

freeze–thaw resistance, it is air that has not been removed by the normal amount of compaction (vibration) used on site.

21.3.4 STEP 4. SELECTION OF WATER/CEMENT RATIO

The w/c ratio is selected using the target mean strength in Table 21.5.

Table 21.4 Estimated Water Contents in kg/m³ (lb/yd³)

Slump mm (in.)	Maximum Aggregate Size mm (in.)				
	9.5 (3/8)	12.5 (1/2)	19 (3/4)	25 (1)	37.5 (1.5)
25–50 (1–2)	207 (350)	199 (335)	190 (315)	179 (300)	166 (275)
75–100 (3–4)	228 (385)	216 (365)	205 (340)	193 (325)	181 (300)
150–175 (6–7)	243 (410)	228 (385)	216 (360)	202 (340)	190 (315)
Approximate amount of entrapped air (%)	3	2.5	2	1.5	1

Data taken from American Concrete Institute Standard 211.1 and reproduced with the permission of the ACI.

Table 21.5 Recommended w/c Ratios

Compressive Strength MPa (psi) Measured on Cylinder in Compression	Water/Cement Ratio
40 (5800)	0.42
35 (5075)	0.47
30 (4350)	0.54
25 (3625)	0.61
20 (2900)	0.69
15 (2175)	0.79

Data taken from American Concrete Institute Standard 211.1 and reproduced with the permission of the ACI.

21.3.5 STEP 5. CALCULATION OF CEMENT CONTENT

This is calculated as the water content divided by the w/c ratio.

21.3.6 STEP 6. ESTIMATION OF COARSE AGGREGATE CONTENT

Table 21.6 gives the volume of dry-rodded coarse aggregate per unit volume of concrete. The mass of coarse aggregate per cubic metre (per cubic yard) is obtained by multiplying this by the dry-rodded bulk density of the coarse aggregate, in kilogramme per cubic metre (lb/yd³). This would be obtained by compacting (dry rodding) a sample of coarse aggregate into a container of known volume, and measuring its mass (see Section 8.4).

21.3.7 STEP 7. ESTIMATION OF FINE AGGREGATE CONTENT

There are two options for this. Estimation by mass or volume:

Mass basis: Table 21.7 is used to estimate the density of the concrete.

The density is the total mass of 1 m³ (yd³). The mass of the cement, water, and coarse aggregate is subtracted from it to give the mass of fine aggregate.

Absolute volume basis: Since the quantities of all the other mix components are known, the fine aggregate volume is obtained by subtraction. For each material, the volume is the mass divided by the absolute density (as defined in Section 8.4).

Table 21.6 Dry-Rodded Volume of Coarse Aggregate Per Unit Volume of Concrete, for Each Size, and Fineness Modulus of the Aggregates

Nominal Maximum Size of Coarse Aggregate mm (in.)	Fineness Modulus of Fine Aggregate			
	2.4	**2.6**	**2.8**	**3.0**
9.5 (3/8)	0.50	0.48	0.46	0.44
12.5 (1/2)	0.59	0.57	0.55	0.53
19 (3/4)	0.66	0.64	0.62	0.60
25 (1)	0.71	0.69	0.67	0.65
37.5 (1.5)	0.75	0.73	0.71	0.69

Data taken from American Concrete Institute Standard 211.1 and reproduced with the permission of the ACI.

Table 21.7 First Estimate of Density of Fresh Concrete

Nominal Aggregate Size mm (in.)	Estimated Density kg/m^3 (lb/yd^3)
9.5 (3/8)	2280 (3840)
12.5 (1/2)	2310 (3890)
19 (3/4)	2345 (3960)
25 (1)	2380 (4010)
37.5 (1.5)	2410 (4070)

Data taken from American Concrete Institute Standard 211.1 and reproduced with the permission of the ACI.

The mass of cement from step 5, in kilogrammes per cubic metre (lb/yd^3), is divided by the absolute density, also in kilogrammes per cubic metre (lb/yd^3) (normally taken as 3150 kg/m^3 = 3.15 kg/L = 5311 lb/yd^3) to give the volume in cubic metres per cubic metre (cubic yards per cubic yard) of concrete.

The coarse aggregate quantity from step 6 is similarly divided by the absolute density.

The water mass from step 3 is divided by the density of water = 1000 kg/m^3 (1686 lb/yd^3)

The volume of entrapped air from Table 21.4 is expressed as a percentage, so it must be divided by 100 to give a fraction of the 1 m^3 (1 yd^3) total volume.

The total volume is 1 m^3 (1 yd^3), so all of these volumes are subtracted from 1 to give the volume of fine aggregate. This is then multiplied by the absolute density to give the mass.

21.4 MIX DESIGN WITH CEMENT REPLACEMENTS

When designing with cement replacements, it is essential to remember that these are generally waste materials and, although they are subject to quality control by the supplier, they may vary considerably, so trial mixes are essential. Their properties are discussed in Chapter 18.

PFA will generally increase workability and reduce strength. A reduction of 3% in the water content may be used for each 10% increase in the proportion of PFA relative to the total cementitious content. For strength, the concept of the efficiency factor may be used. This is a measure of the relative contribution to strength in the mix. It varies generally between 0.2 and 0.45. If a value of 0.3 is used, this means that 10 kg of PFA will be equivalent to 3 kg of cement, when calculating the strength. The efficiency factor concept only works when the PFA content is less than 40% of the total cementitious content. If it is more than 40%, some of it will not react, so it will not contribute to the strength.

GGBS will normally (but not always) act as a water reducing agent. The efficiency factor may be as high as 1 at 28 days. Silica fume substantially decreases workability and increases strength.

21.5 MIX DESIGN FOR AIR-ENTRAINED CONCRETE

21.5.1 EFFECT OF AIR ENTRAINERS

Air entrained mixes are produced with the addition of an air entraining admixture, and are normally used to prevent freeze–thaw damage. The air entrainer will improve the workability and reduce the strength and the density (see Section 24.4).

21.5.2 UK PRACTICE

In UK practice, the mix design process is adjusted as follows:

- It is assumed that there is a 5.5% loss of strength for each 1% by volume of entrained air. The target mean strength is increased to account for this.
- The free water content is reduced by one level of workability in Table 21.2.
- The calculated density is reduced by an amount equal to the saturated surface dry density of the aggregate × the volume fraction of entrained air.

21.5.3 US PRACTICE

In US practice, the water contents in Table 21.4 are reduced as shown in Table 21.8.
The w/c ratios in Table 21.5 are reduced as shown in Table 21.9.
The density in Table 21.7 is reduced as shown in Table 21.10.

21.6 MIX DESIGN FOR SELF-COMPACTING CONCRETE

If enough superplasticiser (or water) is added to a mix, it will flow a lot, and self-compact. The problem is that it will also segregate (i.e., the aggregate will fall to the bottom). The problem can be solved by either having a high powder content (typically 500 kg/m^3 (850 lb/yd^3), as opposed to about 350 kg/m^3 (600 lb/yd^3) in a typical mix), or adding a viscosity modifying admixture (VMA) (see Sections 22.2

Table 21.8 Estimated Water Contents for Air Entrained Concrete in kg/m^3 (lb/yd^3)					
	Maximum Aggregate Size mm (in.)				
Slump mm (in.)	**9.5 (3/8)**	**12.5 (1/2)**	**19 (3/4)**	**25 (1)**	**37.5 (1.5)**
25–50 (1–2)	181 (305)	175 (295)	168 (280)	160 (270)	150 (250)
75–100 (3–4)	202 (340)	193 (325)	184 (305)	175 (295)	165 (275)
150–175 (6–7)	216 (355)	205 (345)	197 (325)	184 (310)	174 (290)
Data taken from American Concrete Institute Standard 211.1 and reproduced with the permission of the ACI.					

Table 21.9 Recommended w/c Ratios for Air Entrained Concrete

Compressive Strength MPa (psi) Measured on Cylinder in Compression	Water/Cement Ratio
35 (5075)	0.39
30 (4350)	0.45
25 (3625)	0.52
20 (2900)	0.60
15 (2175)	0.70

Data taken from American Concrete Institute Standard 211.1 and reproduced with the permission of the ACI.

Table 21.10 First Estimate of Density of Fresh Air Entrained Concrete

Nominal Aggregate Size mm (in.)	Estimated Density kg/m³ (lb/yd³)
9.5 (3/8)	2200 (3710)
12.5 (1/2)	2230 (3760)
19 (3/4)	2275 (3840)
25 (1)	2290 (3850)
37.5 (1.5)	2350 (3970)

Data taken from American Concrete Institute Standard 211.1 and reproduced with the permission of the ACI.

and 24.3). Normally, both of these methods are used. The high powder content also prevents "aggregate interlock" that prevents a mix flowing through narrow spaces.

If a high cement content is used, this will result in high cost and, normally, high strength (assuming a superplasticiser is used with a low water content). Thus, the high powder content will often include a lot of PFA.

There are many published methods for mix design of self-compacting concrete, but a simplified one is as follows:

- Design the concrete mix to achieve the required strength and a slump of 100 mm (4 in.).
- If the cement content is less than 500 kg/m³ (850 lb/yd³), add PFA so that the total cementitious (cement + PFA) is 500 kg/m³ (850 lb/yd³), and the equivalent cement content (= Cement + 0.3 × PFA, assuming an efficiency factor of 0.3) is the same as the cement content from the original mix design.
- Add superplasticiser and VMA in the amounts specified by the manufacturers.

21.7 REDESIGNING MIXES USING TRIAL BATCH DATA

Whichever method of mix design that is used, it is necessary to make trial batches of concrete to confirm and improve the design. When making these batches, it is frequently necessary to adjust the water content in order to get the required workability. Thus, the water content and the w/c ratio will not be

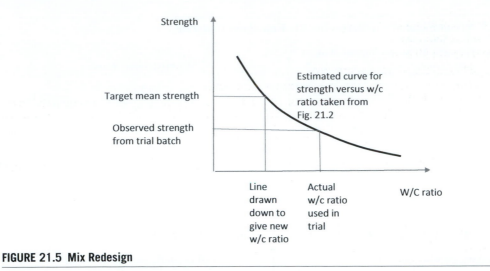

FIGURE 21.5 Mix Redesign

the same as the design mix. A method to improve the mix design based on trial mix data is shown in Fig. 21.5. The actual w/c used and strength observed are plotted, and an estimated line drawn through them. The new w/c ratio is then obtained for the target mean strength.

21.8 US UNITS

In some US literature and websites, the cement content of a mix is quoted in sacks per cubic yard. This is based on traditional sacks of 94 lb (42.5 kg) so 1 sack/yd^3 = 55.6 kg/m^3. Bags of cement are now generally restricted to 25 kg for health and safety reasons.

21.9 CONCLUSIONS

- The workability of concrete is determined by the water content.
- The strength of concrete is determined by the w/c ratio.
- The cement content is obtained by dividing the water content by the w/c ratio.
- US and UK practice use slightly different methods to obtain the aggregate content.
- Cement replacements may be assumed to have an efficiency factor that gives their contribution to strength relative to an equal mass of cement.
- Self-compacting concrete may be made by using a VMA.

TUTORIAL QUESTIONS

1. **a.** Calculate the quantities for a 0.15 m^3 trial concrete mix to the following specification using the UK method:
   Characteristic strength: 35 MPa at 28 days
   Cement: OPC (EN class 42.5)

Percent defectives: 5%

No. of test samples: 5

Target slump: 50 mm

Coarse aggregate: 10 mm uncrushed

Fine aggregate: Uncrushed 50% passing a 600 μm sieve

b. The mix is to be used to make self-compacting concrete, for which a minimum cementitious content of 500 kg/m^3 is required. Assuming that PFA develops 30% of the strength of an equal mass of cement, calculate new cementitious quantities to include PFA as a partial replacement for the cement.

c. Which two types of admixture would normally be used to make the self- compacting concrete?

Solution:

a. Stage 1. Standard deviation = 8 MPa (Fig. 21.1 with five results)

Target mean strength = 35 + 8 × 1.64 = 48.12 MPa (equation (12.3))

Stage 2. Starting point for Fig. 21.3 = 42 MPa (Table 21.1)

Stage 3 and 4. w/c = 0.45

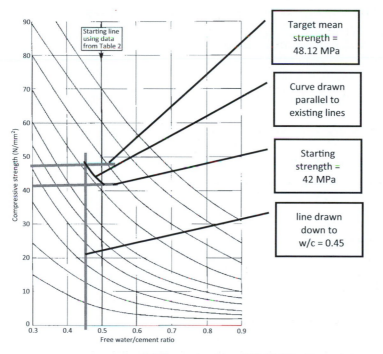

© IHS, reproduced from BRE BR 331 with permission.

Stage 5. Water content = 205 kg/m^3 (Table 21.2)

Stage 6. Cement content = 205/0.45 = 455 kg/m^3

Stage 7. Density = 2340 (Fig. 21.4 using the assumed relative density of 2.6 for uncrushed aggregate)

Stage 8. Aggregate content = 2340 − 205 − 455 = 1680 kg/m³
Stage 9. Percentage fine aggregate = 44% (Fig. 21.2 for 10 mm aggregate and 30–60 mm slump). A line for 50% passing a 600 μm sieve is assumed between 40% and 60% lines.
Stage 10. Fine aggregate content = 1680 × 44% = 739 kg/m³
Coarse aggregate content = 1680 − 739 = 941 kg/m³
Stage 11. The final quantities in kg/m³ are:

	Water	Cement	Fine Aggregate	Coarse Aggregate
Per cubic metre	205	455	739	941
Per 0.15 m³	31	68	111	141

b. To calculate the quantities for the SCC mix two equations are formed.
The total cementitious will be 500 kg/m³, thus:
   Cement + PFA = 500
Also the equivalent cement content must be 455 kg/m³ from the original design, and the assumed efficiency factor is 0.3, thus:
   Cement + 0.3 × PFA = 455
Solving these two equations gives:
   PFA = 65 kg/m³ = 9.75 kg in a 0.15 m³ trial mix
   Cement = 435 kg/m³ = 65 kg in a 0.15 m³ trial mix
c. Superplasticiser and viscosity modifier.

2. Calculate the quantities for the mix from question 1 using the US method. The following additional data should be used:
Cylinder strength: 40 MPa (this is approximately equivalent to a cube strength of 48. See Chapter 22)
Fineness modulus of fine aggregate: 2.8
Dry-rodded density of coarse aggregate: 1600 kg/m³
Absolute density of aggregate: 2600 kg/m³

Solution:

Step 1. The slump is 25–50 mm.
Step 2. The maximum aggregate size is approximately 9.5 mm.
Step 3. The estimated water content is 207 kg/m³ and the entrapped air is 3% (Table 21.4).
Step 4. The w/c = 0.42 (Table 21.5).
Step 5. The cement content = 207/0.42 = 493 kg/m³.
Step 6. Dry-rodded volume of coarse aggregate per unit volume of concrete = 0.46 (Table 21.6).
Thus, mass of coarse aggregate = 0.46 × 1600 = 736 kg/m³.
Step 7. (Using a mass basis.)
Estimated density = 2280 kg/m³ (Table 21.7)
Thus, fine aggregate = 2280 − 207 − 493 − 736 = 844 kg/m³.
(Using an absolute volume basis.)
Volume of water = 207/1000 = 0.207 m³
Volume of cement = 493/3150 = 0.157 m³

Volume of coarse aggregate = 736/2600 = 0.283 m^3
Volume of entrapped air = 0.03 × 1 = 0.03 m^3
Thus, volume of fine aggregate = 1 − 0.207 − 0.157 − 0.283 − 0.03 = 0.323 m^3
Mass of fine aggregate = 0.323 × 2600 = 840 kg/m^3
Thus, the final quantities in kilogramme per cubic metre are:

	Water	Cement	Fine Aggregate	Coarse Aggregate
Mass basis	207	493	844	736
Volume basis	207	493	840	736

3. Calculate the quantities for the following mix using the US method:
Cement: OPC
Target slump: 2 in.
Coarse aggregate: 3/8 in.
Cylinder strength: 5800 psi
Fineness modulus of fine aggregate: 2.8
Dry-rodded density of coarse aggregate: 2700 lb/yd^3
Absolute density of aggregate: 4400 lb/yd^3

Solution:

Step 1. The slump is 1–2 in.
Step 2. The maximum aggregate size is 3/8 in.
Step 3. The estimated water content is 350 lb/yd^3 and the entrapped air is 3% (Table 21.4).
Step 4. The w/c = 0.42 (Table 21.5).
Step 5. The cement content = 350/0.42 = 833 lb/yd^3.
Step 6. Dry-rodded volume of coarse aggregate per unit volume of concrete = 0.46 (Table 21.6).
Thus, mass of coarse aggregate = 0.46 × 2700 = 1242 lb/yd^3.
Step 7. (Using a mass basis.)
Estimated density = 3840 lb/yd^3 (Table 21.7)
Thus, fine aggregate = 3840 − 350 − 833 − 1242 = 1415 lb/yd^3
(Using an absolute volume basis.)
Volume of water = 350/1686 = 0.207 yd^3
Volume of cement = 833/5311 = 0.157 yd^3
Volume of coarse aggregate = 1242/4400 = 0.282 yd^3
Volume of entrapped air = 0.03 × 1 = 0.03 yd^3
Thus, volume of fine aggregate = 1 − 0.207 − 0.157 − 0.282 − 0.03 = 0.324 yd^3
Mass of fine aggregate = 0.324 × 4400 = 1425 lb/yd^3
Thus, the final quantities in lb/yd^3 are:

	Water	Cement	Fine Aggregate	Coarse Aggregate
Mass basis	350	833	1415	1242
Volume basis	350	833	1425	1242

TESTING WET AND HARDENED CONCRETE

CHAPTER OUTLINE

22.1 INTRODUCTION

22.1.1 LABORATORY AND *IN SITU* TESTS

This chapter describes tests that are carried out on concrete at the time of construction. When a concrete mix is being developed, a number of laboratory tests may be carried out and then, when the concrete is made, it will be tested on site. Samples will also be taken on site to be returned to the laboratory to measure the properties of the hardened concrete. Further tests that may be carried out on existing structures to assess their condition are described in Chapter 27. These will measure the effectiveness of the compaction and curing, but it is not common practice to use them at the time of construction.

22.1.2 COMMERCIAL SIGNIFICANCE

Many of these tests are used to determine whether the concrete is acceptable for use in the works. The results coming from them therefore have considerable commercial significance. In particular, failures in compressive strength testing (after considering statistical effects as described in Chapter 13) may result in considerable expenditure to remove the relevant concrete.

22.1.3 SELECTION OF TESTS

There are vast numbers of concrete tests in current use, and only a small number of them are described in this chapter. The aim of this chapter is to introduce the principles of what is being tested. Before carrying out any test, a full description (preferably a standard) should be downloaded and agreed with the client, in order to determine the correct procedure.

22.1.4 ABSOLUTE AND RELATIVE STANDARDS

These tests may either be used to confirm that the concrete has met an absolute standard (such as strength), or used to confirm that successive batches are giving the same result as an initial trial batch that may have been prepared in the lab. The latter approach would be more common for the less established tests, such as those for durability.

22.2 WORKABILITY

22.2.1 DEFINITION

When concrete is poured, work must be done on it to form it into the shape of the formwork, and to remove trapped air from it. Workability may be defined as: "The amount of useful internal work necessary to produce full compaction." "Consistence" is the European word for workability, and is used in European Standards.

22.2.2 SIGNIFICANCE FOR SITE WORK

The most important reason for control of workability is for placing. A concrete with high workability may not require any work to compact it, but a low workability mix will require extensive vibration. If insufficient vibration is used for the workability of the concrete, voidage will occur. This will normally be concentrated at the outer edges of the concrete mass, where the vibration is least effective, and be visible as "honeycombing," and "blowholes" that are seen when the shutters are removed (see Section 26.4.7).

The measurement of workability has been the subject of a considerable amount of recent research because of two increasing trends in construction:

- The use of self-compacting concrete is increasing. The workability tests that were developed for normal concrete do not work for this material, so new ones have been developed.
- The use of concrete pumps and, in particular, the height to which they pump is increasing. Measuring the workability required for effective pumping is complex, and requires viscosity measurements as described in Section 22.2.8.

22.2.3 THE SLUMP TEST

This is by far the most common test for workability. A truncated cone 300 mm (12 in.) high is placed on a flat surface, with the small opening at the top, and concrete is placed in it and compacted. The cone is then lifted, and the extent to which the concrete slumps down is measured. The test is shown in Figs. 22.1 and 22.2.

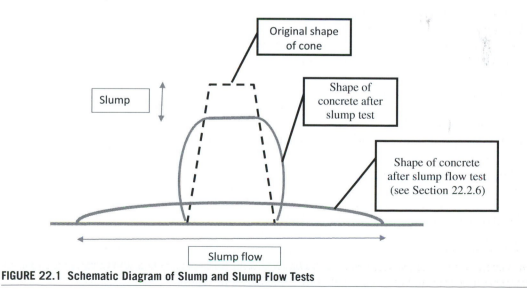

FIGURE 22.1 Schematic Diagram of Slump and Slump Flow Tests

FIGURE 22.2 The Slump Test

After lifting it from the concrete, the cone has been turned upside down to give the height to measure from.

When carrying out this test, check the following:

- Ensure that the base on which the test is to be carried out is flat, level, clean and free from vibration, and of adequate size to stand the cone next to the slumped concrete for measurement.
- If aggregate larger than 40 mm (1.5 in.) has been used, it must be removed before the test.
- Check that the cone is not bent or dented, and the inside surface is clean.
- Check that a length of 16 mm (5/8 in.) diameter steel bar is available for tamping.
- Obtain a representative sample. Do not use the first or last fraction from the truck-mixer.
- Fill the cone, carefully tamping it in three layers. Lift it off vertically. If the concrete collapses to one side, start again.
- If the recorded slump is greater than 250 mm (10 in.) an alternative test method (such as slump flow) should be used.
- Record the result carefully with details of the date, concrete batch, etc.

22.2.4 THE DEGREE OF COMPACTABILITY TEST

This is just one of a large number of tests for workability that have been developed as alternatives to the slump test. It is described here as an example, and also because the apparatus is easy to fabricate and use.

This is a very simple test. The metal container is filled with concrete to the top, taking care not to compact it at all. The surplus concrete is cut away with trowels to give a level top surface and the container is then vibrated until the level of the concrete stops falling.

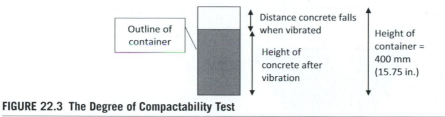

FIGURE 22.3 The Degree of Compactability Test

The degree of compactability is defined as (see Fig. 22.3.)

$$\frac{\text{Height of container}}{\text{Height of concrete after vibration}} \qquad (22.1)$$

Values over 1.4 indicate a concrete that has not compacted when originally placed in the container, and thus has a very low workability.

22.2.5 "SLUMP METERS"

Many ready-mix trucks have slump meters on them that measure the pressure in the hydraulic system rotating the drum. If the mix has a high slump, it will flow more freely and require less power to turn the drum. These meters are affected by the grading of the aggregate and other factors, but can be effective if used with proper precautions.

22.2.6 THE SLUMP FLOW TEST

This is the most common test for self-compacting concrete. It is similar to the slump test, and is shown in Fig. 22.1. It is not necessary to tamp the concrete in the cone. The average diameter of the final spread is recorded, and the time taken to reach a diameter of 500 mm (20 in.) may also be recorded.

22.2.7 THE V FUNNEL TEST

The V funnel is one of the other tests for self-compacting concrete. The funnel (Fig. 22.4) is filled with concrete with the gate closed; the gate is then opened, and the time for the concrete to discharge is measured.

22.2.8 VISCOSITY MEASUREMENTS

The definition of workability is closely related to the physical definition of viscosity (see Fig. 8.1) that refers to the work required to shear a fluid at different shear rates. A viscometer rotates a paddle at different speeds in a concrete sample, and records the torque necessary to drive it. The device resembles a food mixer, and the container for the concrete is typically 300 mm (12 in.) diameter (Fig. 22.5). Computer control is used normally, and the speed of rotation (the shear rate) is increased progressively, and held at different rates, so the torque (shear stress) can be measured. Typical results are shown schematically in Fig. 22.6, and show a curve as the speed is increased, and then when the speed is decreased they approximate to a straight line.

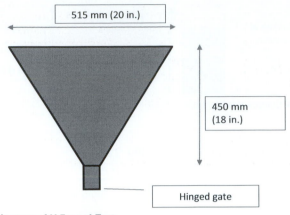

FIGURE 22.4 **Schematic Diagram of V Funnel Test**

The funnel is 75 mm (3 in.) front to back.

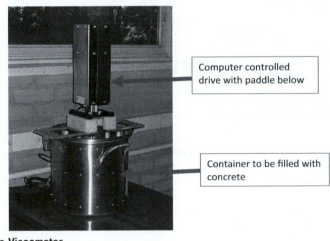

FIGURE 22.5 **Concrete Viscometer**

Referring to Fig. 8.1, concrete is best described as a Bingham fluid because it has a yield stress. This is critical to concrete pumping. If the pump stops for any reason, it will be the yield stress that determines the pumping pressure needed to restart it. If this is too high, the line must be dismantled to clear it. However, the rate of concrete delivery (i.e., the rate of production) is determined by the viscosity.

The enclosed area in Fig. 22.6 indicates the work done on the concrete during the cycle that will make it flow more easily. Before discharging concrete from a truck-mixer that has been standing, it should be mixed at full speed for at least 2 min, in order to overcome this and make it flow.

It may be seen that different mixes may perform better at different shear rates. The tests, such as the slump of the degree of compactability, only measure performance at a single shear rate, so it is possible

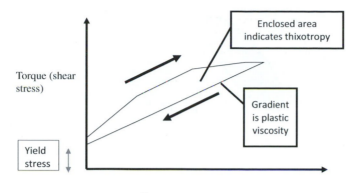

FIGURE 22.6 Schematic Output From Viscometer

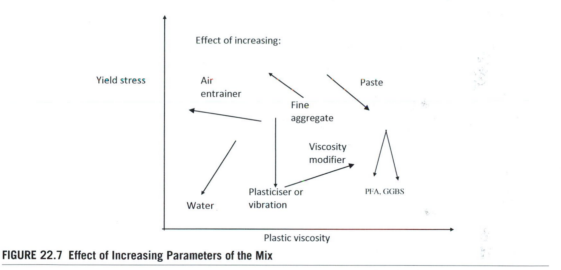

FIGURE 22.7 Effect of Increasing Parameters of the Mix

for one mix to perform better than another in one of these tests, but worse in another. A full understanding of their relative performance can only be seen from viscosity tests.

Figure 22.7 shows schematically the effect of changes to the mix, which may be observed using a viscometer. The effect of vibration is normally to reduce the yield stress, that is, to make the concrete flow with only the effect of gravity. Using both superplasticiser and viscosity modifier reduces yield but (unlike adding water) increases viscosity, and thus prevents segregation. This is self-compacting concrete.

The different types of concrete may be represented as shown in Fig. 22.8. Significantly increasing the paste content of a normal concrete will give a high strength concrete, because the water/cement (w/c) ratio can be reduced. If some of this water is added to this mix, while maintaining the high cement content, this will give a flowing concrete. The self-compacting concrete uses the plasticiser to reduce the yield so it flows, and then the viscosity modifier to increase the viscosity so it does not segregate.

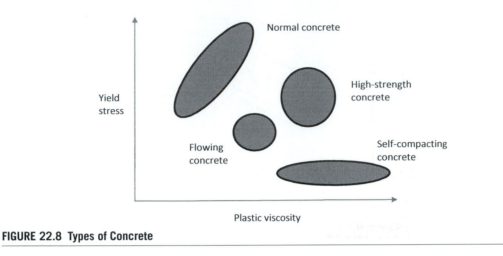

FIGURE 22.8 Types of Concrete

22.2.9 LOSS OF WORKABILITY

A major contribution to the initial loss of workability may be absorption of water by the aggregate. This process will normally be completed by the time concrete is discharged from a truck-mixer.

High temperatures will accelerate the loss of slump with time, due to evaporation and increased hydration rates. In hot climates, it may be necessary to cool the mixing water to control the rate of hydration of the cement. If a water reducer has been used, loss of slump may be quite sudden. If the water temperature exceeds about 70°C (160°F), flash set may occur in which cement substantially hydrates on contact with the water.

Adding water to the mix to regain workability will increase the w/c ratio, and decrease strength and durability.

22.3 BLEEDING AND SEGREGATION
22.3.1 PHYSICAL PROCESSES

Bleeding and segregation occur when the constituents of the mix separate. Bleeding is the loss of water from the mix, and segregation is the loss of aggregate. The consequences are:

* Segregation will reduce the strength both in areas where there is a lack of aggregate, and in areas where there is a surplus of aggregate.
* The lack of aggregate may produce cracking, and the surplus may produce voidage.
* A small amount of bleeding will be harmless because it will reduce the effective water/cement ratio, and provide some water on the surface for curing.
* Larger amounts of water loss will cause plastic settlement and cracking (see Section 23.4.4).
* If the rising bleed water carries cement with it, a dusty, porous surface may result.
* Bleed water trapped under reinforcement will reduce bond and cause voidage.

The process is shown schematically in Fig. 22.9. Note that the total volume remains the same.

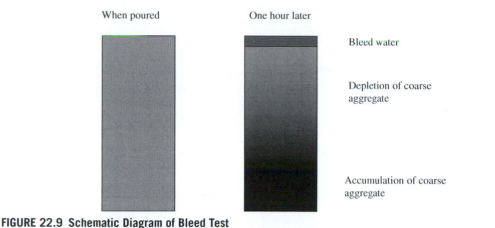

FIGURE 22.9 Schematic Diagram of Bleed Test

22.3.2 MEASUREMENT

Segregation may be observed by inspecting concrete when it is placed and vibrated on a flat surface or, alternatively, by cutting hardened concrete samples, and inspecting the cut surface. Bleeding is measured by placing concrete in a mould and removing and measuring the surface water at various times after casting (Fig. 22.9). Alternatively, the settlement of the concrete surface may be measured. There are two different measurements of bleed that can be made:

- Bleed measurement with the bleed water replaced after measurement. It will normally all be reabsorbed within 24 h.
- Bleed measurement with the water removed and not replaced. This is often a better simulation of site conditions where sun and wind will dry it off.

22.4 AIR CONTENT

22.4.1 TYPES OF AIR VOID

The types of voids in concrete are listed in Table 20.1. Air content tests are intended to measure the content of entrained air because this is essential to give the concrete resistance to freeze–thaw damage.

Air content measurements will measure the potential entrained air (some will be lost during vibration). For an accurate prediction of frost resistance, the air void spacing is measured with a microscope on a cut surface of the hardened concrete.

22.4.2 MEASUREMENT

A pressure type air meter is shown in Fig. 22.10. Concrete is placed in the lower container, and the apparatus is sealed with the water above it. Air is then pumped into the top of the apparatus, and the fall in the level of the water is recorded as the pressure rises.

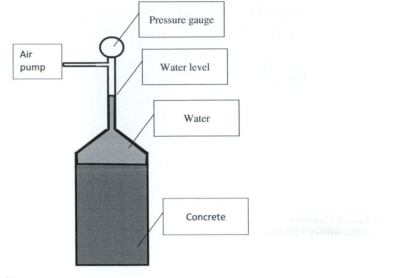

FIGURE 22.10 Air Content Meter

For example, if the original pressure $P_1 = 1$ atm (i.e., open to atmosphere) and a pressure of $P_2 = 2$ atm causes a volume change of 5 mL; from equation (5.3) at constant temperature we know that:

$$P_1V_1 = P_2V_2 \qquad (22.2)$$

where V_1 is the initial volume of the pores, and V_2 is the volume at the higher pressure.

Thus, $V_2/V_1 = 0.5$

Also, $V_1 - V_2 = 5$

Thus, the original entrained air volume $V_1 = 10$ mL.

There is also a second type of meter in which air under pressure above the concrete is released into a known volume at atmospheric pressure. The calculations are similar.

22.5 COMPRESSIVE STRENGTH TESTING

22.5.1 PURPOSE OF THE TEST

When designing concrete structures, a given compressive strength is assumed. The most frequent test used in new construction is intended to measure this strength. This test does not pretend to give an accurate simulation of loading in a structure (e.g., it is "uniaxial" compression, rather than confined "triaxial" compression that will occur in much of a reinforced structure), but has been found over 100 years of use to give a good idea of the "quality" of concrete. One reason why this test has lasted so long, and is widely accepted for contractual purposes, is that, if it is carried out incorrectly,

FIGURE 22.11 Compacting Test Cubes on Site

The smaller 50 mm (2 in.) cubes can only be used for mortar.

it almost always gives a low result, and thus contractors are highly motivated to follow the correct procedure.

The compression test may be carried out on a cube or a cylinder. A cube (often 150 mm on site, but 100 mm in labs) is cast (Fig. 22.11) and tested by turning onto its side and crushing on two opposite as-cast faces. Cylinders are 100 mm or 150 mm diameter, and their height is normally double their diameter. They are tested upright.

Cubes and cylinders are the only samples normally made on sites. They are generally struck from their moulds the day after casting, and sent off to a testing lab within a few days.

22.5.2 SAMPLE PREPARATION

The manufacture of test cubes is often considered to be a low priority task on sites, until some cube failures occur and management wants to know all about it. A small amount of money spent on good practice with cubes will save the massive cost of *in situ* testing after failures. When preparing cubes check the following:

- Check that metal moulds are clean and properly bolted together (Fig. 22.12). Note that moulds may be expensive, but bolts are cheap so always have a box of spare bolts and nuts. Never use wooden moulds or steel moulds held together with tie wire. Plastic cube moulds with a hole in the bottom, so the cube may be removed with compressed air, have been found to be very effective.
- Oil the moulds carefully. Always use proper shutter oil.
- Check that a correct steel tamping bar is available.
- Get a good representative sample (as for the slump test).
- Always compact the concrete in layers.

FIGURE 22.12 Filled Cube and Cylinder Moulds

22.5.3 STORAGE AND DISPATCH

This part of compressive strength testing is often overlooked, and can lead to unnecessary failures.

- Always cover the samples with wet hessian (burlap) covered with polythene, after casting. Note that hessian alone is virtually useless because it has to be continually wetted.
- Always ensure that some form of heating is provided in cold weather, in the area where the samples are stored overnight.
- Try to get the samples collected by the day after casting. If this is not possible, a heated curing tank must be used.
- Always keep good records of the samples, and ensure that they are properly labelled in at least two places.

  Temperature-matched curing, as described in Section 20.2, may also be used.

22.5.4 COMPRESSIVE TESTING

When a sample is tested in uniaxial compression, it will fail in tension due to the lateral strain from Poisson's ratio effects (Fig. 22.13 and also Fig. 3.9).

The observed cube strength will depend on:

- The original concrete (w/c ratio, age, curing conditions).
- Factors in casting and storage (see 22.5.2 and 22.5.3).
- The cleanliness of the platens.
- The alignment of the platens (should be on hemispherical seating that should be checked to make sure it is free to rotate before testing, but locks on loading, see Fig. 22.13). This facility in the

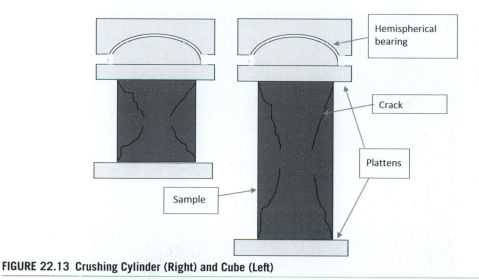

FIGURE 22.13 Crushing Cylinder (Right) and Cube (Left)

machine should not be used to overcome misaligned faces on badly made samples. If the samples are not correctly made, they should not be tested.

- The rate of loading that should be set as specified in the standard. Faster loading gives higher strength (see Fig. 22.14).
- The mode of failure. A failed cube should be two pyramids.

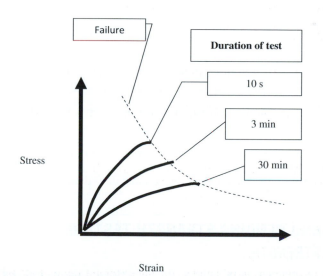

FIGURE 22.14 The Effect of Increasing Load Rates

FIGURE 22.15 Chop Saw for Preparing the Ends of Cylinder Samples for Compression Test

The sample is placed on the bed, and the water supply is turned on to give a good flow to the blade that is then lowered using the handle.

22.5.5 THE DIFFERENCE BETWEEN CYLINDERS AND CUBES

The relative advantages and disadvantages of cubes and cylinders are as follows:

- Cylinders contain more concrete, so they take longer to make and are more difficult to transport.
- A cube is turned onto its side for testing, so accurate cast faces are used for compression. This cannot be done for a cylinder, so the end must be saw cut or capped with a fast-setting capping compound (see Fig. 22.15).
- The samples fail by lateral strain, and this will be restrained by friction between the sample and the machine platen. The effect of this will be less for the cylinder because it is longer.
- A cylinder gives a more accurate representation of a core (see Section 27.3.5).

It may often be assumed that the strength of a cylinder is 20% less than a cube. A more accurate value may be obtained as follows:

$$\text{Cylinder strength} = (0.85 \times \text{Cube strength}) - 1.6\,\text{MPa} \qquad (22.3)$$

If cylinders are made with their height less than double their diameter, their strength will be higher.

22.6 TENSILE AND BENDING STRENGTH TESTING
22.6.1 TENSILE STRENGTH

The tensile strength of concrete is of interest for unreinforced items, but, for almost all structural calculations for reinforced concrete, it is assumed to be zero, so the test is only used for specialised applications.

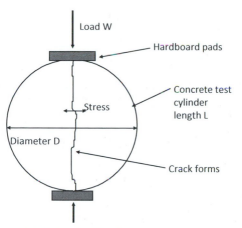

FIGURE 22.16 Schematic Diagram of Cylinder Splitting Test

FIGURE 22.17 Cylinder Sample in Frame for Splitting, and Split Sample After Testing

This is normally measured indirectly by cylinder splitting (see Figs. 22.16 and 22.17). The stress is given by equation (22.4)

$$\text{Stress} = \frac{2W}{\pi LD} \qquad (22.4)$$

22.6.2 FLEXURAL STRENGTH (MODULUS OF RUPTURE)

This is normally measured on $500 \times 100 \times 100$ mm $(20 \times 4 \times 4$ in.) prisms (see Figs. 22.18 and 22.19). The "4-point" loading shown in Fig. 22.18, with two load points on the top of the sample,

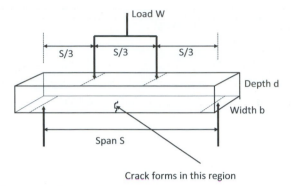

FIGURE 22.18 Schematic Diagram of Flexural Test

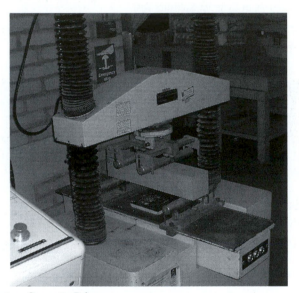

FIGURE 22.19 Flexural Test on Concrete Prism

is used because this gives a region of constant bending moment between them. This makes the test less sensitive to a single aggregate particle or defect that could be below a single loading point, on top of a sample in a "3-point" test. Thus, the 4-point test is used for concrete, while the 3-point test is used for timber or steel (Fig. 33.4).

The stress is given by equation (22.5).

$$\text{Flexural stress} = \frac{WS}{bd^2} \tag{22.5}$$

The modulus of rupture = the flexural stress at failure.

22.7 MEASUREMENT OF MODULUS OF ELASTICITY

22.7.1 TYPES OF MODULUS

The modulus of elasticity may be static (i.e., at rest) or dynamic (i.e., moving, normally vibrating). The static modulus is of considerable interest to structural engineers because it is needed to calculate the deflection of structures under load. Most design codes include methods to estimate the static modulus from the compressive strength, but these may not work for unusual concrete, such as very high strength mixes. The dynamic modulus may be measured with ultrasound (see equation (5.9)).

22.7.2 MEASUREMENT OF STATIC MODULUS

This may be measured by fixing clamps around a cylinder and measuring the distance between them with displacement transducers (see Fig. 3.9) or, alternatively, strain gauges can be glued to the sample. Several transducers must be used, and the average value taken, in case the cylinder compresses more on one side than the other. The reading must be measured on the sample as shown, and not within the compression machine because a machine will deflect at high loads, so any reading will include the effect of the elasticity of the machine components, such as the loading platens.

22.8 DURABILITY TESTS

22.8.1 APPLICATIONS

Clients on major projects are increasingly requiring durability tests at the time of construction; particularly in aggressive environments, such as the Middle East, where the heat and salinity has caused early failure of numerous structures.

22.8.2 ABSORPTION TESTS

Figure 22.20 shows a simple sorptivity test. The water is drawn into the sample by capillary suction, and the flow rate is limited by the permeability. The sample is initially dry, and the weight gain is recorded

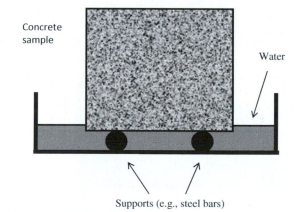

Concrete sample

Water

Supports (e.g., steel bars)

FIGURE 22.20 Schematic Diagram of Sorptivity Test

over a few hours or days. Other tests exposed one face of a sample to water under high pressure, making the results depend more on permeability, and less on capillary suction. A test of this type can give a useful indication of the potential durability.

22.8.3 CHLORIDE MIGRATION TESTS

This test is shown schematically in Fig. 22.21. There are many different forms of it, in different standards. They all use a high voltage to drive chloride into the sample because it takes months to measure significant diffusion, if it is not accelerated in this way. Some of them rely on measuring the total charge passing in a given time (i.e., the average current) to judge the quality of the concrete,

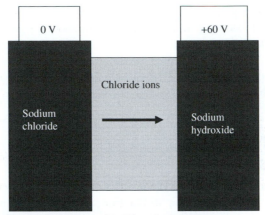

FIGURE 22.21 Schematic Diagram of "Rapid Chloride Migration" Test

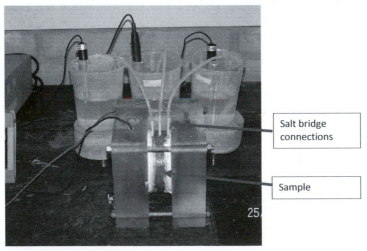

FIGURE 22.22 The Rapid Chloride Migration Test

This sample has additional salt bridge electrical connections to the top, to measure the voltage at different points across it.

while others measure the chloride concentration at different depths in the sample, at the end of the test. The transport process in this test is electromigration (see Section 10.7).

Section 6.9 showed that electrical current in concrete is carried by hydroxyl ions. At the start of this test, all the chloride ions will be near the surface of the sample, and the current flow through the rest of it must be carried by these hydroxyl ions (because the current at all depths must be the same). In Section 18.3, it was noted that pozzolanic cement replacements will deplete the supply of hydroxyl ions. They will therefore limit the current, and thus the flow of chlorides. This test will therefore give misleading results if a pozzolan, particularly an active one, such as silica fume, is added to the concrete. This problem can be detected with the salt bridges shown in Fig. 22.22.

22.8.4 TESTS FOR SULPHATE ATTACK

Sulphate attack can be detected using mortar bar expansion tests, using the same apparatus as for ASR and aggregate shrinkage (see Fig. 19.1).

22.9 CONCLUSIONS

- There are numerous tests for wet and hardened concrete. The relevant standards should always be downloaded before using them.
- Single point tests, such as slump, are useful for site use, but to get a full indication of workability, a viscometer is needed.
- Bleeding involves loss of mix water; segregation normally means loss of aggregate.
- The air content of wet concrete can be measured with pressure meters.
- Proper precautions must be taken with casting, storage, and testing for compressive strength, or a low result is obtained.
- A wide range of durability tests, including adsorption and rapid chloride migration, are available.

TUTORIAL QUESTIONS

1. Rheological tests are carried out on three different concrete mixes, and the results are as follows:

Shear Rate (Relative Units)	Shear Stress (Relative Units)		
	Control Mix	High W/C Mix	Superplasticised Mix
0	0.5	0.4	0.3
0.2	0.62	0.51	0.42
0.4	0.73	0.58	0.53
0.6	0.83	0.62	0.63
0.8	0.91	0.64	0.71
1.0	1.0	0.66	0.80

Discuss the conclusions that may be drawn about the performance of these mixes in the following:
a. A slump test
b. A degree of compactability test
c. A concrete pump
d. Placement from a skip
e. Vibration in a shutter

Solution:

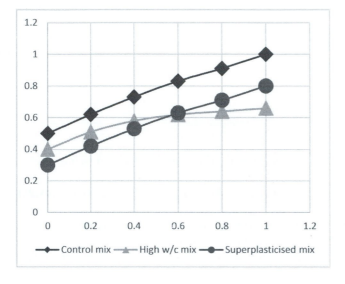

(a, d) Slump test and placement from skip involve low shear rates, so the superplasticised mix will flow most.

(b, e) Vibration (and degree of compactability test that is vibrated) is at medium to high shear rate, so the high w/c mix may be similar, or slightly better, than the superplasticised mix.

(c) Pumping involves high shear rates, so the high w/c mix may pump easiest. It may, however, be less cohesive (indicated by low gradient), and thus segregate. The superplasticised mix has the lowest yield (shear stress at zero shear rate), so it may be the easiest mix to restart pumping if it stops.

2. **a.** What is the difference between air entrained concrete, and foamed concrete?
   b. When the air content of a sample of concrete is measured, using a pressure type air meter, the following observations are made:

   Volume of concrete: 10 L
   Initial pressure: Open to atmosphere
   Pressure change: 1.5 Atm
   Volume change: 80 mL

   What is the percentage of entrained air?

Solution:

a. Air entrained concrete has about 4% air for frost resistance.
   Foamed concrete has about 30–60% air and no coarse aggregate, and is for trench backfill (see Chapter 29).

b. Using the method in Section 22.4.2:
   - $V_2/V_1 = P_1/P_2 = 1/2.5 = 0.4$
   - $V_1 - V_2 = 80$ mL
   - Thus (solving) $V_1 = 133$ mL
   - Thus, air = $133/10{,}000 = 1.33\%$

NOTATION

b	Width (m)
d	Depth (m)
D	Diameter (m)
L	Length (m)
P_1	Initial pressure (Pa)
P_2	Final pressure (Pa)
S	Span (m)
V_1	Initial volume (m^3)
V_2	Final volume (m^3)
W	Load (N)

CREEP, SHRINKAGE, AND CRACKING OF CONCRETE

CHAPTER OUTLINE

23.1 CREEP

23.1.1 TYPES OF CREEP

The following definitions are used for creep.

- *Creep* is long term deformation due to loading.
- *Total creep* is the strain due to loading *and drying*.
- *Basic creep* is the strain due to loading with no loss of moisture.
- *Specific creep* is the creep per unit stress.

Basic creep is almost impossible to measure because it involves keeping a test specimen under load for a long time (often up to 20–30 years), while sealing it to prevent any loss of moisture. Therefore, experimental data generally gives the total creep, but this will depend on the extent and timing of the drying. For most structural purposes, creep is assumed to be proportional to stress, so the specific creep is used.

23.1.2 EFFECTS OF CREEP IN CONSTRUCTION

Creep causes:

- Deflection in structures under continuous loading. This may cause bridges to sag, or cladding systems on buildings to buckle. A tall building may get 50–100 mm shorter during its design life.
- Stress relief that reduces cracking.
- Loss of prestress due to creep of both the concrete and the prestressing steel.

23.1.3 FACTORS AFFECTING CREEP

The factors affecting creep are:

- Water content of the concrete mix. High water contents give high creep.
- Age at load transfer. If structures are permitted to cure for longer before loads are applied, creep will be reduced.
- Section thickness. Thick sections will creep less because moisture movement is reduced.
- Humidity. Creep is higher in humid environments.
- Temperature. Creep increases with temperature.

Most design codes will include a method of calculating specific creep that will take account of many of these factors.

23.2 SHRINKAGE

23.2.1 EFFECTS OF SHRINKAGE IN CONSTRUCTION

Shrinkage causes:

- Cracking – but only if the element is restrained.
- Deflection, normally additional to the creep.

23.2.2 AUTOGENOUS SHRINKAGE

This is the inevitable shrinkage that results from the hydration of cement without additional water, and is typically 40 microstrain after 1 month. It is greatest for mixes with a high cement content, but never sufficient to cause cracking. Stresses from it are rapidly relieved by creep, if no other shrinkage occurs. In wet curing, swelling occurs with similar or greater strains.

23.2.3 THERMAL SHRINKAGE

Concrete is frequently relatively warm, when initial set occurs. This may be due to the heat of hydration, or other effects such as sunlight on the concrete, or the aggregate storage. When it subsequently cools, it will shrink. Typical coefficients of thermal expansion are 5–10 microstrain $°C^{-1}$.

23.2.4 PLASTIC SHRINKAGE

This occurs before final set, and is caused by bleeding. As water is lost from the concrete, its volume decreases. Rapid drying from the surface (i.e., faster than bleeding) will cause substantial plastic shrinkage.

23.2.5 DRYING SHRINKAGE

The effect of early drying is plastic shrinkage. Drying shrinkage is a long-term phenomenon, and occurs when the pore water is lost. Typical values are 500 microstrain at 28 days at 50% RH.

Drying shrinkage is associated more with loss of water from gel pores (formed in the gel during hydration), than with capillary pores that are larger, and are initially occupied by water. Thus, the pastes, which have hydrated more and have a higher proportion of gel pores, will shrink more, for less water loss.

Figure 23.1 shows the effect of various curing environments. Continuous wet curing will cause expansion, continuous drying will cause shrinkage. Sealed curing will just cause autogenous shrinkage. Alternate wetting and drying will cause swelling and shrinkage with each cycle, with a net overall shrinkage or swelling effect depending on the mix.

23.2.6 CARBONATION SHRINKAGE

Carbonation is normally of interest for durability because it causes loss of alkalinity, leading to reinforcement corrosion. It does, however, also cause some shrinkage. This shrinkage is closely related to drying shrinkage, and the combined effect of both will depend on the sequence in which they take place (i.e., carbonation during drying or after it). Typical values for mortar are 800 microstrain at 50% RH.

23.2.7 AGGREGATE SHRINKAGE

Aggregate shrinkage is discussed in Section 19.6.3. Generally, the aggregate will shrink less than the cement paste, so increasing the aggregate content will reduce shrinkage.

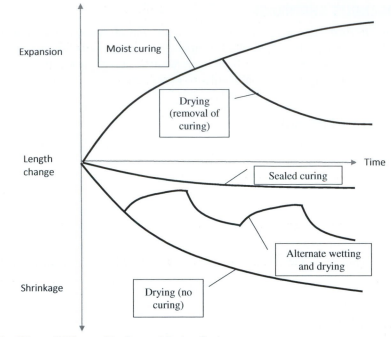

FIGURE 23.1 The Effect of Different Wetting and Drying Cycles

23.3 CRACKING

23.3.1 THE CAUSES OF CRACKING

Cracking occurs when the tensile strain in concrete exceeds its tensile strain capacity. At early ages, when the concrete is weak, this requires far less stress than for mature concrete. Structural cracking occurs when the strains are caused by the loading on the structure. Non-structural cracking is caused by shrinkage. Creep causes "relaxation" of stresses, and will reduce cracking. This is shown schematically in Fig. 23.2.

In the discussion in this chapter, the stress is assumed to be nonstructural. Cracking caused by structural overload would be considered in the topic of structural analysis.

23.3.2 EFFECTS OF CRACKING IN CONSTRUCTION

Cracking may be unacceptable to the client because of its appearance. The significance of the cracks will depend on how easily they may be seen. Thus, while cracks of 0.1 mm may be unacceptable on a major public building at street level, 0.6 mm cracks may not be a problem on parts of a car park that may only be seen from a distance.

Cracks will also cause some loss of durability; in particular, they may form pathways for chlorides to reach the reinforcement. However, defining acceptable widths for durability (i.e., protection of steel) is very difficult because surface width is a poor indicator of crack depth, so corrosion rates do not depend on surface crack width.

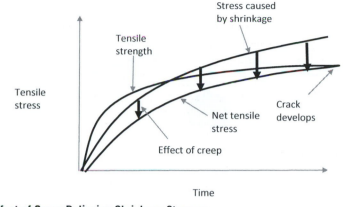

FIGURE 23.2 The Effect of Creep Relieving Shrinkage Stresses

The most important point to note about cracking of concrete is that it occurs in most structures, and is generally harmless because, in structural calculations, the concrete is only assumed to carry compressive loads.

23.3.3 AUTOGENOUS HEALING

Cracks in water-retaining structures will cause leaks. However, for watertightness, a width of 0.2 mm (8×10^{-3} in.) is considered acceptable because, below this width, autogenous healing will probably seal the crack. This is a process in which the water brings cementing minerals into the crack, where they solidify and block it.

23.3.4 DRYING SHRINKAGE CRACKING

Cracking due to drying shrinkage is most common in thin slabs and walls made with mixes with a high water content. It is a long-term effect, and may take weeks or months to appear.

23.3.5 EARLY THERMAL CRACKING

Early thermal cracking may occur from 1 day to 2–3 weeks after casting. It is most common in thick walls and slabs, where the heat of hydration causes significant temperature increases, and subsequent rapid cooling occurs.

23.3.6 PLASTIC SETTLEMENT CRACKING

Plastic settlement cracking is caused by bleeding (see Section 22.3), and is often seen over reinforcing bars, as shown in Fig. 23.3. If there is reinforcement near the top of a reasonably deep section, such as a reinforced footing, this will form a line of cracks above each of the links.

In Fig. 23.4, the concrete settles into the beams due to plastic settlement, but it is restrained at the edges so cracks form.

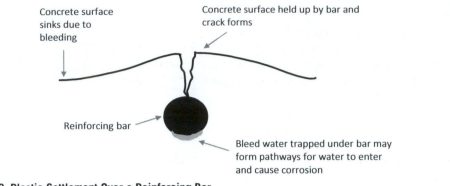

FIGURE 23.3 Plastic Settlement Over a Reinforcing Bar

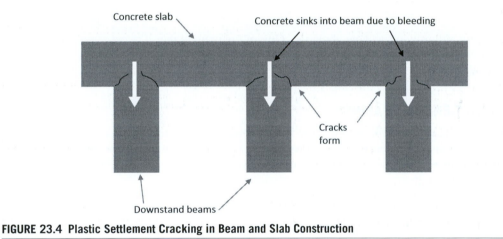

FIGURE 23.4 Plastic Settlement Cracking in Beam and Slab Construction

Plastic cracking may be solved with revibration. It is often considered that this may damage the concrete, but it does not. The immersion vibrators are placed in the concrete as soon as the cracks appear, and this will close the cracks and also disperse any bleed water that has accumulated under the bars. The way to judge whether concrete has set too much for revibration is to inspect the surface after the immersion vibrator is withdrawn. If the surface is flat, without holes in it, then the operation should continue. Simply retroweling the concrete to hide the cracks is not acceptable.

Plastic settlement may be prevented by reducing the water content of the mix, or by using air entrainment.

23.3.7 PLASTIC SHRINKAGE CRACKING

Plastic shrinkage cracking is caused by horizontal shrinkage of slabs. It is common in roads and thin slabs, and may be reduced with improved curing to prevent rapid drying. This keeps the bleed water on the surface, and reduces further bleeding.

23.3.8 CRAZING

Crazing causes a network of fine surface cracks. It occurs when impermeable formwork, such as steel shuttering, is used, that does not absorb any moisture from the wet concrete, and thus has none to release back during setting in order to assist with surface hydration. It may also occur on top surfaces of pours that are overtrowelled. It is most common in mixes with a high cement content, and is reduced with improved curing.

23.3.9 REINFORCEMENT CORROSION CRACKING

When reinforcement corrodes, it will normally form a rust, which occupies more volume than the original steel (see Section 25.3.1 for types of corrosion product). This will cause cracking and spalling of the concrete. The only way to prevent it is to prevent the corrosion of the steel.

23.3.10 ALKALI AGGREGATE REACTION

This will cause map cracking on the surface, characteristically with three cracks radiating from a single point. The causes are discussed in Section 19.6.2.

23.4 PREVENTING PROBLEMS CAUSED BY SHRINKAGE AND CRACKS
23.4.1 CRACK CONTROL STEEL

The purpose of crack control steel is to produce a large number of fine cracks (below the critical widths), rather than a smaller number of larger cracks.

23.4.2 CRACK INDUCERS AND EXPANSION JOINTS

Figures 23.5 and 23.6 show the use of a crack inducer. This is a strip of plastic that is placed into the wet concrete. It is made in two parts, so the upper part can be removed after the concrete has set to make space for a sealant to be placed. By inducing cracks at regular intervals in the slab, the tensile stresses are reduced, and no other cracks form.

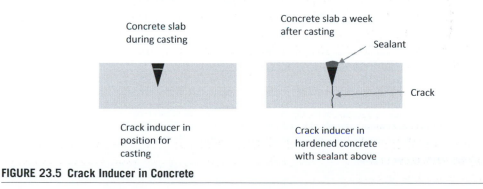

FIGURE 23.5 Crack Inducer in Concrete

FIGURE 23.6 Crack Caused by Crack Inducer

The upper part has not yet been removed to place the sealant.

23.4.3 FILLING CRACKS

Figure 23.7 shows one of many different details that can be used to fill shrinkage cracks. The void is formed between adjacent pours with an inflatable void former that is then extracted, and the wall is left until all of the shrinkage has taken place. The void is then grouted to form a seal that is adequate for a liquid retaining structure.

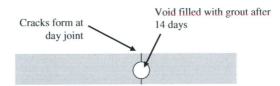

Cracks form at day joint

Void filled with grout after 14 days

Plan view of a section of the top of a wall. A typical wall of this type may be constructed using "hit-and-miss" construction in which alternate panels are cast first and then the in-fills are cast between them.

FIGURE 23.7 Detail Used to Fill Cracks

23.5 CONCLUSIONS

- Basic creep is almost impossible to measure, so total creep that includes drying shrinkage is normally measured.
- Creep can cause deflection of structures, or loss of prestress.
- Almost all concrete will shrink due to effects associated with cooling and drying.
- There are a number of different causes of nonstructural cracking, and it is important to identify the cause, if it is to be prevented.
- In structures, it is generally best to permit cracking to occur in a controlled manner, rather than trying to stop it.

ADMIXTURES FOR CONCRETE

24.1 INTRODUCTION

This chapter describes chemical admixtures that are liquids or powders added in small quantities to a concrete mix. The term "mineral admixtures" is used to describe minerals, such as PFA and GGBS that are used as partial cement replacements. These are described in Section 18.3.

Almost all major projects use admixtures in the concrete, and they are essential to obtain the properties needed for many applications. There are vast numbers of different types of admixtures available; the most significant ones are described in this chapter.

24.2 PLASTICISERS AND SUPERPLASTICISERS
24.2.1 USES OF PLASTICISERS AND SUPERPLASTICISERS

Plasticisers increase the workability of concrete, and are also known as water reducers because they reduce the water content needed for a given workability. Superplasticisers are more powerful, and will give greater increases in workability, but the term "plasticiser" may be taken to include all of them. Significant advances in the technology have meant that there are now large numbers of different plasticisers available, some of which will make the concrete free-flowing. They work by adsorbing onto the cement grains, and dispersing them, so they do not coalesce and inhibit the flow.

The different uses of plasticisers are illustrated in Table 24.1.

The control mix is in the centre of the table, and is a typical concrete with 75 mm slump, and a strength of 35 MPa.

The mix at the top left shows the effect of adding water. The slump is increased, but the water/cement (w/c) ratio is also increased, so the strength is decreased. This problem may be solved as shown in the

Table 24.1 Uses of Plasticisers		
Added Water		**Added Plasticiser**
Cement = 300 kg/m^3, w/c = 0.75, increased slump = 240 mm, reduced strength = 26 MPa		Cement = 300 kg/m^3, w/c = 0.62, increased slump = 240 mm, strength = 35 MPa
	Control Mix	**Added Plasticiser, Reduced Water**
	Cement = 300 kg/m^3, w/c = 0.62, slump = 75 mm, strength = 35 MPa	Cement = 300 kg/m^3, w/c = 0.50, slump = 75 mm, increased strength = 46 MPa
Added Cement, Added Water		**Added Plasticiser, Reduced Cement, Reduced Water**
Increased cement = 370 kg/m^3, w/c = 0.62, slump = 240 mm, strength = 35 MPa		Reduced cement = 250 kg/m^3, w/c = 0.62, slump = 75 mm, strength = 35 MPa

mix in the bottom left, where more cement is added to give the original w/c. However, the higher cement content will increase the cost, and may lead to problems with overheating or cracking.

The mixes in the right-hand column have added plasticiser.

The flowing mix is at the top right. This mix has the same slump as the mix with added water, but it has no increase in water, so the strength is not reduced. This mix will cost less to place, due to the reduced requirement for labour to move it into position and compact it.

The high strength mix is in the middle row on the right. By adding plasticiser and reducing the water, it is possible to increase the strength, without loss of workability. Very high strengths can be achieved with these mixes.

The low cost mix is at the bottom right. By adding plasticiser and reducing both the cement and water contents, it is possible to have the same workability and strength with less cement.

24.2.2 USE WITH CEMENT REPLACEMENTS

When plasticisers are used with replacements, such as PFA or GGBS, the dose should be calculated on total cementitious content. Most plasticisers give similar results to plain cement mixes, but manufacturer's guidance should be checked, and trial mixes carried out.

24.2.3 DURABILITY

Plasticisers can be used to decrease the w/c ratio of a mix, and in doing this they improve durability. The adsorption of the admixture onto the cement grains does not interfere with the adsorption of chlorides, so the apparent diffusion coefficient is reduced, as well as the permeability.

When plasticisers are used to reduce the cement content, without changing the w/c ratio, there is also a small improvement in durability. This occurs because the transport processes, such as chlorides moving in and causing corrosion of the steel, all take place through the cement paste, not the aggregate. Thus, a mix with more aggregate and less cement paste is more durable.

A flowing mix is also likely to be more durable, due to the better compaction and reduction in unwanted air voids.

24.2.4 HARDENED CONCRETE PROPERTIES

Drying shrinkage and creep tend to be slightly higher, possibly up to 20% in the case of some superplasticisers, but using water to increase the workability would have a much larger effect.

24.2.5 MIX DESIGN AND PLASTICISER SELECTION

Admixtures are not a panacea for bad mix design, though, in some cases, they can assist when poor quality aggregates have to be used. If the mix design is not optimised, the admixture performance may be below expectation. For example, if the basic mix has a high plastic viscosity, then water reduction may be less than expected, while a less viscous mix may tend to segregate and bleed more.

High workability superplasticised mixes generally need higher sand levels and good aggregate grading in order to get good flow, without bleed and segregation. A viscosity modifier (see 24.3) may be used at very high workability.

24.2.6 SECONDARY EFFECTS

Studying the heat evolution from a plasticised mix shows that the admixture has a significant effect on the hydration process. This would be expected because the admixture adsorbs onto the surface of the cement grains. However, very little effect on the hardened properties is observed. A few secondary effects become more apparent at high doses, including air entrainment, retardation, and strengths that are less than would be expected for the level of water reduction obtained. If plasticisers are to be used in concrete for waste containment, the adsorption of harmful species on the cementitious matrix may be affected.

24.3 VISCOSITY MODIFYING ADMIXTURES

Self-compacting concretes (SCC) are made with superplasticisers and viscosity modifying admixtures (VMAs). See Fig. 22.7 which shows that, together, they increase the plastic viscosity and reduce the yield. This will give a very high, but slow, slump. These mixes will flow into place without compaction, but will not segregate. Note that the admixtures are not sufficient to make SCC without a sufficiently high powder content, and suitable aggregate grading (see Section 21.6). If the aggregate grading is good, it is also often possible to make SCC with a superplasticiser, without the need for a VMA.

SCC that is also self-supporting for slipforming roads has been developed. This apparent contradiction is possible by exploiting thixotropy that can be increased with special admixtures, and gives a high yield, but low viscosity. This means that mixes flow easily, once they are moving but gel immediately when they stop.

It is not generally recommended to add admixtures at the end of the mixing process; however, this may be done to give a simple demonstration of self-compacting concrete. If a mix is made with a powder content of at least 500 kg/m^3 (850 lb/yd^3) about a 100 mm (4 in.) slump, it may be seen in a laboratory mixer to be a normal looking concrete. If a powerful superplasticiser is added, and mixing is resumed for 1 min, the concrete becomes liquid, with the aggregate all sunk to the bottom. If a VMA is now added, and it is mixed again, the aggregate will reappear at the surface, and a self-compacting mix has been made.

24.4 AIR ENTRAINERS
24.4.1 FROST RESISTANCE

The effect of air entrainers on frost resistance is very significant. A mix that would normally fail after less than 60 freeze–thaw cycles may last for more than 1000, with no damage. The effect is far greater than increasing the strength, or reducing the permeability of a mix. The presence of air bubbles can help to relieve the internal pressures, provided the bubbles are small and closely spaced through the paste phase. A total air content of about 5% is often specified (see Section 22.4).

24.4.2 REDUCED BLEEDING

It may be seen in Fig. 22.7 that air entrainers actually reduce the plastic viscosity; however, it has been observed that they significantly reduce bleeding, and they are used to resolve problems with plastic settlement. As little as 2% additional air can significantly reduce this problem of plastic settlement cracking, a weak top surface, sand runs, and other surface defects caused by poor quality aggregates. Air entrainers may also reduce water loss in pumping, and make a mix easier to pump.

24.4.3 EFFECT OF PFA ON AIR ENTRAINMENT

Residual part-burnt carbon in PFA seriously affects the performance of some air entrainers. It increases the dose required by a factor of between 2 and 5, it can lead to additional air loss on extended mixing, and if the level of residual carbon changes from batch to batch, then control of air entrainment becomes even more difficult. Manufacturers' instructions should be consulted to select a suitable admixture, if PFA is being used (or if it is preblended in the cement).

24.4.4 SECONDARY EFFECTS

Air entrainers usually plasticise, even if not formulated to be dual purpose.

24.5 RETARDERS
24.5.1 THE EFFECT OF RETARDERS

Retarders act chemically on the cement to delay the initial stages of the hydration reaction. It would be expected that this would lead to extended workability retention, as well as a delay in setting and hardening of the mix. In practice, early workability loss is not primarily due to cement hydration, and the only effective way to ensure good workability at a delayed time after mixing is to start with a high initial slump, by using, for instance, a higher plasticiser dose. See Fig. 24.1.

When the main stage of cement hydration does start, any remaining workability is quickly lost, and the concrete will stiffen to the point where a "cold joint" will be formed if further concrete is placed against it. This will mean that there is a poor bond between the newly placed concrete and the previous load discharged into the pour, leading to surface blemishes and, in extreme cases, structural weakness in shear. This is a particular problem that retarders can help with, in hot conditions and/or in large pours, where there may be a significant time delay in progressing the placing of a concrete front. They give

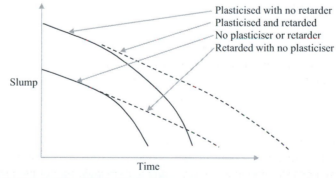

Plasticised with no retarder
Plasticised and retarded
No plasticiser or retarder
Retarded with no plasticiser

Slump

Time

FIGURE 24.1 Schematic Graph of the Effect of Retarders

their main benefit as the concrete approaches initial set. At this point, the retarded mix will hold some workability (0–30 mm slump) for longer, and delay the onset of setting. This level of workability is usually insufficient for primary compaction, but is enough to give good intermixing under vibration at the interface, when fresh, more workability concrete is being laid against an older mix.

24.5.2 SECONDARY EFFECTS

Retarders are based on sugars that will "kill" a mix, and prevent hydration. At low doses, they only retard, but care must be taken not to overdose, or the mix will not set.

Retarders will delay early strength gain, but it should recover within 28 days. Retarders will often also plasticise, and affect bleeding and segregation.

Retarders are very temperature sensitive. A mix that is intended to be "4-h retarded" may not set overnight, if the temperature drops.

24.6 ACCELERATORS

24.6.1 USES FOR ACCELERATORS

Accelerators can be set accelerators that reduce the stiffening time of the mix, or strength accelerators that give higher early age strengths. By 28 days, there is usually little improvement over a control mix.

Using water reducers and accelerators, it is possible to replace concrete road surfaces during overnight closures, and open them to traffic within 6 h of pouring (with strengths of 7 MPa (1000 psi)).

Accelerators are essential for concreting at very low temperatures. They will accelerate hydration, and also accelerate the development of the heat of hydration, and may thus prevent freezing.

24.6.2 CHLORIDE-BASED ACCELERATORS

These provide both set and strength acceleration, their action is proportional to dose over a wide range, and the performance is unbeatable. Unfortunately, chloride ions help to promote corrosion of steel reinforcement, and so chloride accelerators are unacceptable in all concrete that contains any form of embedded steel. The extensive use of calcium chloride accelerator, and the widespread problems caused by the resulting corrosion or reinforcement, made many engineers reluctant to use admixtures at all for many years, and significantly delayed their widespread introduction and use.

24.6.3 CHLORIDE-FREE ACCELERATORS

Performance is never as good as from chloride-based accelerators, and they tend to be relatively expensive in use; however, they are the only choice for reinforced concrete.

24.6.4 SUPERPLASTICISERS AS STRENGTH ACCELERATORS

At temperatures above about 15°C (60°F), the most effective strength accelerator is by far a superplasticiser used as a high range water reducer. By 24 h at 20°C (70°F), strengths will be significantly higher than from even a chloride-based accelerator, and additionally, the strength improvement will be maintained at later ages. Superplasticisers also respond well to accelerated high temperature curing.

24.7 OTHER ADMIXTURES

24.7.1 FOAMING AGENTS

"Foamed concrete" is made by adding foaming agent into a mixed mortar in a truck-mixer. Foamed concrete is a nonstructural void filler that can be dug out with excavators. It is used for trench back-fill and similar applications. The air content is over 50%, compared with about 5% in air entrained concrete.

24.7.2 SHRINKAGE COMPENSATORS

These are used in grouts, when complete void filling is required. If they are not used, the shrinkage during setting will leave a small air space above the grout. They are metal powders, either zinc or aluminium. These give off small amounts of gas, when they react with the mix water. Unlike expansive cements, the pressure of expansion is low and will not cause damage.

24.7.3 CORROSION INHIBITORS

Calcium nitrite is used as an anode inhibitor. Other admixtures can inhibit the cathode. Corrosion can also be reduced with amine/esters that reduce the permeability, but they also tend to reduce the strength.

24.7.4 ALKALI AGGREGATE REACTION INHIBITORS

Lithium-based compounds are used to reduce AAR.

24.8 USING ADMIXTURES ON SITE

24.8.1 TIME OF ADDITION

Plasticisers are more effective if they are added slowly after mixing. For example, all of the increase in slump may be lost after 2 h, if they are added during mixing, but if they are added in small doses every 15 min for 2 h after mixing, the workability will steadily increase. The air content of concrete is also significantly reduced, if it spends a long time in a truck-mixer.

Thus, if the site is a long way from the batching plant, it is theoretically best to add admixtures immediately before discharge. However, this involves much greater risk of operator error, and may not be permitted by some quality assurance (QA) schemes. Special methods are needed for adding admixtures in a truck-mixer because, if they are simply poured in from the top, they may not flow down into the main part of the load.

24.8.2 USING MORE THAN ONE ADMIXTURE

The order and timing of admixture addition can be critical. Trial batches should be used to establish the best procedure. Never premix admixtures before adding them to the concrete, but it often works well to add them to the mix water.

24.9 CONCLUSIONS

- Plasticisers and superplasticisers can be used to improve workability, reduce the w/c ratio, or decrease the cementitious content without reducing strength.
- If superplasticisers are used to reduce the w/c ratio, they will increase strength and durability.
- Air entrainers are mainly used for frost resistance, but they will also reduce segregation and bleed.
- Retarders should be used with care, but are effective at reducing cold joints in large pours in hot climates.
- Chloride-based accelerators should not be used in reinforced concrete.
- If superplasticisers are used to reduce the w/c ratio, they will often increase early strength as effectively as an accelerator.
- Foaming agents entrain significantly more air than air entrainers.

TUTORIAL QUESTIONS

1. Describe how admixtures should be used in concrete, in the following circumstances:
   a. A high strength base, 4 m deep by 10 m square, for a large machine.
   b. A slab for an outdoor parking area.
   c. A superflat floor in which the concrete is to be placed with minimal vibration.
   d. A building in a hot country, where the delivery time from the ready-mix plant is over an hour.
   e. A precasting yard where the moulds must be used on every working day.

   Solution:

   a. Use a superplasticiser to reduce cement content in order to reduce heat (a cement replacement could be used for this).
   b. Air entrainer for frost resistance.
   c. A superplasticiser to give high workability, and possibly a VMA to prevent segregation.
   d. A retarder to delay final set.
   e. Could use an accelerator, but a superplasticiser to give low w/c would also get high early strength, and would give much better long term properties.

2. Describe how concrete admixtures should be used in the following circumstances:
   a. In a concrete road.
   b. In a heavily reinforced beam.
   c. In rapid construction of a multistorey frame.
   d. In the production of a large number of unreinforced precast units.

   Solution:

   a. An air entrainer for frost resistance and a plasticiser for workability/economy.
   b. A superplasticiser to make the concrete flow around congested reinforcement. In extreme cases, it may be necessary to reduce the size of the coarse aggregate.
   c. A chloride-free accelerator may be used in cold conditions, but normally a superplasticiser would be used to give higher strengths at all ages.
   d. Calcium chloride accelerator might be used because there is no reinforcement but it but may cause efflorescence, and will also cause corrosion in metal forms and tools. A superplasticiser may be preferable.

DURABILITY OF CONCRETE STRUCTURES

CHAPTER OUTLINE

25.1 INTRODUCTION

25.1.1 TYPES OF DETERIORATION

In this chapter, the different processes that limit the durability of concrete structures are discussed. It is very important to understand the causes, so that the useful life of a structure may be predicted, effective repairs may be carried out, and future designs and methods improved. There are many different processes that may cause deterioration, including sulphate attack, alkali–silica reaction, and frost damage, but more than half of all materials related failure of concrete is caused by reinforcement corrosion.

25.1.2 THE IMPORTANCE OF THE TRANSPORT PROCESSES

The common characteristic with all of these processes is that they require something to be transported into the concrete from outside. Even alkali–silica reaction, a reaction between components that are already present inside a structure, requires water for the reaction to take place. Transport may take place in sound or cracked concretes but, even where cracks are present, nothing will move without a transport process.

25.1.3 TRANSPORT IN THE COVER LAYER

The key to durability of a concrete structure is to limit the transport in the cover layer. Unfortunately, as shown in Fig. 25.1, this is likely to be the poorest quality concrete. It will be the most likely part to suffer from poor curing and, if it is a horizontal surface, as shown, it will have bleed water passing through it during the initial set.

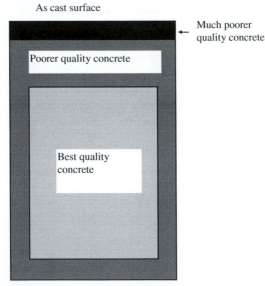

FIGURE 25.1 Cover Concrete

25.2 TRANSPORT PROCESSES IN CONCRETE

The four main transport processes are listed in Table 25.1, and the processes that inhibit or increase them are listed in Table 25.2.

Before considering the processes in detail, the exact nature of what is being transported must be defined. Damage may be caused to concrete by water itself, or chemicals dissolved in water. The water itself will move with the ions in it (as discussed in Chapter 9), or the ions may move through the water (as discussed in Chapter 10). Thus, the transport processes may cause damage both by movement of water, or by ionic movement in the water.

25.2.1 PRESSURE DRIVEN FLOW

"Permeability" is defined as the property of concrete that measures how fast a fluid will flow through concrete, when pressure is applied. Figures 25.2 and 25.3 show examples of pressure-driven flow in structures.

Figure 25.2 shows the effect of pressure-driven flow on a tunnel with a concrete lining. These processes will be of greatest concern, if the groundwater has salt in it. Most concrete tunnel segments are reinforced to prevent breakage during installation, and the chloride will make this corrode and cause spalling. Adsorption onto the hydrated cement matrix will significantly reduce the flow of chlorides, and the water reaching the steel will probably have a very low concentration throughout the design life.

Figure 25.3 shows the wicking process. It is characterised by a line of concrete spalling about a metre above the ground, where the last of the moisture evaporates, leaving the salt behind.

Table 25.1 The Key Transport Processes

Process/Parameter	Cause	Reference Section
Permeability	Pressure gradient	9.3
Diffusion	Concentration gradient	10.4
Thermal migration	Temperature gradient	9.4
Electromigration	Voltage gradient	10.6

Table 25.2 Processes That Inhibit or Increase Transport

Process	Effect	Reference Section
Adsorption	Inhibits transport	10.5
Capillary suction	Increases transport	9.5
Osmosis	Increases transport	9.6

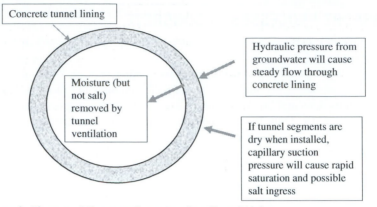

FIGURE 25.2 Schematic Diagram of Transport Processes in a Tunnel Lining

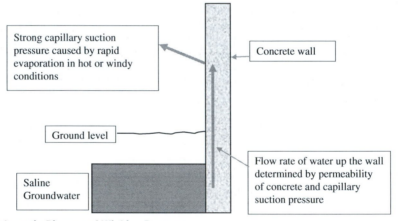

FIGURE 25.3 Schematic Diagram of Wicking Process

25.2.2 DIFFUSION

Diffusion is a process by which an ion can pass through saturated concrete, without any flow of water. Moisture diffusion will take place in a gas when the concentration of water vapour is higher in one region than another. This mechanism will enable water to travel through the pores of unsaturated concrete.

Figure 25.4 shows an example of how diffusion may work in combination with other processes to transport chloride to the reinforcing steel.

25.2.3 ELECTROMIGRATION

Electromigration occurs when and an electric field (voltage difference) is present. This may be derived from an external source, such as leakage from a direct current power supply, but is also frequently

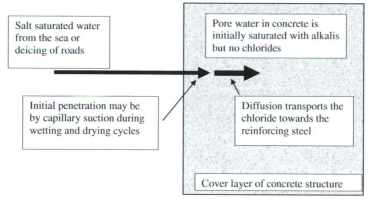

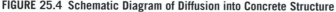

FIGURE 25.4 Schematic Diagram of Diffusion into Concrete Structure

caused by the electrical potential of pitting corrosion on reinforcing steel. If an electric field is applied across the concrete, the negative ions will move towards the positive electrode.

Electromigration will also have a significant effect whenever charged ions are diffusing into a structure. If, for example, chloride ions (that have a negative charge) are diffusing through the cover layer, this will create an electric current. However, an electric current cannot flow unless it is going around a circuit, otherwise electric charge builds up. This will have the effect of reducing the diffusion, and also causing other ions, such as hydroxyl ions (OH^-), to migrate in order to dissipate the charge build-up.

Electromigration can be measured from the electrical resistance of the concrete because it is the only mechanism by which concrete can conduct electricity. Because it is conducted in this way, rather than with electrons, as in a metal, direct current will carry chloride ions into the concrete, or possibly out of it. This mechanism is used in the desalination process in which a positive voltage is applied to the surface of a concrete structure, in order to extract the chlorides from it. This process has the added advantage of applying a negative voltage to the reinforcing steel that will directly inhibit corrosion, in the same way as cathodic protection (see Chapter 31). Unfortunately, it also has the disadvantage of removing other ions from the concrete, a fact that may cause problems such as ASR.

25.2.4 THERMAL GRADIENT

The most obvious situation when this process may occur is when a concrete structure that has been contaminated with deicing salt heats up in sunlight. The salt-saturated water in the surface pores will migrate rapidly into the structure. Even if it does not reach the steel by this mechanism, the salt may diffuse the remaining distance.

25.2.5 CONTROLLING PARAMETERS FOR CONCRETE DURABILITY

Having defined the transport processes, the key questions are how they may be controlled, and how they will affect the durability of a structure. In Fig. 25.5, the left-hand column shows a number of factors we may expect to affect the properties of concrete. They do not, however, directly affect the

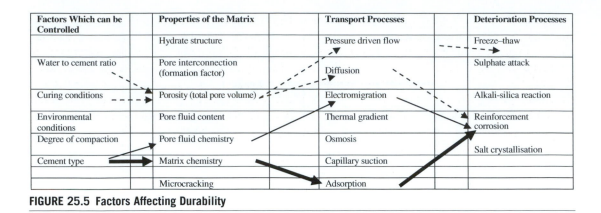

Factors Which can be Controlled	Properties of the Matrix		Transport Processes		Deterioration Processes
	Hydrate structure		Pressure driven flow		Freeze–thaw
Water to cement ratio	Pore interconnection (formation factor)		Diffusion		Sulphate attack
Curing conditions	Porosity (total pore volume)		Electromigration		Alkali–silica reaction
Environmental conditions	Pore fluid content		Thermal gradient		Reinforcement corrosion
Degree of compaction	Pore fluid chemistry		Osmosis		
Cement type	Matrix chemistry		Capillary suction		Salt crystallisation
	Microcracking		Adsorption		

FIGURE 25.5 Factors Affecting Durability

transport properties, and the next column shows the actual internal properties that they are likely to affect. These include microcracks, the chemistry, and also the "formation factor" for the pores, which is a measure of how many direct paths there are through the pore structure. The third column shows the transport properties, and the final column shows the deterioration processes that we want to inhibit. Thus, we change something in the left hand column in the hope that it will affect the next two columns, and finally get a result in the last column. The complexity of this situation explains why it is so difficult to achieve durability in a structure. For structural analysis, the relationship between what we do and the results that we get is defined by quite precise equations. For durability, even moving from one column to the next in the table is unfortunately only possible by using experimental data that is often difficult to interpret. There are, however, numerous significant relationships on the table that may be exploited.

The dashed line arrows start from the water to cementitious ratio, and the curing conditions. If the water content is kept low, this will reduce the number of capillary pores that contain it, and the overall porosity will be reduced. Similarly, good water-retaining curing will promote full hydration, and the resulting products will fill many of the pores and reduce the porosity (see Chapter 20). Following across the table shows that this will reduce the pressure-driven flow, and the diffusion. The pressure-driven flow causes frost attack (in combination with capillary suction), and diffusion is a key mechanism in reinforcement corrosion by carbonation, or chloride ingress.

The thin solid arrows show that the cement type will be the major factor determining pore fluid chemistry. In particular, the use of a pozzolanic material, such as pulverised fuel ash or silica fume, will reduce the amount of hydroxyl ions in solution. These are the main charge carriers, and therefore the electromigration will be greatly reduced. The key effect of this is to reduce corrosion of reinforcement because this depends on the electromigration of these negatively charged ions from the cathode to the anode, so they can combine with positive metal ions.

The thick solid arrows show that the cement type will also be the key factor in determining the chemistry of the cement matrix that forms the structure of the hydrated paste. In particular, if a sulphate-resisting cement is used, there will be few aluminates in it. This will, in turn, severely limit the ability of the matrix to adsorb chloride ions, and they will thus remain free to cause corrosion.

There are very many other links in the table; indeed, almost every factor in each column affects every factor in the next one to some extent, and exploring each of them reveals methods that offer the potential to be exploited to improve durability.

It may be seen, from the discussion, that reducing transport processes will normally improve durability. An exception to this is the deterioration of saturated concrete in fire. The Channel Tunnel linking England with France experienced a severe fire, and photographs of the damage remarkably reveal apparently undamaged reinforcing steel with virtually no remaining concrete. In this incident, the exceptionally low permeability of the concrete prevented the escape of steam from the pores of the saturated lining segments, and caused them to literally explode. This phenomenon may be demonstrated by placing very low permeability concrete in a microwave oven. The only solution to this that is known to the author is to mix the concrete with polypropylene fibres that melt at high temperatures, and provide pathways for the steam to escape.

When considering Fig. 25.5, it must also be observed that building structures with concrete with low transport properties is of no use at all, if the depth of the cover layer is not maintained. If the reinforcement is just a few millimetres below the surface, nothing will protect it from the external environment.

25.3 **CORROSION OF REINFORCEMENT**

25.3.1 **GENERAL DESCRIPTION**

The corrosion of steel is considered in detail in Chapter 31. Steel reinforcement embedded within concrete will not corrode, due to the formation of a protective iron oxide film that passivates the steel in the strongly alkaline conditions of the concrete pore fluid. However, due to a number of causes, this protective film may be destroyed or rendered ineffective. The most significant of these are:

- Carbonation (neutralisation of the alkaline pore fluid).
- Chloride ions.

The corrosion process is described by equations (25.1) and (25.2).

$$Fe \rightarrow 2e^- + Fe^{++} \qquad \text{Anode} \tag{25.1}$$

$$1/2\,O_2 + H_2O + 2e^- \leftrightarrow 2(OH^-) \qquad \text{Cathode} \tag{25.2}$$

If there is oxygen present, the products of these reactions may then combine as in equations (25.3) and (25.4) to form "red rust."

$$Fe^{++} + 2(OH^-) \rightarrow Fe(OH)_2 \tag{25.3}$$

$$4Fe(OH)_2 + 2H_2O + O_2 \rightarrow 4Fe(OH)_3 \tag{25.4}$$

Or, if the oxygen supply is limited, they may form black rust, as shown in equation (25.5).

$$3Fe^{++} + 8OH^- \rightarrow Fe_3O_4 + 2e^- + 4H_2O \tag{25.5}$$

In general, red rust occupies a greater volume than the original materials, and this causes spalling. Black rust does not cause spalling, and is therefore difficult to detect. Red rust is characteristic of carbonation, and black rust of chloride ingress, in situations such as prestressing ducts, where the air is excluded. If exposed to the air, black rust will generally oxidise to red rust within an hour. The colour difference is very clearly visible.

25.3.2 CARBONATION

As CO_2 enters concrete from the atmosphere, it reacts with the lime contained in the pore fluid, and reduces its alkalinity to a point where the passive oxide layer on the steel surface can no longer be supported. The reaction is shown in equation (7.9). The interface between carbonated and noncarbonated concrete is abrupt, but fairly uniform. Consequently, corrosion due to carbonation is generally characterised by a widespread surface rusting (red rust), even though it may occur in patches of different intensity, reflecting local variations in steel, and concrete characteristics, and cover depth (Fig. 25.6).

A key secondary effect of carbonation is that it reduces adsorption of chlorides. As the pH is reduced, the chloroaluminates that form to bind the chlorides to the matrix become unstable, and the capacity factor is reduced. This makes the combination of carbonation and chloride exposure cause significantly more damage to highway structures, than the two processes would have caused individually.

The rate of carbonation is influenced largely by water/cement (w/c) ratio, initial curing and environmental conditions, and increases with increasing w/c ratio. It is highest at relative humidities, in the range of 60–70%. The chemical reaction produces water and, if the humidity is too high for this to escape, the process will slow down.

Carbonation generally progresses at a rate proportional to the square root of time. Thus, if a building is 5 years old, and has only carbonated to 5 mm depth, it may only be expected to carbonate to 10 mm after 25 years.

FIGURE 25.6 Carbonated Concrete on a 50-Year-Old Chapel at Coventry Cathedral

The concrete has carbonated, causing the steel to corrode and this has resulted in spalling. The stone cladding panel has been taken off to repair it. The slenderness of the concrete column has limited the amount of cover that could be provided.

25.3.3 CHLORIDE INGRESS

Chlorides also break down the passive layer on steel. The main sources of chlorides attacking concrete structures are from deicing salts run-off and spray, and also from the marine environment. These can affect bridge decks, piers, and beams. The key reactions are shown in equations (25.6) and (25.7).

$$Fe^{++} + 2Cl^- \rightarrow FeCl_2 \tag{25.6}$$

$$FeCl_2 + 2H_2O \rightarrow Fe(OH)_2 + 2HCl \tag{25.7}$$

It may be seen that the chloride does not form part of the rust, and is released to participate in further reactions.

In marked contrast to corrosion due to carbonation, chloride-induced corrosion tends to cause severe localised damage in the form of pitting. This process is shown in Fig. 25.7, and can produce a rapid local loss in reinforcement cross-section. Corrosion measurement techniques tend only to establish average corrosion intensity. Thus, the measurement of corrosion induced by chlorides is potentially unsafe, since local corrosion may be more intense that the measured average value indicates.

25.3.4 INFLUENCE OF LOCAL ENVIRONMENT OF STEEL

Figure 25.8 shows a typical arrangement of reinforcement in a large structure. The outer layer of bars has greater access to oxygen from the air, so it will tend to form a cathode, and the corrosion damage will occur on the inner layer that forms an anode.

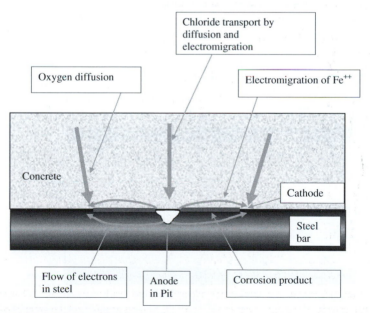

FIGURE 25.7 Pitting of Steel Reinforcement in Concrete

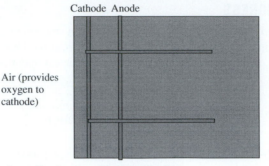

FIGURE 25.8 Anode and Cathode on Two Layers of Reinforcing Bars

25.3.5 INFLUENCE OF POTENTIAL

In the corrosion process, the anode and cathode are normally close together, but they may be far apart, for example, on marine structures where the anode may be in the tidal zone, and the cathode under water. Typical current flow and potential distributions are shown in Fig. 25.9.

Figure 25.9 shows two different examples of corrosion of steel reinforcement. In the upper example, the structure is in air. The current flow between the anode and the cathode will take place by electromigration of ions in the concrete and, due to the relatively high resistivity, the distance between them will be limited. In the lower example, the sample is in seawater that has a low resistivity so the anode and cathode can be further apart. This means that the cathode can be spread over a much larger area of the steel surface, giving far higher corrosion rates.

Figure 25.10 shows an example where severe corrosion may occur. When semisubmerged concrete oil production platforms are used, it is common practice to use the hollow legs for oil storage. In this instance, the oxygen supply to the cathode comes from the oil that may be several metres below the anode, on the splash zone.

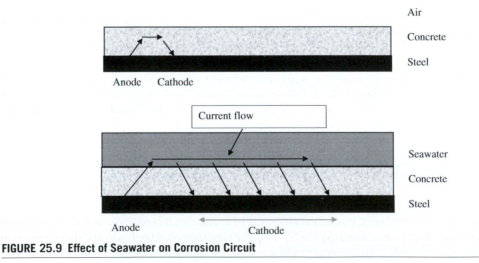

FIGURE 25.9 Effect of Seawater on Corrosion Circuit

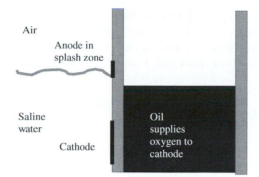

FIGURE 25.10 Corrosion in an Oil Retaining Structure

Other consequences of the electrical nature of the corrosion process are as follows:

- Reducing the area of the anode (e.g., by coating part of the corroding steel) may increase corrosion elsewhere. This is a common problem in repairs where severe corrosion occurs around the edge of a patch repair. However, the cathode typically has an area 10 times larger than the anode, so, within a restricted hole in epoxy coating, there is unlikely to be much anodic corrosion.
- Corroding areas may be located by measuring an increased anodic potential (see Section 27.4.4).
- Application of a positive potential to the surface of the concrete will stop the corrosion process (cathodic protection, see Chapter 31).
- Stray currents from welding, dc. conductor rails or contact between different metals may produce rapid corrosion by creating an anodic region.
- Using a cementitious material with a high resistivity, such as a pozzolanic mix, will decrease corrosion. This would affect both examples in Fig. 25.9.

25.4 SULPHATE ATTACK

Sulphate attack occurs in concrete when sulphates from the surrounding environment react with the calcium hydroxide, and calcium aluminate hydrate in hardened cement paste. The products of the reactions – gypsum and calcium sulphoaluminate – have considerably greater volume than the compounds that they replace, so the reactions lead to expansion and disruption of the concrete. Common sources of sulphate for attack on concrete are:

- Groundwater.
- Sulphate-rich soils.
- Seawater.
- Demolition hardcore that contains gypsum plaster.

The usual method of prevention of sulphate attack is to use sulphate resisting cement, or additions/replacements to the cement (particularly, GGBS). If the structure has not been adequately protected in this way, any repair must effectively prevent further penetration of sulphates.

Thaumasite formation is a different type of sulphate attack. This has caused problems in a number of structures, even when sulphate resisting cement has been used, because it does not need a high alumina content to progress. It has been observed in concrete footings, and depends on ground conditions. The process causes the concrete to disintegrate and, even before this happens, it reduces chloride binding.

25.5 ALKALI–SILICA REACTION

This affects the aggregate and is discussed in Section 19.6.2.

25.6 FROST ATTACK

The high incidence of freezing and thawing cycles, common to many temperate climates, creates an environment destructive to concrete, if it is not correctly designed for such conditions. The extent of damage by freezing is dependent on many factors, such as environmental conditions (rate of temperature change and humidity), and the nature of the concrete itself, in particular, its porosity and permeability. Concrete can be damaged by freezing at two stages; either soon after casting, while the concrete strength is insufficient to resist freezing stresses, or when it has hardened. It is the latter that is of concern, since the former can be minimised by avoiding cold weather concreting.

Continuous freezing of a structure that may occur in polar regions, or in facilities such as liquefied natural gas storage, does not normally damage the concrete. Damage only occurs when there are freeze–thaw cycles.

Early attempts to explain the mechanisms of freezing and thawing in concrete were based on the fact that water expands by 9% on freezing. If a closed vessel is more that 91.7% full of water, it will be stressed on freezing, fact that led to the hypothesis that materials have critical saturation point of 91.7%. In fact, the situation is far more complex because water in very fine capillaries cannot freeze, but the simple model serves to explain most of the observed effects.

Frost attack is normally prevented by using an air entraining admixture

25.7 SALT CRYSTALLISATION

This process has damaged many structures in the Middle East. The salts are taken into the concrete by absorption of saline water. The water then evaporates, leaving the salt (Fig. 25.11). As salt accumulates, it builds up an internal pressure within the sample, which eventually causes spalling. Figure 25.3 shows an example of where this may occur (in addition to reinforcement corrosion caused by the chlorides).

25.8 DELAYED ETTRINGITE FORMATION

Ettringite is one of the minerals that forms the hydrated cement gel but, if it forms long after the concrete has set, it causes expansive stresses and cracking. Fortunately, it is restricted to components where high temperature curing has been used, normally precast.

FIGURE 25.11 Salt Accumulation on Concrete Masonry Blocks in a Parking Garage

Water dripping from the air conditioning units has mixed with salt brought in on car tires, and been drawn up by capillary suction. A short distance up, it dries out, leaving the salt behind in a clearly defined band.

25.9 DURABILITY MODELLING

There is an increasing requirement from construction clients for designers to model the durability of structures, in order to confirm the design life. Traditionally, this has been done by considering just one deterioration process, for example, the diffusion of chlorides into the structure. Diffusion may be calculated from an integrated form of equation (10.2), and this may be used to calculate the time taken for the chloride concentration at the cover depth to reach the critical value for initiation of corrosion. However, it may be seen from this chapter that there are many transport and other processes taking place at the same time, so considering one in isolation is not sufficient. Finite element modelling, which is now commonly used for structural analysis, can be used to solve this problem. Small elements of the concrete are considered for very short times, and the effect of every relevant process is calculated. The software repeats the calculations for every element and time step to give a model of the whole process. The results are then considered statistically, as discussed in Section 13.6.

25.10 CONCLUSIONS FOR CORROSION AND CORROSION PROTECTION

25.10.1 WHY SHOULD REINFORCED CONCRETE STRUCTURES FAIL?

The corrosion process is a chemical reaction between oxygen (from the air), water, and the metal (steel).

- The air and water can move easily through the concrete to the steel.

25.10.2 THE MAIN REASON WHY CONCRETE STRUCTURES DON'T FAIL

The main products of the reaction between cement and water are:

- Calcium silicate hydrate (CSH gel) – this is the main structural part.
- Lime (calcium hydroxide) – this provides alkalinity that promotes the formation of the passive film that protects the steel.

25.10.3 THE MAIN REASONS WHY THE PROTECTION FAILS

- Chlorides from seawater and deicing salts.
- Carbonation from atmospheric carbon dioxide.
- These enter the concrete using the transport processes.

25.10.4 HOW TO REDUCE CARBONATION

Reduce the transport of carbon dioxide by:

- Using a carbonation resisting coating, or
- Using concrete with lower transport rates.

25.10.5 HOW TO REDUCE TRANSPORT RATES IN THE COVER LAYER

- Increase the depth of cover, or
- Reduce the porosity (i.e., the volume of voids) by:
  - Reducing the water/cementitious ratio
  - Using pulverised fuel ash, or blast furnace slag (refines the porosity)
  - Locally reducing the w/c ratio with controlled permeability formwork (see Section 26.2.4)
  - Good compaction
  - Good curing
- Or increase adsorption.

25.10.6 HOW TO PROMOTE THE ADSORPTION OF CHLORIDES

- Do not use sulphate-resisting Portland cement – it has a lower aluminate content.
- Do use pulverised fuel ash, or blast-furnace slag. They have additional adsorption sites.

25.10.7 WHAT CAUSES THE ADSORPTION TO BE REDUCED

The reaction between lime and carbon dioxide to produce carbonates. This reduces the pH and makes chloroaluminates unstable.

These processes are very important. Few reinforced structures, in regions where chlorides are present, would be in serviceable condition today if the alkaline protection and adsorption of chlorides did not occur in the concrete.

TUTORIAL QUESTIONS

1. State, with reasons, which one or more of the four main transport processes for ions in concrete will be most significant in the following circumstances. In each case also describe the effects that will be caused by the ions in the concrete.
   a. The parapet of a road bridge subject to frequent applications of deicing salt.
   b. A concrete slab placed directly over hardcore in moist conditions.
   c. External cladding panels on a building.
   d. A tank containing seawater.
   e. A bridge beam under a direct current power rail.

Solution:

   a. Absorption (capillary suction) and diffusion will transport chlorides into the concrete and cause corrosion of the reinforcement, causing rust staining, and loss of strength. Carbon dioxide from the atmosphere may also enter by diffusion, causing corrosion directly, and reducing chloride binding.
   b. Diffusion, absorption, and (if head of water present) pressure-driven flow will transport sulphates into concrete from hardcore, if the membrane under the slab is absent or failed. Sulphates will cause expansion and subsequent total failure. Carbonation or chemical ingress into the top of the slab is also possible.
   c. Carbonation by pressure-driven flow of air, and solution diffusion in pore water will cause corrosion in areas of low cover. Acid attack from diffusion of atmospheric pollutants is also possible.
   d. Pressure-driven flow of seawater will cause corrosion due to chlorides on steel, and sulphate attack on concrete.
   e. Electromigration of chlorides due to induced electrostatic fields in reinforcement may occur, but polarisation of steel will probably cause more corrosion damage.

2. Discuss the processes that will cause deterioration of a reinforced concrete retaining wall in a marine structure. Identify the processes that are controlled by the transport properties of the concrete, describe the most significant transport properties for each process, and explain why they control the deterioration.

Solution:

Mechanical processes:
- Abrasion
- Impact
- Structural failure

Processes controlled by transport properties:
- Chlorides. Chlorides from salt generally enter the concrete by capillary suction of water in which they are dissolved. This process requires wetting and drying (e.g., in tidal zone). When they are in the concrete, they will diffuse inwards, if the pores are full of water. The tide may also cause a pressure difference across the cover zone that will cause pressure-driven flow,

controlled by the permeability. All movement will be restricted by adsorption onto aluminate phases of cement or PFA to form chloroaluminates. When they reach the steel, they will break down the passive alkaline film, and generally cause pitting.

- Sulphates. These enter in the same way as chlorides but, because they cause deterioration of the concrete itself, their progress is helped by cracking and loss of surface. They react with the aluminates to form expansive compounds.
- Freeze–thaw. This will often be prevented in a marine environment because the sea rarely freezes but, if it does occur, the chloride may actually make it worse. This is because the simple concept of water freezing in pores is not a true description of the process. Note that it only happens if there are freeze–thaw cycles.
- Salt crystallisation. This will occur in a hot environment, with little rainfall. The salts are drawn up through concrete structures by capillary suction, and then the water is lost by evaporation. This will cause progressive build-up of salt crystals in the pores, and subsequent cracking.
- ASR. ASR can be stopped if water is excluded, but this would almost certainly not be possible on a marine structure.

CHAPTER

PRODUCTION OF DURABLE CONCRETE

26

26.1 INTRODUCTION

This chapter outlines methods that may be used to apply the theory discussed in previous chapters, and produce durable concrete.

Durability is critically important for sustainability. There is no point in constructing a bridge with sustainable materials if it requires constant maintenance, or even early reconstruction. The environmental impact of these activities will be far greater than any saving made in the original construction.

275

It is now common to specify a design life of 120 years for concrete structures. This is a good practice. It is possible to argue that, transport technology will change significantly within 120 years, so the type of structure now being built will no longer be required. However, the additional cost or using a lower water/cement (w/c) ratio and a bit more cover is very low. compared with the risk of having to reconstruct a failed road bridge.

There is a school of thought that concrete is unsuited basically to very severe exposure to aggressive environments. This philosophy has led to, for example, the "boxing in" of beams on highway bridges, and providing membranes around buried structures. However, even with these precautions to reduce the risk of failure, the requirement for a durable concrete structure should not be reduced, so the structure lasts even if the protection fails.

The requirement for durability must be embedded throughout the construction process, starting with the specification and design, and continuing through the construction itself. If there is a failure of durability, disputes arise frequently, in which attempts are made to blame the specifier, the designer, or the construction contractor. In many instances, however, the responsibility lies with all parties, and their failure to work together and communicate.

Durable concrete is "high quality" concrete, and quality depends, more than anything else, on personnel. If good durability is to be achieved, the designer and the contractor must employ adequately qualified and, above all, motivated personnel. Without this, theoretical knowledge is of little use.

26.2 DESIGN FOR DURABILITY
26.2.1 ACHIEVING ADEQUATE COVER

Ensure that the detailing is adequate to achieve the specified cover. Many design packages now give large-scale views of the reinforcement. Make sure these are used, and that the reinforcement can be bent to the necessary radii to give the required cover (see Fig. 26.1).

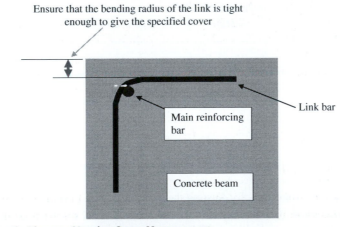

FIGURE 26.1 Schematic Diagram Showing Cover Measurement

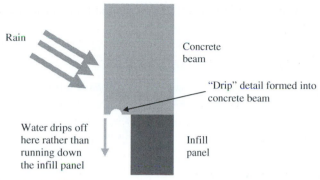

FIGURE 26.2 Design Detail to Shed Water

FIGURE 26.3 Poor Detail of Precast Sill That Does not Shed Water

It results in unsightly staining and potential durability issues.

26.2.2 KEEPING RAINWATER OFF SURFACES

Avoid water running down vertical surfaces – provide "drips" to keep it clear (Figs. 26.2 and 26.3).

26.2.3 AVOIDING CONGESTED REINFORCEMENT

Avoid highly congested reinforcement that may cause voidage. Reinforcement couplers may help. These are mechanical devices that work with screw threads, formed on the ends of reinforcing bars, or clamping or poured bonding systems, and avoid the need for the lap lengths normally required in reinforcement detailing. A high workability or self-compacting mix with a small aggregate may also overcome the problem.

26.2.4 CONTROLLED PERMEABILITY FORMWORK

Consider specifying controlled permeability formwork. This is a fabric that is stretched across the surface of the formwork. When the concrete is wet, it will provide channels for the surface water to flow away, locally reducing the w/c ratio in the cover layer. Some water is also retained in the fabric, and is then available to assist with early curing. Both of these processes improve durability.

26.2.5 ADMIXTURES AND COATINGS

Durability enhancing admixtures are discussed in Chapter 24. Various coating systems are available, ranging from silanes and siloxanes that are sprayed onto the surface, and are intended to block the pores, to resins such as acrylics and polyurethanes that form a complete waterproof seal, unless they are damaged. Evidence to demonstrate their effectiveness should be studied with care.

26.2.6 NON-FERROUS REINFORCEMENT

Under severe conditions, consider using coated/stainless or fiber-reinforced polymer reinforcement. Epoxy coated bar costs approximately 50–100% more than plain bar, but may last significantly longer. However, if the coating may be damaged if it is not handled carefully (e.g., lifting with chains, rather than soft slings). Galvanised bar is a similar price, and has been used successfully on some projects, despite poor performance in some reported laboratory tests. Stainless steel bar costs approximately 100–200% more than plain bar, and may be durable (depending on the grade), but may have poor ductility. Fiber-reinforced polymer bars have good durability, but must be made to their final shape because they cannot be bent.

26.2.7 THE COST OF DURABILITY

The costs for concrete are typically 45% formwork, 28% reinforcement, 12% placing, and just 15% for the concrete itself, so specifying a cheap mix will not save much money, but reducing the w/c ratio may double the useful life of the structure.

26.3 SPECIFICATION FOR DURABILITY
26.3.1 TYPES OF SPECIFICATION FOR CONCRETE

Performance and method specifications are discussed in Section 14.2.2. They are used in the three main types of specification.

26.3.1.1 Designed mix

This is a performance specification, the producer selects the mix proportions (designs the mix). The performance is usually specified as a 28 day cube strength, but may include durability tests such as those described in Section 22.8. An element of method specification is also included normally, in that the specification may include a minimum cement content, or maximum w/c ratio, in order to achieve durability.

26.3.1.2 Prescribed mix

This is a method specification. The purchaser specifies the mix proportions, and is responsible for ensuring that the mix has the required performance.

26.3.1.3 Designated mix

A designated mix is a performance specified mix in which it is intended that the purchaser is guaranteed a given level of durability. The purchaser specifies the required strength, and also the severity of exposure to which the concrete will be exposed. The supplier must then adhere to requirements for cement type and minimum cement content laid down in the standard for the given exposure, as well as achieving the required strength. Quality assurance is often also specified (see Section 14.6).

26.3.2 GUIDANCE ON SPECIFICATION

The key difference between a designed mix and a designated mix is that the element of method specification in a designated mix is taken from the standards, rather than being particular ideas from an individual designer. This will generally reduce cost and improve quality because the concrete suppliers will be familiar with the standards, but may have difficulty with complex specifications that they have not seen before. Also, it is expected that standards will be updated with the latest information on the requirements for durability.

Many problems can arise on site, if too many different mixes are specified. Even if it involves over-specifying the concrete for some elements of the works, it is always better to keep the number of mixes to a minimum, to avoid possible mistakes.

26.4 PLACING DURABLE CONCRETE

26.4.1 CONCRETE SUPPLY

- Be very careful about adding water to the mix on site. Sometimes, this may be necessary to give adequate workability for placement, but it will increase the w/c and thus reduce strength and durability.
- Cement replacements, such as PFA and GGBS, require better curing, so it is essential to find out if they have been used in the mix.
- If a load looks wrong, reject it. The client's engineer may effectively do this by insisting that the contactor makes test samples from it, at which point the contractor will normally send it back.

26.4.2 OBJECTIVES FOR PLACING OPERATIONS

The objectives of concrete placing procedures are to produce a homogeneous mass, with the same properties throughout.

- Nothing else must be in with the concrete (air or debris).
- Nothing must be lost (leakage or breakage, when striking the shutters).
- Nothing must sink to the bottom or float to the top (segregation and bleeding).

FIGURE 26.4 Defects in Concrete Surface

26.4.3 PREPARATION

- Ensure that the shutters are secure, and will not leak. The extent of grout leakage through minor gaps will depend on the workability of the mix. If it occurs, it will result in "honeycombing," in which the coarse aggregate is visible on the concrete surface (Fig. 26.4).
- Remove all debris, loose tying wire, etc.
- In narrow walls use a tremie pipe to prevent segregation due to "ricochet" from reinforcement or shuttering (Fig. 26.5). If this is not used, some concrete will get caught up in the steel near the top of the pour, and have set by the time the concrete is placed around it, and trap air voids within the pour.
- Ensure that the concrete supply is sufficient to achieve a vertical rise of 2 m/h, otherwise cold joints and colour variations may occur (see Section 24.5 for the use of retarders, if this is unavoidable).

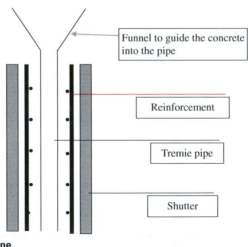

FIGURE 26.5 Use of Tremie Pipe

26.4.4 **SHUTTER OILS AND RELEASE AGENTS**

All shutters should be treated with an oil or release agent, otherwise they will probably take the surface off the concrete when removed. A release agent is a surface retarder that stops the outer surface from setting before the shutter is removed. Shutter oils are mineral oils, which physically prevent the concrete from adhering to the shutters. Release agents have the advantage that, unlike oils, they dry on the shutter, so they are less likely to get onto the reinforcement. Never use any oil that is not sold for the purpose, and never dilute release oils with other oil, such as fuel oils.

Sealing a wooden shutter (e.g., with varnish) is not a good idea because it will generally give a less durable concrete surface (a porous surface works like controlled permeability formwork (see Section 26.2.4)).

26.4.5 **SPACERS**

Durability is substantially dependent on reinforcement corrosion, and this is dependent on cover. Cover is generally achieved with spacers that are fixed to the steel, and rest against the shutters. Spacers are generally made of plastic, and clip onto the steel, but may also be made of other materials, and fixed with tying wire.

- Make sure that the spacers are the correct size, and are fixed in the correct place (normally the links, not the main steel).
- Use a minimum of one spacer per square metre of shutter.
- If the spacers are fixed with tying wire, keep it away from the shutter. Stainless steel tying wire is sometimes specified to avoid surface staining from this.
- Spacers may also be used for other purposes, such as holding wall shutters apart where this is not achieved with standard tie systems. Concrete spacers may be cast on site, and used as shown in Fig. 26.6. If this concrete is not as good as the concrete in the rest of the pour, it may severely affect the durability. It must therefore be cast and cured very carefully.

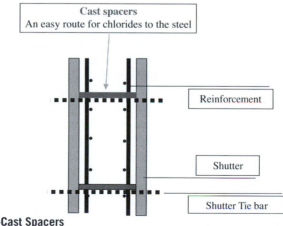

FIGURE 26.6 Use of Site-Cast Spacers

Skip suspended from crane above

FIGURE 26.7 Concrete Placement From a Skip for a Below Ground Parking Garage

26.4.6 PLACING CONCRETE

Concrete may be placed directly from a truck-mixer or with dumpers, skips (see Fig. 26.7), or pumps. These are discussed in books on construction methods. Always consider pumping; this often gives better quality because very poor mixes will not pump.

26.4.7 COMPACTION

Before compaction, concrete with 75 mm slump may contain 5% entrapped air, but at 25 mm slump, it may contain as much as 20%. If the air is not removed by proper compaction, the presence of these large voids will reduce the strength of the concrete, more than 5% loss of strength for each 1% of air. It will also increase the permeability, and hence reduce the durability and protection to the reinforcement, as well as reduce the bond between the concrete and the reinforcement. An additional consequence will be visual blemishes, such as blowholes (small air bubbles trapped against the shutter), and honeycombing (see Fig. 26.4).

The most common method for compaction of concrete is with internal (immersion) vibrators. These work with an eccentric weight that is rotated in the tip. The power for this normally comes from a motor (air/petrol/electric), via a flexible shaft. The alternative is electric immersion vibrators, with motors in the tip (that are very effective).

When using internal vibrators:

- Concrete should be placed in layers approximately 600 mm deep, and each layer should be vibrated before the next one is placed.
- Use vibrators that are in good condition, and powerful/big enough for the job.

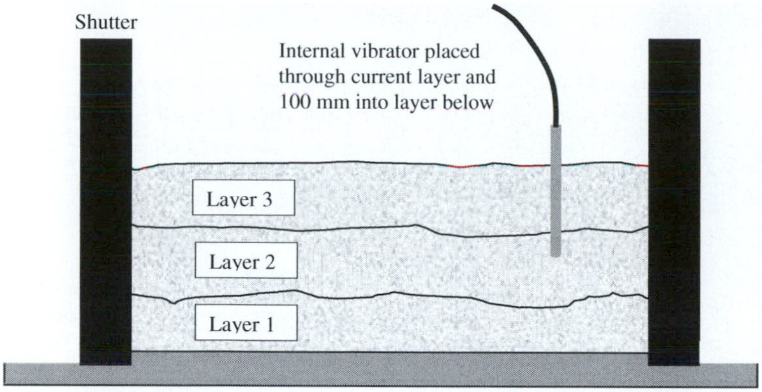

FIGURE 26.8 Sequence of Vibration

- Make sure that the vibrators are ready (and petrol units will start) *before* the concrete arrives. Each layer of the pour must be vibrated before the next layer is placed. Filling a deep shutter, and then starting vibration will result in extensive air voidage.
- Watch where the vibrators are put. They can move steel, and damage spacers and shutters very quickly.
- Never leave the vibrators lying in the concrete, keep them moving. If they are left in one place for more than about 20 s, they will not expel any more air, but will drive away the coarse aggregate.
- Use vibrators about every 500 mm across the surface of the pour (this is for a 60 mm vibrator, closer for less powerful one).
- Always lower the vibrators vertically to the bottom of the fresh concrete, and about 100 mm into previous layers (see Fig. 26.8).
- Never use the vibrators to move concrete around the pour. This will cause segregation.
- Continue vibration until all of the air is expelled, not just until the top of the pour is flat.
- Undervibration is common, overvibration is rare, so, if in doubt, continue vibration.
- Revibrate the top 75–100 mm after 3–4 h, if plastic settlement has occurred (see Section 23.3.6)

  Other methods of vibration are:

- Beam vibrators, drawn across the surface of slabs (Fig. 26.9). Internal vibrators should be used at the edges of a slab when a beam vibrator is used.
- Shutter vibrators, fixed to the back of shutters. These apply considerable stresses to the forms, so they must be fixed securely, and the forms must be robust.

26.5 CURING

This is the last item to be considered, but arguably the most important. The purpose of curing is discussed in Chapter 20.

FIGURE 26.9 Placing Concrete Slab

Some methods of curing are:

- Wrap elements (e.g., columns) in polythene after removing shutters.
- Spray with curing membrane after removing shutters.
- Cover slabs with polythene (and pour ground slabs on polythene).
- For heat retention, use polystyrene on the back of shutters (especially, steel ones).
- Simply leaving shutters in place for a few extra days (especially wooden ones).
- 50 mm of sand can work on slabs.
- Ponding (i.e., forming a pool on the concrete surface) is by far the most effective.

Note about curing:

- Curing is not usually priced as a separate item – this does not mean that no money may be spent on it.
- Make sure that curing is applied as soon as possible. A few hours may make a substantial difference.
- Spray-on curing membranes are not very effective, and in windy conditions they are often useless. On difficult areas (such as columns) they may, however, be the only option.
- Remember that PFA, GGBS and, especially, CSF need *much* better curing (often 5 days, rather than 3 days).
- Permitting the bleed water to dry off will encourage more bleeding, and plastic cracking.
- Slabs on ground should have a polythene sheet placed under them, to prevent excessive water absorption by dry soils.

26.6 CONCLUSIONS

- Good design, specification, placing, and curing are all essential to achieve good durability.
- Design details are important. The design should ensure that the cover is achievable, and water is shed from surfaces where possible.

- The use of designated mix specifications that use standard provisions to achieve durability is recommended.
- Concrete pours should be checked carefully to ensure that adequate spacers have been used.
- Vibrators should be used carefully to ensure full compaction.
- Curing is critical for both heat retention, and water retention.

TUTORIAL QUESTIONS

1. Explain the consequences of the following, in reinforced concrete construction:
   a. Too much cover to the reinforcement.
   b. Insufficient cover to the reinforcement.

Solution:

   a. Makes the structure inefficient, for example, the span of bridges would be reduced.
   b. Will reduce durability by decreasing transport times for harmful species to the steel.

2. Describe curing methods which would be suitable for the following:
   a. A road slab.
   b. A thin concrete wall in cold weather.
   c. A thick concrete wall in cold weather.
   d. A concrete beam in a heated precasting shed.

Solution:

   a. Heat retention is not needed because early strength is not needed. Water retention may be achieved with a spray-on membrane, but windy conditions will make this difficult. Once the concrete has set, tarpaulins, etc. may be used. If tarpaulins are held off the surface, a "wind tunnel" effect must be avoided.
   b. In extreme conditions, it may be necessary to apply heat in cold weather. If not, thick straw mattresses or recycled carpet, etc. may be used; these will also give water retention. Leaving the shutters in place for a few days will help. If they are metal, fix polystyrene on the back. This may be followed up with a spray-on membrane when they are removed.
   c. Heat retention is only needed to reduce temperature gradients. External heating should not be used. Water retention as (b)
   d. Water retention with a membrane (works OK out of the wind). Steam curing for rapid work.

3. Describe the correct procedure for compaction of concrete in the following pours, indicating the different types of plant that might be used:
   a. A road slab.
   b. A heavily reinforced beam.

Solution:

   a. Road slab: possible to use internal vibrators, but beam vibrator is more likely.
   b. RC beam: could use internal vibrators – should be powerful, electric, or petrol units would give more power at a smaller diameter than compressed air units. Shutter vibrators could be used, but formwork must be very secure for these.

4. Describe the effect of the following on the corrosion of reinforcement in concrete:
   a. Construction with increased depth of cover.
   b. Construction with an increased w/c ratio.
   c. Applying a silane coating to the surface of the concrete.

Solution:

   a. This will delay the onset of corrosion due to carbonation or chlorides. In the case of carbonation, the relationship is quadratic, that is, doubling depth increases life by factor of 4.
   b. Increasing w/c will increase rates for all transport processes, and will therefore cause corrosion.
   c. Silanes may reduce permeability, and thus inhibit carbonation and chloride corrosion.

CHAPTER

ASSESSMENT OF CONCRETE STRUCTURES

27

27.1 INTRODUCTION

Assessment of concrete structures is a rapidly growing industry, and employs large numbers of engineers. Some of the reasons for carrying out assessments are as follows:

1. To determine the nature of repairs. In this instance, there are always visible signs of deterioration, for example, cracking.

2. To decide the value of a building, when it is being sold or insured.
3. To determine the expected remaining useful life of a building, when refurbishment or change of use is being considered.
4. To determine the ability of the structure to take an increased loading. This may be required for an old structure, when a change of use or additional construction is being considered, but it is often for new structures where the concrete must be tested to see if it has gained sufficient strength to support additional lifts, or falsework removal. There have been some serious failures caused by inadequate testing of the early *in situ* strength of concrete, particularly when the temperature has been lower than anticipated.
5. Routine safety inspection (e.g., road bridges).

As with the tests on new concrete, described in Chapter 22, there are vast numbers of concrete tests in current use, and only a small number of them are described in this chapter. Before carrying out any test, a full description (preferably a standard) should be downloaded, and agreed with the client in order to determine the correct procedure.

27.2 PLANNING THE TEST PROGRAMME
27.2.1 PRELIMINARY DESKTOP STUDY

Before any site work is started, all available documentation should be studied. Of greatest interest are mix designs, cube test results, and sources of materials. Unfortunately, for many structures, virtually no records have been retained from the original design and construction.

27.2.2 INITIAL VISUAL SURVEY

It is always worth spending sufficient time for a detailed record of all cracks, spalling, segregation, or movement (often most evident at door frames) to be made. A good pair of binoculars is essential for external surveys. At this stage, the structure should be measured for the preparation of layout drawings, if the originals have been lost.

27.2.3 PLANNING THE INVESTIGATION

There are many restraints on the investigation; these include cost, time, access, safety, and damage caused by the testing (e.g., cutting cores). The following principles apply:

- Do not be overambitious. A few tests clearly analysed and reported on time are far more use than a mass of data.
- Ensure that adequate control areas are used. Almost all of the tests are comparative, so there must be some results from sound areas to compare with the suspect areas.
- Where possible, carry out the tests on an accurately measured grid. This will remove bias from the data. If a grid is not used, the tests will tend to be in easily accessible locations that may have also been accessible for maintenance, and will thus be in better condition, or more exposed, leading to a worse condition.

- Try not to rely on the results of a single test. Many of the tests can give quite misleading results, for reasons that may not be apparent at the time of testing. Several test methods should be used, and the correlation between them should be checked.

Decisions about structural assessment are often made in the absence of any clear idea of the objectives. If the manager of a major building is considering a repair programme, he will commission a "thorough" investigation in order to support his case for the budget to carry out the repair. They will then have 500 observations of carbonation depth, when 10 would have been sufficient. Similarly, the author has observed that a committee faced with a major problem will sometimes order further tests just to be seen to be doing something.

27.3 TEST METHODS FOR STRENGTH
27.3.1 ULTRASONIC PULSE VELOCITY TESTING

Ultrasound is very high frequency sound (outside the audible range). Unlike sound in the audible range, it transmits very poorly in air but relatively well through dense materials. Both sound and ultrasound are pressure waves, and their velocity will depend on the Young's modulus of the material, through which they are travelling (see equation (5.9)

The apparatus consists of an ultrasonic emitter and a receiver, and an electronic circuit that records the time taken for the pulse to travel from one to the other (Fig. 27.1).

The general procedure for use is to establish acoustic contact between the concrete and the emitter and receiver, and record the transit time for the pulses. The distance is then measured, and the velocity calculated.

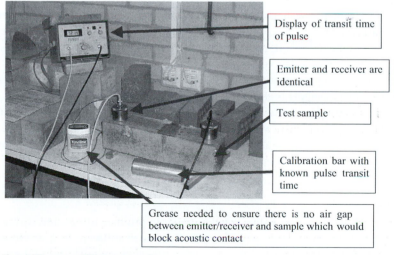

Display of transit time of pulse

Emitter and receiver are identical

Test sample

Calibration bar with known pulse transit time

Grease needed to ensure there is no air gap between emitter/receiver and sample which would block acoustic contact

FIGURE 27.1 Indirect Ultrasound Measurement

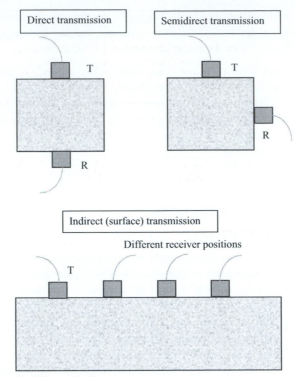

FIGURE 27.2 Configurations for Ultrasonic Measurements

T, transmitter; R, receiver.

The three basic geometric arrangements for the test are shown in Fig. 27.2. It is far preferable to use direct transmission (e.g., through a wall), but this is often the most difficult geometry for measuring the path length. For the indirect method, the path length is curved, so several readings must be taken and plotted, and the gradient used as the velocity.

Because it cannot travel through air, the pulse travels around cracks and other voids (unless they are full of water). Thus, unexpectedly high transit times may indicate cracks.

Figure 27.3 shows the effect of an interface in the structure. This may, for example, be where an overlay has been placed on a road, or where delamination has occurred in a bridge deck. The pulse will travel along the interface (and may also be reflected from it). The resulting data obtained is shown schematically in Fig. 27.4. The gradient changes when the pulse travels along the interface.

27.3.2 IMPACT ECHO

This involves striking the concrete with a small hammer to produce ultrasound waves, recording the echo, and analysing it in great detail with complex computer software to give strengths and layer thicknesses.

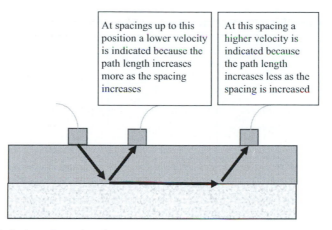

FIGURE 27.3 Ultrasonic Paths at Layer Interface

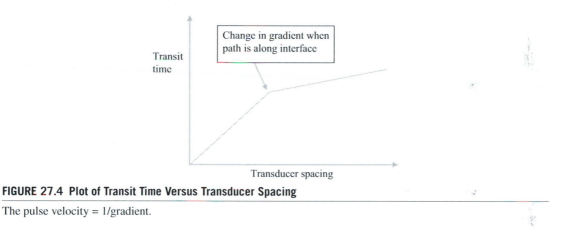

FIGURE 27.4 Plot of Transit Time Versus Transducer Spacing

The pulse velocity = 1/gradient.

27.3.3 **REBOUND HAMMER TESTS**

The rebound hammer (often a "Schmidt" hammer) measures the rebound of a weight from the back of an "anvil" pushed against the concrete surface. The method may be compared to dropping a large ball-bearing onto a concrete surface. If the surface is very rigid (i.e., high modulus), the ball bearing would be expected to bounce back to a good height, but it would bounce less from a less rigid surface.

If a good number of observations are made over a grid, the rebound hammer will give a reasonable indication of relative strength (derived from the Young's modulus). A good use for it is for deciding where to take cores, and extrapolating core test results. Note that a hard surface layer (e.g., from carbonation) may prevent a rebound hammer from detecting the effect of high alumina cement.

Table 27.1 Typical Core Strengths as Percentages of Strength at Bottom of Pour

Position in Pour	Wall	Beam	Column	Slab
Top	45	60	75	80
3/4	55	65	90	90
Mid	70	75	95	95
1/4	85	88	95	98
Bottom	100	100	100	100

27.3.4 CORE TESTS

Cores are cut with diamond tipped core drills.

- Table 27.1 shows typical strengths. Do not expect to get the target mean. The results will vary significantly because of the effects of better consolidation at the base of a pour, bleed water rising through the upper parts, and imperfect curing of the top surface.
- Cores are tested on end, so the ends must be ground, cut, or capped before testing. The quality of the result is only as good as the quality of this work.
- Always use plenty of water or the core bit will be damaged (see Fig. 27.5).
- If carrying out chemical tests on the core, remember that the cutting water will have washed out some chlorides from near the core surface. In the lab, this may be avoided by using oil as the cutting fluid.
- Use a cover meter (see Section 27.4.1) to locate the steel, and try to avoid it. Cutting the steel is bad for the structure, slow, wears out the coring bit, and produces a core which cannot be tested.

Figure 27.6 shows a typical lightweight coring machine. It is effectively a large drill with a special coupling to the core bit to introduce the water. Figure 27.7 shows a core recovered from a road slab. This was an experimental slab with a thin layer of concrete over a grout base, with no aggregate.

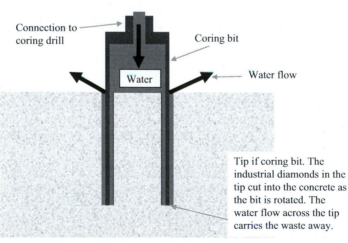

FIGURE 27.5 Cross-Section Showing Water Flow Around Coring Bit

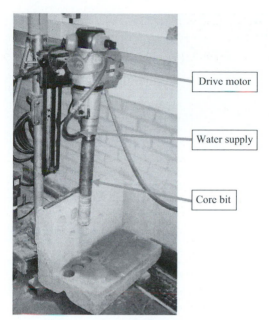

FIGURE 27.6 Coring Machine

FIGURE 27.7 Core Recovered From Road Slab

FIGURE 27.8 Pull-Out test

The stud is being pulled out using a screw thread, and a torque wrench. This measures the force needed to pull it out.

27.3.5 PULL-OUT TESTS

There are many of these including the Windsor Probe. In them, a stud is cast, drilled, or shot fired into the surface, and the force required to pull it out is measured (see Figs. 27.8 and 27.9). This can give a reasonable indication of strength (better than rebound hammers, but not nearly as good as cores).

27.3.6 LOAD TESTS ON STRUCTURES

These are very expensive, but can yield a lot of information. The load applied should be the working load plus a suitable safety factor. A number of different methods may be used to load structures, heavy vehicles may be driven onto bridges, buildings may have temporary tanks of water on the floors, or stressed cables through to ground level.

In a load test, the deflection is measured throughout loading *and unloading*. One of the most important measurements is to check whether the loading results in any permanent deformation, that is, whether it has been loaded beyond its elastic limit, as defined in Section 2.5 (see Fig. 27.10).

27.4 TEST METHODS FOR DURABILITY
27.4.1 MEASUREMENT OF COVER

There are a number of different electronic devices on the market that claim to measure cover to the reinforcement and bar diameter. They all work quite well at locating reinforcement, are reasonably accurate at cover depth, but are less reliable when measuring bar diameter. They do not work in areas of highly congested reinforcement, or if there is other metal in the concrete, for example, conduits, or excessive amounts of tying wire.

FIGURE 27.9 Test Cube After Pull-Out Test

The stud was installed in a drilled hole and has been pulled out, and this has cracked the cube.

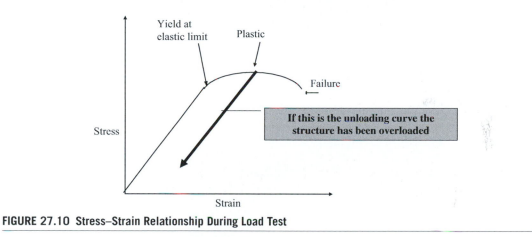

FIGURE 27.10 Stress–Strain Relationship During Load Test

27.4.2 MEASUREMENT OF CRACK WIDTH

The width of cracks on the surface of concrete is measured with optical micrometres, or feeler gauges. The result of a single observation is not much use because the surface width of a crack does not correlate with the depth. If structural movement is suspected, more interesting results may be obtained by monitoring crack movement over time, by installing strain gauges across a crack, or filling it with a rigid filler, and checking to see if new cracks appear as movement continues.

27.4.3 **INITIAL SURFACE ABSORPTION TEST (ISAT)**

The processes that permit the water to flow into a concrete surface (capillary suction, pressure driven flow, see Chapter 9) are also the processes that control some aspects of durability, and the results of this test correlate quite well with measurements of chloride ingress, corrosion, freeze/thaw, etc.

The equipment for the ISAT is shown in Figs. 27.11 and 27.12. The rate of adsorption of water into a small area of the surface of concrete is measured as movement of water in a capillary. The test procedure is:

1. Secure the Perspex cap to the concrete surface.
2. Fill the apparatus with water, with the tap open so all air is expelled, and it flows out through the capillary tube. This is the start time for the test.

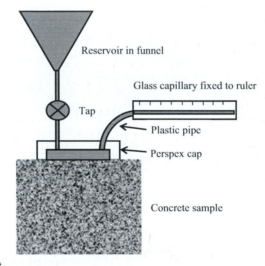

FIGURE 27.11 The ISAT Test

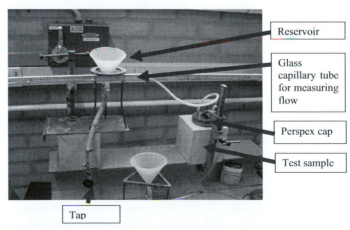

FIGURE 27.12 ISAT Test Apparatus

3. At intervals up to 2 h, close the tap. The flow of water into the concrete surface will cause the glass capillary to empty slowly. Measuring the movement of the meniscus along it, in a set time (typically 30 s), will give the rate of flow into the concrete. A low rate of flow indicates a durable concrete.
4. Reopen the tap after each reading.

The rate decreases with time, and a curve of the form:

$$\text{Flow} = at^{-n} \qquad (27.1)$$

where t is time and a and n are constants for each run.

This may be fitted to the results, using the trendline function on a spreadsheet.

From the constants, the ISA_{10} (10 min) value is obtained at $t = 600$ s. This is more accurate that simply making an observation at 10 min because it includes all the data.

The difficulty with the test is that it is substantially affected by the initial moisture content of the concrete. For this reason, exterior tests are not permitted within 2 days of rain. Even with this constraint, results vary from winter to summer, etc. This may be solved by placing some indicating silica gel desiccant into the ISAT cap, placing it on the concrete surface, and applying a vacuum to it. The vacuum will dry the concrete surface, and when the desiccant changes colour, the concrete is ready to test.

The standard states that the water level in the reservoir should be a fixed height above the concrete. However, the capillary suction pressure is substantially greater than the pressure head, so the test actually works well even if the reservoir is below the concrete.

The test is almost nondestructive but, when testing structures, it is necessary to drill fixing holes in the concrete in order to clamp the cap to the surface.

27.4.4 FIGG TEST

The arrangement for the Figg air index test is shown in Fig. 27.13. A 10 mm diameter hole is drilled into the concrete, and the top of it is sealed with sealant. A hypodermic needle is put through the seal,

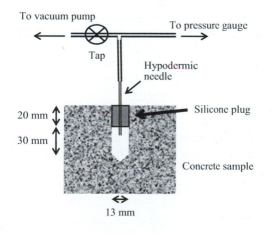

FIGURE 27.13 The Figg Test

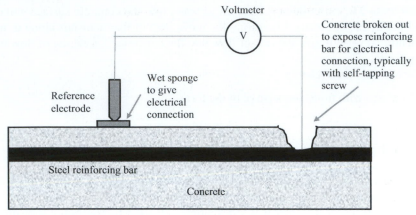

FIGURE 27.14 Circuit for Potential Mapping

and the space below it is evacuated. The time taken for the vacuum to decay, due to pressure-driven flow through the concrete, is measured.

The only transport parameter that controls this test is the permeability (it is pressure-driven flow). This test is affected by the moisture in the concrete and, as with the ISAT, results may be improved by vacuum drying.

27.4.5 POTENTIAL MAPPING

The relationship between the rest potential of steel reinforcement and the corrosion rate is discussed in Chapter 31. The arrangement for potential mapping is shown in Fig. 27.14. The electrode is used to establish electrical contact with the concrete surface, with a known contact potential. The electrode may be copper/copper sulphate, standard calomel or silver/silver chloride, and there is a constant voltage difference between results from each type.

The results may be interpreted using Table 27.2 to give the probability of corrosion, but are best used comparatively to identify areas for further investigation. The data in Table 27.2 should only be used as a very rough guide. It is a good idea to use a contour plot of the surface of the concrete, to show up areas at high risk.

Table 27.2 Interpretation of Measures Rest Potential

Measured Potential (Calomel Standard Electrode) of Concrete Surface (mV)	Statistical Risk of Corrosion (%)
More negative than −350	90
From −200 to −350	50
Less negative than −200	10

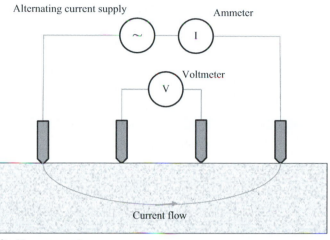

FIGURE 27.15 Resistivity Measurements

Potential mapping is quick, nondestructive, and cheap. Large areas may be covered. However, the results must be treated with caution. For example, carbonation increases concrete resistivity, and therefore increases apparent rest potential. The results only indicate that corrosion is possible, and give no information about corrosion rates.

27.4.6 RESISTIVITY MEASUREMENTS

Figure 27.15 shows an arrangement for measuring the electrical resistivity of concrete. Four electrodes are used because the two that are used to apply the current cannot be used to measure an accurate voltage. An alternating current is used because a direct current would cause the ions in the concrete to migrate toward the electrodes, and the measured current would progressively reduce, as the charge built up (see Section 6.9).

For corrosion to proceed, a current must flow through the concrete, and therefore a high resistivity indicates low corrosion rates. Table 27.3 gives guidance on rates. This is a good test, and should be used when a potential survey is being carried out. However, it will be affected by carbonation, and the use of pozzolanic cement replacements that deplete the charge carriers, as discussed in Section 22.8.3.

Table 27.3 Interpretation of Resistivity Results	
Resistivity (Ωm)	**Likelihood of Significant Corrosion**
<50	Very high
50–100	High
100–200	Low/moderate
>200	Low

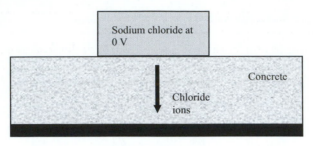

Reinforcing bar at +60 V

FIGURE 27.16 Rapid Chloride Measurements on *in situ* Concrete

The charge passing in 6 h gives an indication of concrete durability.

27.4.7 LINEAR POLARISATION

This is described in Chapter 31. It is the only way of measuring actual corrosion rates of reinforcing steel.

27.4.8 CHLORIDE MIGRATION TESTS

Figure 27.16 shows the *in situ* version of the test in Fig. 22.21. It is less frequently used, and has the same limitations. It would also cause concern about the structure by promoting corrosion, but this is unlikely in just 6 h of operation.

27.4.9 THE PHENOLPHTHALEIN TEST

Phenolphthalein is an indicator solution that is used to show carbonation. It goes purple at pH less than 10, and remains clear above 10. Thus, if a freshly broken surface is sprayed with it, the carbonation depth may be measured. This is a fairly basic but exceptionally cheap and useful test. Theoretically, the steel will corrode at some pH levels that still show purple, but generally the carbonation front is well defined.

27.4.10 OTHER CHEMICAL TESTS

These are generally carried out on samples returned to a laboratory. They may either be wet chemical tests, X-ray diffraction, or X-ray fluorescence. They may be used to detect chlorides, sulphates, original cement types and contents. The most reliable tests are for chlorides, sulphates, and HAC.

27.4.11 OTHER TESTS

There is an almost endless range of other tests:

Radar may be used to assess the condition of reinforcement in concrete.

Thermal imaging (see Section 4.9) may be used to identify hot or cold regions on the surface of a structure. If one part of a concrete surface heats up faster than the rest, when exposed to sunlight, this could indicate delamination that is preventing heat loss back into the structure.

27.5 PRESENTING THE RESULTS

The results of an assessment of a structure can have significant financial implications, and may also guide decisions regarding the safety of continued use. Generally, the different properties of concrete will correlate, so an area of high strength would be expected to show good results for the durability tests. These relationships should be checked, and any anomalies investigated. The comments in Section 16.6.4 on presentation of results from a research programme also apply to a structural assessment. In order to remove bias, it is technically correct to decide on the method of presentation before the data is obtained. This should include the methods used to dismiss any anomalous data.

27.6 CONCLUSIONS

- Careful initial assessment and good planning are essential for effective testing of structures.
- Ultrasound can give a good indication of strength.
- Core tests give the most reliable indicators of strength, but the results will vary considerably, depending on the position with a pour, and will not normally reach the target mean strength.
- Load tests are expensive, but can give very useful results.
- Reinforcement cover meters can be confused by other metal components in concrete.
- Transport tests, such as the ISAT, can give a good indication of the durability of a structure.
- Potential mapping is quick to carry out, and can give an indication of areas for further investigation.
- Resistivity measurements are useful, and should be included with a potential survey.
- To avoid bias, the method of presentation of results should be decided before the tests are carried out.

TUTORIAL QUESTIONS

1. A concrete structure has been built, and a few months after the concrete has been cast, 10 cores are taken from it. The cores are tested in compression, and they show four results below the characteristic strength for the concrete. Describe how a decision should be made about the benefit of further testing.

    Solution:

    The following questions should be answered before a decision is made:
    a. Is the stated characteristic strength required, or has it been specified in order to achieve durability?
    b. Were the bad cores taken from near the top of wall/column pours? This would give low results. (See Table 27.1.)
    c. Were any unusual failures reported?
    d. Did any of the cores have steel or debris in them?
    e. What was the spread of the results? Are there figures available to show the expected standard deviation for sets of cores taken in this way?
    f. Are the areas from which the bad cores came on the critical path for construction? Is there time for further tests? What could/would be done, if low strengths were proven?
    g. Were all of the failed cores from the same part of the structure, or were they randomly distributed?
    h. Does the organisation that took and tested the cores have a good reputation and accreditation?

2. A major fire has occurred in a reinforced concrete structure, and you are required to determine the extent of the loss of durability of the structure, in the event of subsequent exposure to chlorides and sulphates. Describe how an experimental programme to do this should be designed.

Solution:

The aim of the programme will be to measure the transport properties of the concrete, and compare it with unaffected, "good" concrete.

In situ flow tests. ISAT, FIGG, etc. These can be used to measure transport of gas and water. Preconditioning of the concrete is important. They have other limitations, including depth of measurement for ISAT, and effect of drilling for Figg.

Electrical tests. Potential survey is unlikely to work very well because the fire will have calcined the surface (see Section 7.9), and may have caused carbonation. In any event, it is not a good predictor of future corrosion.

Linear polarisation needs cast-in probes and, even if there are any, these may have been debonded by the heat. However, even this measures current corrosion rate – not future trends.

Resistivity is probably best of the electrical tests, but is also affected by surface calcination.

Lab tests. Permeability tests on cores would probably be the best. Absorption tests (or even ISAT) could work well because the surface layer could be cut off.

For sulphates, an expansion test would be useful on a prism cut from a core (see Section 22.8.4).

The programme. Always compare with the control area (half the tests should be control).

Plan the programme in advance.

Check for agreement between the different tests.

MORTARS AND GROUTS

CHAPTER OUTLINE

28.1 INTRODUCTION

This chapter discusses mixes made with cement that contain no coarse aggregate. They are used for a wide variety of purposes where the large aggregate particles cannot be placed. Some of the strength, some of the water demand, much of the stability, and some of the mixing of concrete is provided by the coarse aggregate. Mixes without coarse aggregate will therefore:

- Generally, have lower strengths.
- Generally, have lower water/cement (w/c) ratios.
- Have high shrinkage; this makes them unsuitable for use in thick sections.
- Often require high shear mixing.

Types of masonry and methods of masonry construction are discussed in Chapter 34. The discussion in this chapter is about the mortar; however, the high priority given to the appearance of the finished work applies as much to the mortar as to the masonry units (bricks or blocks).

Grouts are used widely as void fillers in civil engineering structures. They are often placed in critical locations where failure will damage the entire structure, so they must be accurately prepared and placed.

Grouts are normally cementitious, in that they are made with cement and cement replacements. However, there are other types in which the matrix is a polymer, such as in epoxy grouts. These are discussed in Chapter 35.

28.2 MASONRY MORTARS

28.2.1 MORTAR SUPPLY

Mortar is normally mixed on site in small rotating drum mixers so quality control is difficult. On larger construction sites, the mortar is often held in silos. These may be "dry silos" that will hold around 30 tonnes of mortar fully mixed, but without the water. An automated system dispenses small batches of the powder and water, and mixes them ready for use. Alternatively, mortar may be delivered fully mixed into "ready mix" silos from which it is dispensed as required.

28.2.2 MATERIALS USED TO MAKE MORTAR

Masonry mortars are made with the following materials:

Cement. Masonry cement is supplied for use in mortar and contains an air entrainer to give improved workability, and subsequently frost resistance after hydration.

Cement replacements. GGBS may be used in mortar. PFA will give it a dark colour that may be unattractive. Natural pozzolans have been used traditionally.

Sand. Bricklayers will prefer a well-rounded sand to a crushed sand with angular particles.

Air entrainer. This should be added separately if a masonry cement is not being used.

Retarder. When delivered to ready mix silos, mortar is retarded normally for 36 or 72 h.

Plasticiser. Special plasticisers are formulated for masonry mortars. Detergents sold for dishwashing will work, but their use should be discouraged; particularly, cheaper ones that contain salt as an extender.

Pigment. A wide variety of colours are available, and are often specified by architects. Pigments are often premixed into the lime and sand, before delivery to site in order to give a consistent colour.

Water. The water is added at the mixer to give the workability required by the bricklayers.

28.2.3 TYPES OF MORTAR

Cement lime mortar may be made at ratios of 1:0.25:3 (cement:lime:sand) for higher strength and durability down to 1:3:12 for a weaker mix capable of accommodating more movement. Cement lime mortar should not normally be used below the damp proof course (dpc).

Masonry cement mortar is used where the sand is suitable to give sufficient workability without lime or plasticiser. Mix ratios vary from 1:2.5 (cement:sand) for a rich mix for structural brickwork, such as manholes in roads, down to 1:7 for more movement.

Plasticised mortar is made with ratios of 1:3 to 1:8. These will set faster than cement lime mortars, and thus, reduce the risk of frost attack during construction.

Lime mortar is a traditional building material, and is now only normally used for restoration, and repair of old buildings. It was generally made with a mix ratio of about 1:3 (lime:sand), and sets by carbonation. Pozzolans such as volcanic ash or ground clay bricks, and tiles were often added to give a higher strength.

28.2.4 THE REQUIREMENTS OF MORTAR WHEN IT IS WET

- It should be sufficiently workable to spread easily. Lime is the most common component, which works as a plasticiser.
- It should be sufficiently cohesive to stay on a trowel, and then adhere to a masonry unit (brick or block) during laying.
- It should remain useable for 2 h after mixing (note that, unlike most concrete, mixing is not continued until use). Porous bricks draw water out of mortars
- It should set fast enough to permit walls to be raised 1 m each day, and to give some resistance to overnight frost (although protection should be used). Accelerators rarely work due to the thin sections. Do not use calcium chloride, as it leaches as well as causing corrosion. Work should normally stop below 3°C.

28.2.5 THE REQUIREMENTS OF MORTAR AFTER IT HAS SET

- It should be strong enough to carry the loads that are applied on it by the masonry units. Apart from the w/c ratio, the sand grading has a substantial effect on strength.
- It should be flexible enough to take up movement in the structure. All movement caused by thermal effects and moisture movement (often high in new bricks) should be taken up in the mortar, not the masonry units, because they will crack. Therefore, as a working rule, the mortar must not be stronger than the units that it is bonding.
- It should resist water penetration. Note that some water penetration into a wall cavity is harmless because it is removed by tray dpcs and weep holes (see Chapter 34)
- It should have an attractive uniform appearance. Uniformity requires accurate proportioning for mixing. Adding more water will give generally a lighter colour to the set mortar and widely varying water contents can ruin the appearance of brickwork.
- It should be durable. The main durability considerations are:
  - Frost resistance. This is easily achieved with air entrainment and, if it fails, requires repointing.
  - Sulphate resistance. Some bricks leach sulphates, and if sulphate resisting cement is not used the consequences are almost impossible to rectify. Below the dpc sulphates in groundwater may be a problem.
  - Protection of embedded steel, that is, reinforcement (mesh) and wall ties. This requires resistance to carbonation.

28.3 RENDERING

This has two main functions:

- To improve appearance (often on block walls).
- To resist water penetration, often on solid walls. In countries such as the United Kingdom with a large stock of old masonry properties, up to 30% were built before cavity walls were introduced, so they have solid walls that rely on the render to prevent moisture ingress.

The materials used are similar to building mortars. Good adhesion and low shrinkage are essential. The render should not be stronger than the substrate. A very coarse sand gives poor adhesion, while a very fine sand gives high water demand, and thus high shrinkage. A first coat is used to even out the water absorption of the substrate before the outer durable coat is applied. The render should be cured by protecting it from strong sunlight and drying winds.

28.4 CEMENTITIOUS GROUTS

28.4.1 GROUTING MATERIALS

Cementitious grouts often have a high content of cement replacements such as PFA, which will reduce the rate of generation of heat of hydration and also improve workability. They will also normally contain admixtures including plasticisers and shrinkage compensating admixtures (see Section 24.7.2).

28.4.2 GROUT MIXING AND PLACING

Grouts require high-shear mixing because they lack the coarse aggregate that assists with the mixing in concrete. This is achieved usually by pumping it as shown in Figs. 28.1 and 28.2.

Water is adder first to the mixer tank, and the pump started with the diverter valve set to return the flow to the tank. The powders (typically, cement and PFA) are then added, and the shearing action of the pump mixes them very effectively with the water. Once the mixing is completed, the diverter valve is set to discharge to the works. This uses normally a small diameter flexible pipe. If grouts are to be prepared in the laboratory, high shear mixing is also essential.

Grouts are not normally vibrated, but the need to expel all the air is just as critical as it is for concrete, and failure to do it effectively has led to some widespread structural failures (particularly in tendon ducts as described in section 28.4.3).

28.4.3 GROUTING IN STRUCTURES

Figures 28.3 and 28.4 show a very typical application for grout in construction. The voids are formed around the bolts when they are cast into the concrete base, to permit some movement to allow for tolerances when positioning the steel column. When the column is fixed into position, it is supported on shims

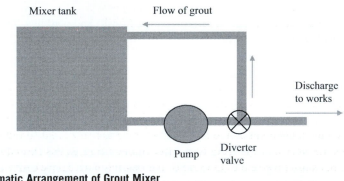

FIGURE 28.1 Schematic Arrangement of Grout Mixer

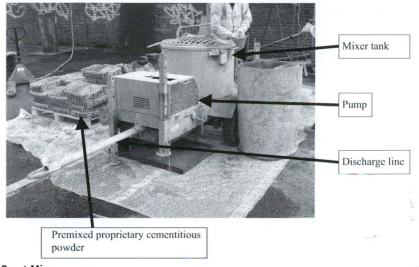

FIGURE 28.2 Grout Mixer

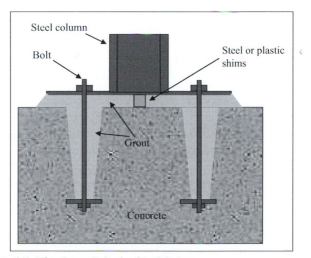

FIGURE 28.3 Arrangement of Holding Down Bolts for Steel Column

so the bolts may be tightened, and the frame assembled. Once it is correctly positioned, grouting may follow. It may be seen that if the voids are not properly cleaned out, and the grout is not properly mixed and placed, there may be air pockets left around the bolts. These may fill with water and cause corrosion.

Figure 28.5 shows the arrangement of a roller bearing for a bridge deck. The grout may be considered an insignificant part of the works, but grout failure under bridge bearings has caused significant damage to bridges. The Fig. 28.5 shows a space to the side of the bearing where a jack may be used to take the load to permit maintenance of the bearing, or replacement of the grout. This provision is recommended.

FIGURE 28.4 Holding Down Bolts With the Expanded Polystyrene Void Formers Still in Position Around Them

The void formers must be removed before the baseplate is positioned on the bolts.

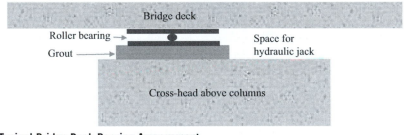

FIGURE 28.5 Typical Bridge Deck Bearing Arrangement

Figure 28.6 shows a typical arrangement of a prestressing tendon duct in a bridge deck. This is typically a 75 mm (3 in.) steel or plastic duct containing a number of 16–20 mm (5/8–¾ in.) diameter cables. It may be seen that if it is not fully grouted, water may gather in it. If deicing salt is used on the bridge (or carried onto it from other parts of the road by the traffic) this water will contain salt. Due to the exclusion of air, the corrosion product will be black rust, which does not cause expansion so there will be no external signs of the corrosion (see Section 25.3). The duct will prevent potential survey and other methods of investigation described in Section 27.4 from working. There have been a number of serious failures of bridges of this type.

In order to overcome this problem, very strict grouting regimens have been introduced. A number of vent pipes are cast into the slab. Grout is injected from one end, and must appear at each of the vents in turn. The construction method will stipulate that a good volume of grout that is free of air must flow from each vent before it is plugged off.

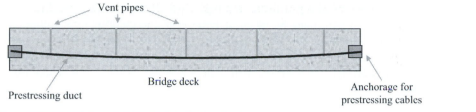

FIGURE 28.6 Arrangement of Prestressing Duct in Bridge Deck

In some critical installations, such as nuclear pressure vessels, the tendons may be left ungrouted, so they can be removed for inspection; but this is not practical in road bridges where they would fill with water. External tendons have been used on some bridges to avoid this problem.

28.4.4 PREPLACED AGGREGATE CONCRETE

Preplaced aggregate concrete is used in places where it would not be possible to place normal concrete at an adequate rate to fully fill the required volume. The coarse aggregate is placed carefully, and compacted into position. A sanded grout is then injected under pressure to surround it. In order to avoid trapped air pockets, the grout would normally be injected from the bottom of the pour.

28.4.5 GEOTECHNICAL GROUTING

Geotechnical grouting is a major subject area for geotechnical engineers. The grout may be used to stabilise bad or contaminated ground, or it may be used to seal it against water flow. A grout curtain is often installed to seal cracks in the rock around a dam, using special grouting tunnels that penetrate considerable distances to either side. Anti-washout admixtures are used, so the grout remains stable if there is groundwater in the cracks.

28.4.6 WASTE CONTAINMENT

Cementitious grouts are very effective for waste containment due to their high capacity for adsorption (see Section 10.5). This will mean that, provided the pH remains high, harmful species will be bound onto the cement matrix, even if they are not retained physically by a structure with low transport (the main transport processes are diffusion and pressure driven flow). Grouts may be injected to prevent groundwater contamination from existing waste deposits. In the nuclear industry, cement grouts are very effective for the long-term containment or radioactive waste because the capacity factors for the most harmful species such as plutonium are very high.

28.5 CEMENTITIOUS REPAIR MORTARS

The most common application for repair mortars is the repair of spalling caused by reinforcement corrosion in concrete structures with chloride ingress. This is a major industry with large repair contracts in progress worldwide. Proprietary products containing cements, cement replacements, and plasticisers, and shrinkage compensating admixtures are supplied preblended.

Figure 28.7 shows a typical repair detail and Figs. 28.8–28.11 show a typical repair sequence on a critical structure. Great care must be taken to remove all the chloride contaminated and spalling concrete, particularly behind the bar. The reinforcing bar is then shotblasted clean, and inspected to ensure that the loss of section due to corrosion is not critical for the structure. The grout is then poured in as shown or, alternatively, it may be pumped in at the bottom of the repair. The grout must have a

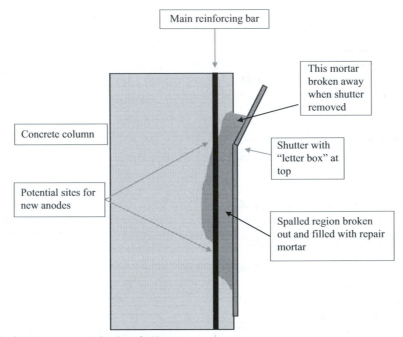

Main reinforcing bar

This mortar broken away when shutter removed

Concrete column

Shutter with "letter box" at top

Potential sites for new anodes

Spalled region broken out and filled with repair mortar

FIGURE 28.7 Typical Arrangement for Repair Mortar

FIGURE 28.8 Repair Sequence 1

Jacks are positioned to carry the load during the repair.

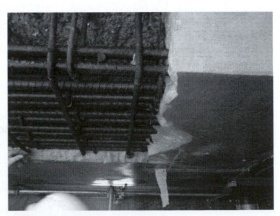

FIGURE 28.9 Repair Sequence 2

Chloride contaminated concrete is removed by hydrodemolition and the reinforcing bars are cleaned.

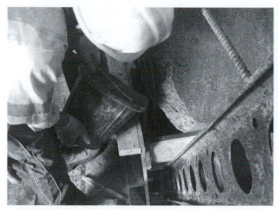

FIGURE 28.10 Repair Sequence 3

The shutter has been placed and the grout is being poured into it.

FIGURE 28.11 Repair Sequence 4

A completed repair.

similar coefficient of thermal expansion to the parent concrete to avoid debonding during hot or cold weather, so cementitious grouts are generally preferred to epoxies (see Section 35.6.3). A problem with patch repairs is that the new grout will promote the formation of a strong cathode, and this may lead to "incipient anodes" at the edges of the repair causing rapid corrosion. Cathodic protection may be necessary to stop this (see Chapter 31). Good curing is essential for repairs; air curing is seldom adequate.

28.6 FLOOR SCREEDS

A floor screed is a layer, typically 50 mm (2 in.) thick, that is placed on top of a cast slab, and possibly some insulation, to provide a flat floor to take finishes such as tiles or carpets. Traditionally, trowelled sand/cement screeds have been used. Gypsum/anhydrite (see Section 7.10) screeds have been found to be very effective because they can be foamed, and made flowing, so they are self-levelling, and they also provide the necessary insulation without the need for a separate insulating layer. Precautions should be taken to prevent sulphate attack from the gypsum on the concrete slab.

28.7 CONCLUSIONS

- Mortars and grouts contain no coarse aggregate, and have many applications in construction.
- The appearance of mortar is often critical in brickwork, and depends on accurate mix proportioning.
- Rendering is needed, if masonry is required to be weather-proof.
- Grouts require high shear mixing because they have no coarse aggregate.
- Structural grouting is often critical for load bearing.
- Grouts are effective for waste containment due to their high adsorption.
- Cementitious repair materials require careful application to avoid the formation of incipient anodes.
- Gypsum/anhydrite floor screeds can be foamed to make them self-levelling and insulating.

SPECIAL CONCRETES

CHAPTER OUTLINE

29.1 INTRODUCTION

This chapter discusses ways in which the various aspects of concrete technology that have been introduced in previous chapters may be used to meet the requirements of specific applications.

29.2 LOW COST CONCRETE

29.2.1 CEMENT REPLACEMENTS

Replacing cement with PFA will normally reduce the cost, and improve durability. The cost of GGBS may be higher, depending on location, but will almost always be less then cement. It is possible to make mixes with CSF replacement that achieve the same strength, at a higher water/cement (w/c) ratio and thus less cement. This is not recommended, as it will reduce durability. (See Chapter 18).

29.2.2 WATER REDUCERS

Water reducers (plasticisers) may be used to reduce the cement content, while maintaining the w/c ratio. This is recommended and may help the durability. (See Chapter 24)

29.2.3 LOW STRENGTH MIXES

Low strength mixes are required for applications, such as trench backfill, where re-excavation may be needed. A hydraulic excavator can dig out concrete with strengths up to 1–2 MPa (150–300 psi). Making these mixes with ordinary cement with a high w/c ratio will waste money, and may also segregate. Foamed concrete can be used but, despite having a very high air content, it is still quite expensive (see Section 24.7.1). Low cost controlled low strength materials (flowable fills) can be made for this application with high percentages of PFA, or with types of mineral waste, such as steel slags and waste gypsum that are unsuitable for structural concrete.

29.3 CONCRETE WITH REDUCED ENVIRONMENTAL IMPACT

29.3.1 REDUCING CARBON FOOTPRINT

Cement replacements such as PFA or GGBS save 850 kg of CO_2 for every tonne of Portland cement saved. Using magnesia-based cements in place of current calcium–silica-based cements has the potential for large savings. Recycled aggregate, from demolished concrete or industrial mineral waste, will save

CO_2, provided haulage distances are kept short. The carbon footprint from road haulage can be significant, so locally sourced materials should be used, if possible.

29.3.2 REDUCING FLOODING AND RUN-OFF

No-fines concrete is highly permeable, and can reduce or eliminate run-off from roads, and prevent flooding and pollution. It is also known as "pervious" concrete. See Fig. 29.1.

29.3.3 REDUCING HEAT IN CITIES

A bright white concrete surface with GGBS will reduce the "heat island" effect of cities in hot climates.

29.3.4 REDUCING ROAD NOISE

Concrete road surfaces can cause significant noise pollution, both to drivers, and nearby residents. Some surface texture is needed for skid resistance, but any grooves should be kept at random spacings to avoid resonances. Good results for noise reduction have been obtained with exposed aggregate surfaces (similar to Fig. 29.2).

29.4 LOW DENSITY CONCRETE
29.4.1 NO-FINES CONCRETE

With no-fines aggregate, this has a lower density than normal concrete, and can be produced at strengths high enough for structural use (see Fig. 29.1).

FIGURE 29.1 Pervious (No-Fines) Concrete Being Discharged From a Truck-Mixer

The coarse aggregate is all clearly visible in the flowing concrete.

FIGURE 29.2 Exposed Aggregate Concrete

This has also been divided into panels, as an architectural feature.

29.4.2 LIGHTWEIGHT AGGREGATE CONCRETE

The strength and density of this concrete will depend on the aggregate used. Sintered PFA and slag will give structural strengths, with modest reductions in density. As the density decreases, the thermal conductivity decreases, and the mixes become good insulators (see Section 19.5.1). With very lightweight aggregates, concrete can be made that will float in water.

29.4.3 FOAMED CONCRETE

This is a very low strength nonstructural mix (see Section 24.7.1).

29.5 HIGH-DENSITY CONCRETE

High-density concrete is used for radiation shielding. It is made with high density aggregate. A dense rock aggregate, such as granite, may be sufficient for thick radiation shielding walls (see Section 19.5.2).

29.6 UNDERWATER CONCRETE
29.6.1 ORDINARY CONCRETE

Any concrete will set under water (and cure very well), provided the cement does not wash out. In very still water, with careful placement using a tremie pipe, this may be sufficient (see Section 26.4.3). Hessian (burlap) sacks and other forms of containment can be used in rivers.

29.6.2 CONDENSED SILICA FUME

CSF concrete is very cohesive, and resistant to washout. This is a good application for this material because of the good curing conditions (see Section 18.3.4).

29.6.3 ANTI-WASHOUT ADMIXTURES

These are made specifically for underwater placement, and may be used in combination with CSF.

29.7 ULTRA-HIGH STRENGTH CONCRETE
29.7.1 HIGH-RANGE WATER REDUCERS

Powerful superplasticisers may be used to achieve w/c ratios below 0.3, and very high strengths. These mixes have very good durability, and permit fast construction of very tall buildings. However, the modulus of elasticity may not increase as much as the strength. They will have high heat of hydration, and require very good curing. A strong aggregate is required.

29.7.2 CONDENSED SILICA FUME

CSF can be used with superplasticisers to achieve very high strengths, but should be used with great care. Compressive strengths of 100 MPa (14.5 ksi) are quite easily achieved with CSF and superplasticiser. These mixes may use fibre reinforcement to give tensile strength.

29.7.3 VACUUM DEWATERING

This technique involves removing water by vacuum from a slab, after casting. It is labour intensive, but will give a good, strong, durable concrete, without the cost of superplasticisers.

29.7.4 MACRO DEFECT FREE MIXES

These are highly specialised mixes made with a cement–polymer matrix that may be cured under pressure, to ensure that they have no porosity. They have very high strengths in compression, and good strength in tension. It is possible to make "concrete springs" with them. However, they have been available for several decades, and have yet to find a significant application.

29.8 ULTRA-DURABLE CONCRETE
29.8.1 ULTRAHIGH-STRENGTH CONCRETE WITH GOOD CURING

Normally, high strength concrete with a w/c ratio below about 0.33 will have low transport properties and, thus, good durability. Specific deterioration mechanisms, such as sulphate attack or freeze–thaw, must also be addressed, as noted in Chapter 25.

29.8.2 CONTROLLED PERMEABILITY FORMWORK

This will locally reduce the w/c ratio near the concrete surface (see Section 26.2.4).

29.8.3 COATED, STAINLESS, OR NON-METALLIC REINFORCEMENT

These will inhibit reinforcement corrosion, but good quality concrete and well-managed placement is still essential (see Section 26.2.6).

29.9 CONCRETE WITH GOOD APPEARANCE (ARCHITECTURAL CONCRETE)

29.9.1 SURFACE FEATURES

- A plain concrete surface will always have small visible blemishes, such as marks at the edges of the shutter surfacing, and the wall ties. By far, the best way to make this finish attractive is to form features in it, to draw the eye from noticing the blemishes (this also applies to masonry and other materials).
- A ribbed finish can be formed with wooden laths on the shutter. Different spacings may be used to give the impression of stonework (see Figs 29.2 and 29.3).
- A board finish can be formed with rough sawn boards as shuttering. However, this textured surface may collect dirt and moss.
- A brush finish on a slab may be made with a stiff brush, before the concrete sets.
- Various types of special imprints may be used to give the impression of bricks and blocks on slabs.

FIGURE 29.3 Ribbed Concrete

The ribs are almost sufficient to divert the eye from the severe cracking (probably plastic settlement – see Section 23.3.6), and the inadequate vibration on the right hand side.

29.9.2 EXPOSED AGGREGATE

Exposing the coarse aggregate of concrete is a common method to improve appearance (Fig. 29.2). This may be achieved with a surface retarder on the shutter that then permits the fine material at the surface to be washed off, when the shutter is removed. Alternatively, the surface may be removed with bush hammers or needle guns, but this is labour intensive. On horizontal surfaces, the fine material may be washed off, or a decorative aggregate placed into it, before final set.

29.9.3 PIGMENTS

A wide variety of different colour pigments are available for concrete. These should generally be used with a white cement because the grey colour of normal cement will hide most other colours.

29.10 FAST SETTING CONCRETE

29.10.1 ACCELERATING ADMIXTURES

Chloride free admixtures should be used for reinforced concrete, but calcium chloride can be used if there is no risk of corrosion (see Section 24.6).

29.10.2 SUPERPLASTICISERS TO REDUCE THE WATER CONTENT

This method is often more effective than accelerators.

29.10.3 HIGH TEMPERATURE CURING

This is used in precasting works, but can cause delayed Ettringite formation, at very high temperatures. Autoclaved concrete (high temperature and pressure) is used for rapid production (see Section 20.2).

29.11 CONCRETE WITHOUT FORMWORK

Shotcrete (also known as sprayed concrete, or gunite) is a semidry mix that is sprayed onto a surface, and is consolidated onto it by the impact, and requires no formwork. The nozzle from which it is sprayed may be held by an operative or, on larger projects, such as tunnel linings, it is on a robot arm. The water may be added at the mixer (wet process), or at the nozzle (dry process). The mix normally only contains a small coarse aggregate, or may be a mortar and is often fibre reinforced.

29.12 SELF-COMPACTING CONCRETE

This material is expensive, due to the cost of the admixtures, high cementitious content, and often better aggregate grading. However, using it will reduce labour costs, reduce noise, reduce the risk of vibration-related health hazards (vibration white finger), and give a better surface finish.

29.13 ROLLER-COMPACTED CONCRETE

For the construction of road bases and dams, a mix with a low water content, and thus low workability, is used; that is compacted with vibrating rollers. Because the water content is low, the cement content can be low, while maintaining an adequate w/c ratio (see Fig. 17.4).

29.14 CONCLUSIONS

Concrete can be made to meet a range of different requirements.

It is possible to produce concrete that will be suitable for almost any environment.

TUTORIAL QUESTION

1. Describe the advantages and disadvantages of using high strength concrete in a tall office block.

   Solution:

   Advantages of high strength concrete:

   - High durability (e.g., abrasion and carbonation resistance).
   - High strength to weight ratio, particularly for lightweight HSC.
   - High early strength allows faster construction, and formwork reuse.
   - Increased elastic modulus and lower creep.
   - Very tall buildings become feasible.
   - Lower column costs for concrete and steel.
   - Smaller columns allow for more floor space.
   - Box girder and solid girder bridge spans may be increased, and designs simplified.

   Disadvantages of high strength concrete:

   - Increased cost per unit volume.
   - More stringent quality control of materials and construction needed.
   - CSF mixes may self-desiccate, giving failure at day joints.
   - Workability difficult to define, and often declines rapidly with time, after mixing.
   - Timing of concrete delivery, and addition of admixtures becomes critical.
   - High heat evolution may necessitate use of low heat binders (e.g., PFA), and cooling measures.
   - Stiffness (modulus) does not increase in proportion to strength.
   - More than 28 days may be required to reach specified strength.
   - Structural members may exhibit brittle failure.
   - Use of high strength concrete is not covered by design codes.

STEEL

30.1 INTRODUCTION

Steel is one of the most important materials used in construction (Fig. 30.1). The crystalline structure and the effect of grain boundaries in steel was discussed in Chapter 2. In Chapter 6, it was noted that metals have free electrons in them that can conduct electricity (and heat), and in Chapter 7 the structure of atoms with fixed nuclei and electrons in a cloud around them was discussed. In this chapter, the microstructure and manufacture of steel is outlined, in order to explain the different types of steel, and the effect of heating (that occurs in welding). Steel corrosion is discussed in Chapter 31.

30.2 IRON–CARBON COMPOUNDS
30.2.1 CARBON CONTENTS

Iron is an element with the chemical symbol Fe. Steel and cast iron are described as "ferrous" metals, and are made from iron with different carbon contents (Table 30.1). Increasing the carbon content of steel improves yield strength, and ultimate tensile strength, but reduces ductility and toughness. Thus, a construction steel will not fail in impact, but a tool steel will be hard enough to maintain a sharp edge. Cast iron has high carbon content, and is therefore normally hard and brittle, and it is used for heavy items – such as big machine bases, and tunnel lining segments (see Fig. 30.2). This rule is, however, by no means universal. Variations in manufacturing process and minor alloy compounds can produce malleable cast irons that are soft and ductile.

The low carbon content of about 0.04% is used for sheet or strip steels, intended to be shaped by extensive cold deformation. Structural steels rarely contain more than about 0.25% carbon; increasing the carbon content above this brings problems with welding.

FIGURE 30.1 A Substantial Steel Frame for a Library That has High Loadings From the Books

Table 30.1 Iron–Carbon Compounds

Carbon Content (%)	Material
0.04–0.3	Low carbon mild and high yield steels. Used in construction.
0.3–0.7	Medium carbon. Used in nuts and bolts and machine parts.
0.7–1.7	High carbon and tool steels.
1.8–4	Cast irons (Figure 30.2). Used in low strength applications.

FIGURE 30.2 Students Admiring the World's First Iron Bridge, Which was Built in 1779, Spanning the River Severn in England

Iron was displaced by steel for most structural applications, about 100 years later.

30.2.2 CARBON IN THE MICROSTRUCTURE

The structure of the crystals of steel, when it is slowly cooled to room temperature, is shown in Fig. 30.3. This material is soft and ductile, despite being crystalline. It is called body centred cubic because the carbon atom is in the centre of the cube. Only a small amount of carbon may be held in it.

There is another type of crystal that has a different structure (see Fig. 30.4). It is called face centred cubic and, as can be seen, it can contain more carbon. It is formed at elevated temperatures, and may be present in steel that has been cooled rapidly.

The bulk of the carbon in steel may be held as iron carbide (Fe_3C) that is a hard, brittle material.

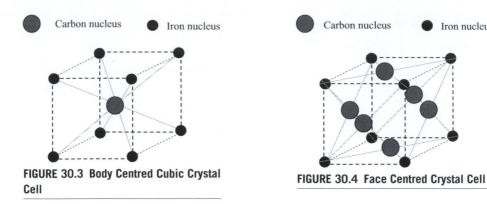

FIGURE 30.3 Body Centred Cubic Crystal Cell

FIGURE 30.4 Face Centred Crystal Cell

The nomenclature is:

Ferrite or αFe	This is the body centred cubic iron that is formed on slow cooling, and may contain less than 0.01% carbon at room temperature.
Austenite or γFe	This is the face centred cubic iron that is formed at high temperatures, and may contain up to 1.7% carbon.
Cementite	This is iron carbide (Fe_3C) that contains about 6.7% carbon.
Pearlite	This is the laminar mixture of ferrite and cementite, and has an average carbon content of about 0.78%.

Figure 30.5 shows the compounds in slowly cooled, low carbon steels. It may be seen that complex changes take place as the material is heated, and these can be exploited to achieve different properties. If the steel is cooled rapidly, or has minor alloying elements in it, the transitions will not take place as it is cooled. Thus, for example, some stainless steels contain austenite at room temperature.

30.3 CONTROL OF GRAIN SIZE

30.3.1 EFFECT OF GRAIN SIZE

The effect of grain size is discussed in Section 3.2.2. It may be seen that decreasing the grain size will increase the strength, by hindering the movement of dislocations. The grain size can be changed by altering the history of rolling and heating treatments.

30.3.2 CONTROL BY HEATING

Heat treatment (whether deliberate or accidental) affects the properties of steels substantially. In general, heating followed by rapid cooling (quenching) will harden them, but heating followed by slow cooling (annealing) may soften them.

Steels that are cooled slowly in furnaces are known as fully annealed steels. If steel is cooled in air, it will "undercool," that is, the transitions in Fig. 30.5 will not be completed. This is called normalising,

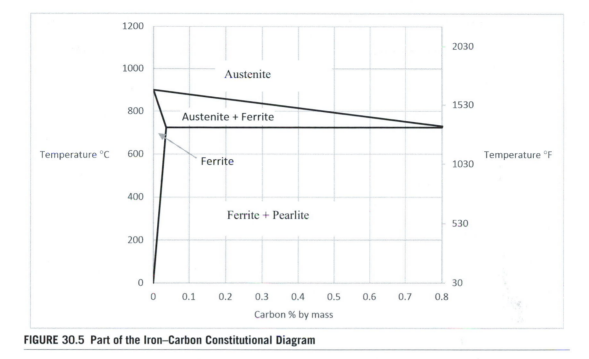

FIGURE 30.5 Part of the Iron–Carbon Constitutional Diagram

and the temperature from which cooling starts is called the normalising temperature. The process will increase the proportion of ferrite, and reduce the grain size of the ferrite, and make finer lamellae in the pearlite. These changes give higher yield strength, and better ductility, and toughness.

30.3.3 RAPIDLY COOLED STEELS

Rapid cooling (e.g., quenching in cold water) gives fine grain structure and products not found with slower cooling. This will give a hard, brittle steel. If a tool, such as a chisel or a screwdriver, is damaged and the end is reformed, it should be heated to red heat, and then quenched to harden it. While structural steels are not normally quenched, very rapid cooling, resulting in brittleness, can occur after welding. Steel compositions with poor hardening characteristics must therefore be used for welding.

30.3.4 CONTROL BY WORKING

By cold working steel, that is, compressing or twisting it beyond its yield point, it is possible to achieve considerable increases in strength, by reducing the grain size. This is the basis for the production of some high tensile steels.

Structural steels are produced by rolling ingots, and this rolling process affects the grain size. If the rolling is carried out at temperatures well within the austenite range, the grains reform as fast as they are broken up. They are generally rolled at temperatures just above the ferrite + austenite range. This gives an ordinary normalised microstructure. If rolling is carried out in the ferrite + austenite range, the ferrite and austenite are each rolled out along the rolling direction and, on cooling, long bands of pearlite will form. This does not cause severe harm.

Low carbon steels may be rolled at lower temperatures, but the presence of pearlite makes it difficult. At temperatures above 650°C (1200°F), the ferrite grains will reform but the carbide laths in the pearlite are broken up.

If rolling is carried out at room temperature, none of the grains can reform, and the strength will increase. The steel becomes less ductile, and would eventually split. Other methods of deformation, for example, twisting reinforcing bar, have the same effect.

Reheating to between 650°C (1200°F) and 723°C (1330°F) will reform the ferrite grains, but not the carbide. Greater deformation before recrystallisation and lower treatment temperatures will give finer grain sizes. This is called subcritical annealing because austenite is not formed. The resulting microstructure has good ductility; sheet steel for pressing for car body panels is used in this state.

30.3.5 CONTROL BY ALLOYING

Small additions of aluminium, vanadium, niobium, or other elements combined with manganese help to control grain size. These high strength, low alloy steels have high strength and toughness, combined with ease of welding. These steels are used for special applications, such as pressure vessels, where the cost is justified.

30.4 MANUFACTURING AND FORMING PROCESSES

30.4.1 THE PROCESSES

- Rolling: almost all construction steel is rolled. In this process, a length of steel is passed between a number of pairs of heavy rollers that progressively form it into the required shape.
- Forging, in which a heated ingot is mechanically worked into shape (ingots can weigh over 100 tonnes).
- Extrusion, used for special shapes, for example, seamless tube.
- Cold rolling: thin plate, down to 1 mm, is produced by cold rolling that will reduce the grain size, and also gives a smoother surface.

30.4.2 ROLLED STEEL SECTIONS

Figure 30.6 shows some common rolled sections. Rolled steel column sections have approximately equal height and width, to give efficient performance to prevent buckling in columns. If these are used for piling, they have webs and flanges of equal thickness, and are designated HP shape in the United States. Universal beam sections (designated W shape in the United States) are deeper, to give efficient performance in bending, but rolled steel column sections can be used for beams, if height is restricted. Rolled hollow section (structural tubing) sections can be round (as shown), or rectangular. Round sections have become popular for exposed steelwork, as the software to design and fabricate the complex shapes needed at connections has become widely available. Rolled steel angle (designated L shape in the United States) is typically used for lighter components, such as cladding rails. Very large numbers of other sections, such as channels, tees, and sheet piling are produced.

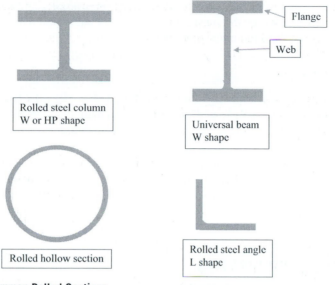

FIGURE 30.6 Some Common Rolled Sections

30.4.3 PLATE GIRDERS

These are larger beams, made by welding steel plate into beam sections. Box girders are similarly made, by welding plate into large hollow sections. They are normally welded together at a fabrication yard, and then transported to site.

30.5 STEEL GRADES

30.5.1 EUROPEAN GRADES

The first character of the code in European grade designations is the application letter, for example, S for structural, and B for reinforcing bar.

For some application designations, another letter is included before the property value. This number is used to indicate any special requirements or conditions.

The next three digits give the minimum yield strength, in MegaPascal.

Additional symbols may then follow for impact testing strengths, testing temperatures, delivery condition, such as annealed, or normalised, or other specific requirements.

Thus, for example, a grade S355K2W is a structural steel with a yield stress of 355 MPa. The K2 is a grade of toughness, and the W indicates it is a weathering steel (see Section 31.6.2).

30.5.2 US GRADES

The ASTM specification starts with the letter A to indicate it is a ferrous metal followed by a number that has been serially assigned to refer to a detailed specification in the standards covering composition and properties. The number itself is arbitrary, and does not relate to any property. For some types, this is followed by a grade, which is the yield stress in ksi. Thus, for example, A529Gr.50 is a carbon steel with a yield stress of 50 ksi. Within the standard for A529 steel additional requirements are specified, such as the ultimate strength and minimum elongation.

30.5.3 STEEL SUPPLY

Structural steel is ordered by size and weight, for example, a 457 × 152 × 82 UB is 457 mm high, 152 mm wide universal beam, and has a mass of 82 kg/m run. Steel manufacturers roll substantial quantities at a time, before changing to a different size, so, unless a large amount is required, steel is normally bought from a stockholder.

Reinforcing steel is either mild, or high yield. Mild steel bars are smooth and round, and designated by the letter R (e.g., R10 or R16 for 10 or 16 mm diameter). High yield bars are normally rolled with "ribs" on the surface to improve their bond to the concrete, and are designated by the letter Y (e.g., Y16).

30.6 MECHANICAL PROPERTIES
30.6.1 STRESS–STRAIN RELATIONSHIPS

Figure 30.7 shows the stress–strain relationship for mild steel from Fig. 3.1, with the curve for a length of prestressing wire added. This is produced by cold drawing hot-rolled rods that are subsequently stress-relieved, by heating to about 350°C (660°F) for a short time. The cold drawing (stretching) significantly reduces the grain size, and increases the strength. The two bars looked similar, and had similar diameters, but it may be seen that their strengths were very different. The prestressing wire is not ductile, and would snap if significantly bent, and cannot be welded because the heating would greatly reduce the strength.

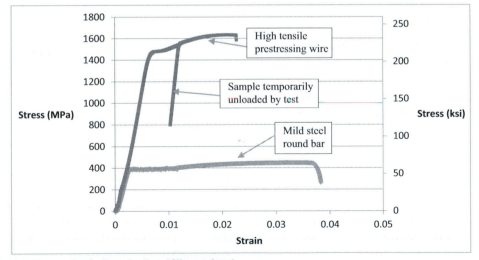

FIGURE 30.7 Stress–Strain Data for Two Different Steels

The trace for the high tensile wire includes a section of temporary unloading that was programmed into the test machine to occur after yield. It may be seen that this is almost parallel to the original loading (it is not quite parallel due to slippage in the jaws, during the original loading. This trace could not be obtained with a displacement transducer, as shown in Fig. 3.2 because it would have been damaged when the bar failed).

30.6.2 THE 0.2% PROOF STRESS

If the material exhibits plastic deformation (yields), and does not return to its original shape when unloaded, this is clearly unacceptable for most construction applications. Thus, for a brittle material (e.g., concrete), strength is defined from the stress at fracture, but for a ductile material the yield stress must be used. For some steels, the exact yield strength on curved, like Fig. 30.7, is not very clear, and the strength is defined from the 0.2% proof stress (see Figs 30.8 and 30.9). This exploits the performance shown in the unloading in Fig. 30.7, and is the stress required to give 0.2% permanent strain. This is a good test for commercial applications. The sample is loaded to the specified 0.2% proof stress, and then unloaded. The permanent extension is measured, and, if this is a strain of less than 0.2%, it has met the specification.

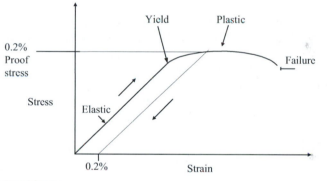

FIGURE 30.8 The 0.2% Proof Stress

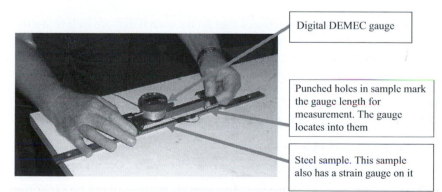

FIGURE 30.9 Measuring Extension of a Steel Sample With a "Digital DEMEC Gauge"

Table 30.2 Ductile–Brittle Transition Temperatures of a Low-Carbon Steel

Grain Size (μm)	Strength MPa (ksi)	Brittle Transition Temperature °C (°F)
25	255 (37)	0 (32)
10	400 (58)	−40 (−40)
5	560 (72)	−60 (−76)

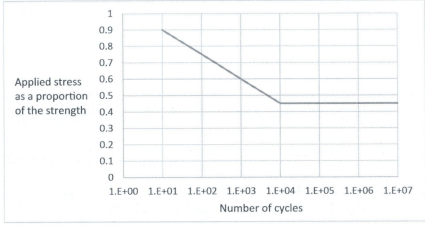

FIGURE 30.10 Schematic Graph of Fatigue Failure Loads

30.6.3 PERFORMANCE AT LOW TEMPERATURES

At low temperatures, steel may lose its ductility and become brittle, leading to structural failures. Table 30.2 shows typical temperatures for low carbon steel. It may be seen that it is important to have a small grain size.

30.6.4 FATIGUE

Steel will fail, if it is subject to a large number of load cycles, even if they are well below the strength for a single cycle (see Section 2.8). Figure 30.10 shows a typical relationship between the failure stress due to fatigue, and the number of cycles. The stress that the steel can take will progressively decrease, until it reaches a lower limit, that may be used in the design for materials subject to large numbers of cycles.

30.7 STEEL FOR DIFFERENT APPLICATIONS

30.7.1 STRUCTURAL STEELS

The properties required of structural steels are:

*   Strength. This is traditionally specified as a characteristic value for the 0.2% proof stress.
*   Ductility to give impact resistance. Ductility increases with reducing carbon content.
*   Weldability (see 30.8).

30.7.2 **REINFORCING STEELS**

Reinforcing steels are tested for strength, and must also comply with the requirements of a "rebend" test, to ensure that they retain their strength when bent to shape. This limits the carbon content.

30.7.3 **PRESTRESSING STEELS**

Prestressing steels (high tensile steels) are not bent, so they can have higher carbon contents than normal reinforcement, and have higher strengths. This limits the ductility, but is necessary to avoid loss of prestress due to creep.

30.8 **JOINTS IN STEEL**
30.8.1 **WELDING**

Welding is a complex subject, and specialist literature should be downloaded before attempting critical work. The principle of welding is that a small amount of new metal is melted onto the surfaces to be jointed, and the surface of these is heated sufficiently to melt into the joint. The main methods are:

- Gas welding. In order to produce a hot enough flame, a combustible gas (e.g., acetylene) is burnt with oxygen. This method is not used for major welding jobs, but has the advantage that the torch will also cut the metal.
- Arc welding. In this method a high electric current is passed from the electrode (the new metal for the weld) to the parent metal. The electrode is coated with a "flux" that helps the weld formation, and prevents contact with air that would cause oxide and nitride formation.
- Inert gas shielded arc welding. This method uses a supply of inert gas (often argon) to keep the air off the weld, so no flux is needed.

The carbon content of a steel must be restricted, if it is to be welded, otherwise it will be damaged by the process. In practice, the limitation is applied to the "carbon equivalent value" that is calculated from the carbon, manganese, chromium, molybdenum, vanadium, nickel, and copper contents.
When welding

- Do not look directly at a welding process (especially, electric arc). It may damage your eyes.
- Always allow for the effect of heating and uncontrolled cooling of the parent metal. For example, if a high yield reinforcing bar is welded, the effect of the cold working will be lost – and with it much of the strength. This heating will also often cause distortion.
- Check the welding rods. If they have become damp, the flux will be damaged. Use the correct rods for the steel (e.g., stainless).
- Remember that the welding process cuts into the parent metal and, if done incorrectly, may cause substantial loss of section.

30.8.2 **BOLTED JOINTS**

Bolted joints are common in steelwork (Fig. 30.11), and care should be taken to ensure that the correct bolts are used. Standard "black" bolts are used for normal connections, and generally carry loads in shear, but are generally specified by their tensile strength. Care should be taken that the specified bolts (and nuts) are not mixed with ungraded bolts that may be on site for other purposes. High strength friction grip bolts (and nuts) are made from higher-grade steel, and are torqued to transfer loads by friction between the steel members.

FIGURE 30.11 A Typical Bolted Joint in a Steel Frame

30.8.3 **RIVETS**

Rivets are no longer used for new steel fabrication, but may be seen in large numbers in older structures (Fig. 30.12). They were placed by spreading the end of the hot rivet by hammering, resulting in dome shaped ends. From the number that may still be seen, it is evident that their durability is good.

FIGURE 30.12 Bolts and Rivets on the Brooklyn Bridge in New York

30.9 CONCLUSIONS

- Different types of iron and steel are made by varying the carbon content.
- The strength of steel is increased by decreasing the grain size with rapid cooling, or cold working.
- Most structural steelwork is made with standard sizes of rolled steel sections. Larger sections are welded together, using steel plate.
- The 0.2% proof stress may be used to grade steel because the yield stress is often not clearly defined.
- The selection of steel for different applications requires both strength, and ductility.
- A welded joint is formed by melting the weld metal using an electric arc, or gas flame.

TUTORIAL QUESTIONS

1. The following figure shows observations made when a 100 mm length of 10 mm diameter steel bar was loaded in tension. The equation given is for a trendline that has been fitted to the straight portion of the graph.

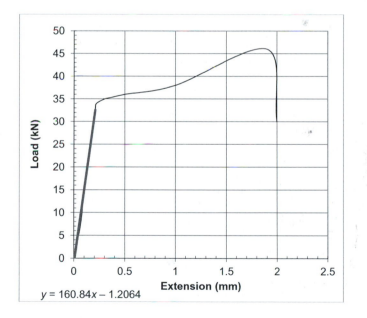

$y = 160.84x - 1.2064$

Calculate the following:

a. The cross-sectional area of the bar, in m^2.

b. The Young's modulus, in GPa.

c. The 0.2% proof stress, in MPa.

d. The estimated yield stress, in MPa.

e. The ultimate stress, in MPa.

Solution:

a. Cross-section area = $\pi \times 0.01 \times 0.01/4 = 7.85 \times 10^{-5}$ m^2

b. Gradient = 161 kN/mm (from equation on graph)

$E = 161 \times 10^6 \times 0.1/7.85 \times 10^{-5} = 205$ GPa　　　　　　　　　　　(2.8)

The student may alternatively obtain this from measuring the gradient on the graph.

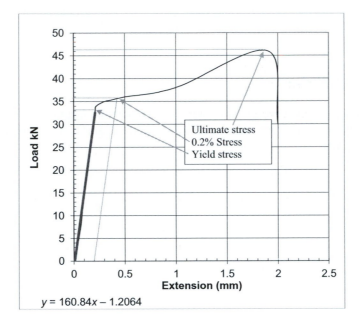

$y = 160.84x - 1.2064$

c. 0.2% of 100 mm is 0.2 mm.

Thus, the line on the graph is at 0.2 mm greater extension than the data.

Load = 36 kN (from graph)

Thus, stress = $36 \times 10^3/7.85 \times 10^{-5} = 458$ MPa　　　　　　　　　　(2.3)

d. Load = 33 kN (from graph)

Thus, stress = $33 \times 10^3/7.85 \times 10^{-5} = 420$ MPa　　　　　　　　　　(2.3)

e. Load = 46 kN at failure

Thus, stress = $46 \times 10^3/7.85 \times 10^{-5} = 586$ MPa　　　　　　　　　　(2.3)

2. A flat steel bar measuring 50 mm by 20 mm by 3 m long is placed in tension, supporting a load of 5 tonnes.

a. If the Young's modulus of the bar is 220 GPa, what is the length of the bar when supporting the load?

b. If the Poisson's ratio of the steel is 0.35, what are the dimensions of the steel section when supporting the load?

c. If the 0.2% proof stress of the steel is 200 MPa, what load would be required to give an irreversible extension of 0.2%? (Give your answer in tonnes.)

d. If the load calculated in (c) is applied, and then reduced by half, what is the final length (with the load still applied)?

Solution:

a. Area $= 50 \times 10^{-3} \times 20 \times 10^{-3} = 10^{-3} \, \text{m}^2$

Load $= 5 \text{ tonnes} = 5 \times 10^4 \, \text{N}$ (assume $g = 10$)   (2.2)

Stress $= 5 \times 10^4 / 10^{-3} = 5 \times 10^7 \, \text{Pa}$ (2.3)

Strain $= 5 \times 10^7 / 2.2 \times 10^{11} = 2.27 \times 10^{-4}$ (2.7)

Extension $= 2.27 \times 10^{-4} \times 3 = 6.81 \times 10^{-4} \, \text{m}$ (2.5)

Length $= 3 + 6.81 \times 10^{-4} = 3.000681 \, \text{m}$

b. Horizontal strain $= 2.27 \times 10^{-4} \times 0.35 = 7.94 \times 10^{-5}$ (2.9)

Extension along $= 50 - (50 \times 7.94 \times 10^{-5}) = 49.996 \, \text{mm}$ (2.5)

Extension across $= 20 - (20 \times 7.94 \times 10^{-5}) = 19.998 \, \text{mm}$ (2.5)

c. $200 \text{ MPa} \times 10^{-3} \, \text{m}^2 = 2 \times 10^5 \, \text{N} = 20 \text{ tonnes}$ (2.3)

d. See the following figure.

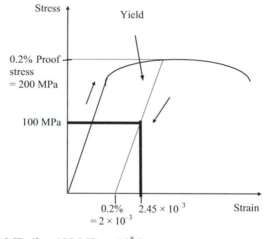

Half the stress $= 200 \text{ MPa}/2 = 100 \text{ MPa} = 10^8 \, \text{Pa}$

Elastic strain $= 10^8 / 2.2 \times 10^{11} = 4.5 \times 10^{-4}$ (2.7)

Total strain $= 0.2\% + 4.5 \times 10^{-4} = 2 \times 10^{-3} + 4.5 \times 10^{-4} = 2.45 \times 10^{-3}$

Total extension $= 3 \times 2.45 \times 10^{-3} = 7.35 \times 10^{-3} \, \text{m}$ (2.5)

New length $= 3 + 7.35 \times 10^{-3} = 3.00735 \, \text{m}$

3. The following figure shows observations that were made when a 4 in. length of 3/8 in. diameter steel bar was loaded in tension. The equation given is for a trendline that has been fitted to the straight portion of the graph.

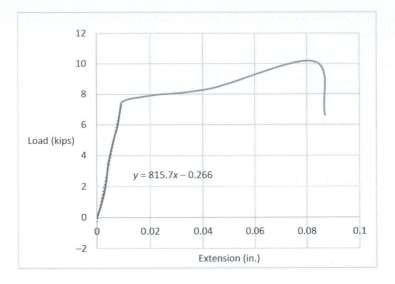

Calculate the following:

a. The cross-sectional area of the bar.
b. The Young's modulus.
c. The 0.2% proof stress.
d. The estimated yield stress.
e. The ultimate stress.

Solution:

a. Cross-section area = $\pi \times 0.375 \times 0.375/4 = 0.110$ in.^{2}
b. Gradient = 815.7 kips/in. (from equation on graph)
    $E = 815.7 \times 10^3 \times 4/0.11 = 29,661$ ksi   (2.8)
    The student may alternatively obtain this from measuring the gradient on the graph.
c. 0.2% of 4 in. is 0.008 in.
    Thus, the line in the above figure will be at 0.008 in. greater extension than the data.
    Load = 7.9 kips (see example in 1(b))
    Thus, stress = $7.9 \times 10^3/0.11 = 71.8$ ksi   (2.3)
d. Load = 7.6 kips (from graph)
    Thus, stress = $7.6 \times 10^3/0.11 = 69.1$ ksi   (2.3)
e. Load = 10.2 kips at failure
    Thus stress = $10.2 \times 10^3/0.11 = 92.7$ ksi   (2.3)

4. A flat steel bar measuring 2 in. by ¾ in by 10 ft. long is placed in tension, supporting a load of 11200 lb.
    a. If the Young's modulus of the bar is 30,000 ksi, what is the length of the bar when supporting the load?
    b. If the 0.2% proof stress of the steel is 30 ksi, what load would be required to give an irreversible extension of 0.2%?

c. If the load calculated in (b) is applied, and then reduced by half, what is the final length (with the load still applied)?

Solution:

a. Area $= 2 \times 0.75 = 1.5$ in.^{2}
Stress $= 11,200/1.5 = 7,467$ psi .... (2.3)
Strain $= 7,460/3 \times 10^7 = 2.49 \times 10^{-4}$ .... (2.7)
Extension $= 2.49 \times 10^{-4} \times 10 = 2.49 \times 10^{-3}$ ft. .... (2.5)
Length $= 10 + 2.49 \times 10^{-3} = 10.00249$ ft.
b. $30 \times 10^3 \times 1.5 = 45,000$ lb
c. See 2(d) as an example of this analysis.
Half the stress $= 30$ ksi/2 $= 15$ ksi
Elastic strain $= 15 \times 10^3/3 \times 10^7 = 5 \times 10^{-4}$ .... (2.7)
Total strain $= 0.2\% + 5 \times 10^{-4} = 2 \times 10^{-3} + 5 \times 10^{-4} = 2.5 \times 10^{-3}$
Total extension $= 10 \times 2.5 \times 10^{-3} = 0.025$ ft. .... (2.5)
New length $= 10 + 0.025 = 10.025$ ft.

CORROSION

31

CHAPTER OUTLINE

31.1 INTRODUCTION

Corrosion of metals is a primary concern for engineers. The cost of reinforcement corrosion to the world economy was discussed in Section 17.10. In order to reduce this damage, it is essential to understand the electrical nature of the process, and students should be familiar with the description of the electrical properties in Chapter 6, before studying this chapter. The theory of corrosion of steel in concrete is discussed in this chapter, but the practical consequences are explained further in Chapter 25. The detailed analysis in this chapter is used to explain the test method called linear polarisation, the only nondestructive method that can give a true indication of corrosion rates. Rest potential and resistivity measurements are described in Section 27.4.5, and can indicate the probability of corrosion occurring.

31.2 ELECTROLYTIC CORROSION
31.2.1 CORROSION OF A SINGLE ELECTRODE

When a metal is placed in water, there is a tendency for it to dissolve (ionise) in the solution. Equation (31.1) shows this for iron or steel.

$$Fe \rightarrow Fe^{++} + 2e^-$$
(31.1)

where e^- is the electron which remains in the metal (Fig. 31.1).

Positive metal ions are released into the solution, and the process continues until sufficient negative charge has built up on the metal to stop the net flow (because opposite charges attract – see Section 6.2). In this condition, metal ions are being lost from, and gained by, the metal at an equal rate. Table 31.1 gives the potential that is reached by a number of metals.

Each of the ions lost from the metal carries a fixed charge, so the rate of loss of metal may be expressed as the anode current I_a, where a denotes anode because this is an anodic process. I_{a-} is the current caused by the flow of metal ions back into the solid (known as the exchange current). The voltage at which the net flow stops, that is, $I_a = I_{a-}$ is the anodic rest potential (electrode potential), and is denoted V_{ao}.

The current will depend exponentially on the difference between the potential and the rest potential:

$$I_a = I_{ao} e^{[(V - V_{a0})/B_1']}$$
(31.2)

where V is the voltage across the anode and B_1' is a constant for all samples.

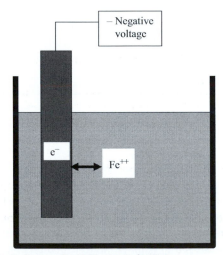

FIGURE 31.1 Corrosion from a Single Iron or Steel Electrode in Water

Table 31.1 Electrode Potentials Relative to a Standard Hydrogen Electrode

Metal	Electrode Potential (V)
Magnesium	−2.4
Aluminium	−1.7
Zinc	−0.76
Chromium	−0.65
Iron (ferrous)	−0.44
Nickel	−0.23
Tin	−0.14
Lead	−0.12
Hydrogen (reference)	0.00
Copper (cupric)	+0.34
Silver	+0.80
Gold	+1.4

Similarly for the exchange current:

$$I_{a^-} = I_{a0} e^{[(V_{a0}-V)/B_1']}$$ (31.3)

These two equations are plotted in Fig. 31.2, using typical values for the constants.

It may be seen that at voltages well above V_{a0} the exchange current is negligible, and the voltage may be expressed by rearranging equation (31.2):

$$V = V_{a0} + B_1 \mathrm{Log}\left(\frac{I_a}{I_{a0}}\right)$$ (31.4)

where $B_1 = B_1' \mathrm{Ln}(10)$.

This has been expressed as a log to base 10 to follow convention. See Section 1.5 for the notation for logarithms.

The process will stop, unless there is some mechanism available to overcome this potential barrier and clear the "log jam" of electrons. Possible mechanisms include applied voltages, connection to dissimilar metals, or exposure to oxygen or acids and are discussed in Sections 31.2.3–31.2.6.

31.2.2 SLOW CORROSION IN PURE WATER

If steel is immersed in pure (oxygen free) water, very slow corrosion takes place (Fig. 31.3).

The small amount that does take place is caused but the pH of water being 7, not infinite, this means there are 10^{-7} grams of hydrogen ions per litre in neutral water (see Section 7.6). They are the product of the equilibrium of the reaction:

$$H_2O \leftrightarrow H^+ + OH^-$$ (31.5)

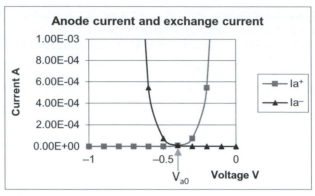

FIGURE 31.2 Anode Currents

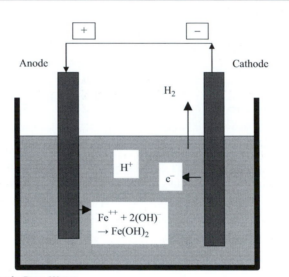

FIGURE 31.3 Slow Corrosion in Pure Water

in which the OH⁻ is a hydroxyl ion that may combine with the iron ions in solution:

$$Fe^{++} + 2(OH)^- \rightarrow Fe(OH)_2 \tag{31.6}$$

The product is ferrous hydroxide that is a green precipitate.

The reaction of the hydrogen ions with the electrons in the metal:

$$2H^+ + 2e^- \rightarrow H^2 \uparrow \tag{31.7}$$

is known as the cathodic reaction, and the dissolution of the metal ions:

$$Fe \rightarrow Fe^{++} + 2e^- \tag{31.8}$$

is the anodic reaction.

These two reactions cannot take place in the same location because for the cathodic reaction to proceed, the metal must have a negative potential relative to the solution (or the H^+ would never approach it) and similarly, the anode must be at a positive potential, relative to the solution, or the Fe^{++} ions would not escape. Thus, the cathode is at a negative potential relative to the anode. Minor variations in surface condition or environment can achieve this.

This reaction is very slow, and is normally insignificant, compared to the other mechanisms listed in Sections 31.2.3–31.2.6. However, in some deep nuclear disposal facilities, it is the only long-term mechanism, and the evolved hydrogen is a major concern.

31.2.3 APPLIED VOLTAGES

If a positive voltage is applied to the metal, it will make it "anodic." This will remove the surplus electrons, and permit the corrosion current to flow, releasing metal ions into solution (Fig. 31.4). This may happen in buildings due to stray currents. These currents must be direct current, not alternating current, and could come, for example, from welding, or charging circuits for batteries.

In order to stop corrosion, a negative voltage may be applied. This is the process of cathodic protection that is widely used to prevent corrosion (see Section 31.6.4).

Small positive applied voltages will cause significant corrosion. Figures 31.5–31.7 show a laboratory corrosion experiment, and it may be seen that the loss of section in the bars is very rapid.

31.2.4 CONNECTING TO DIFFERENT METALS

If two different metals are in a solution with an electrical connection between them, a current flows as shown in Fig. 31.8. The metal with the lower electrode potential in Table 31.1 will corrode, and the other metal will be protected from corrosion. This is how a battery works. The examples of copper and zinc are given in Fig. 31.8 because they have significantly different electrode potentials, and the reaction will therefore be fast enough to produce visible hydrogen bubbles.

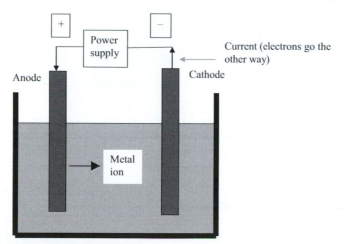

FIGURE 31.4 Corrosion Caused by Applied Voltage

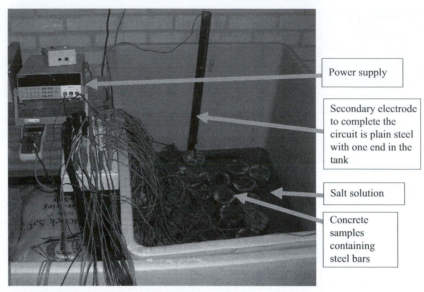

FIGURE 31.5 Corrosion Samples with Applied Voltage

The applied voltage is measured with a reference electrode that is dipped into the salt solution (see Section 27.4.5).

FIGURE 31.6 Corrosion Samples After Just 2 Months at +100 mV

If the copper is replaced with steel, the current will flow in the direction shown, but will be less. In this way, the zinc becomes a "sacrificial anode," and dissolves while protecting the steel (see Section 31.6.5).

Unintentional circuits can arise when different metals come into contact to give bimetallic corrosion. The metal with the lowest potential in Table 31.1 will always corrode. If copper nails are used

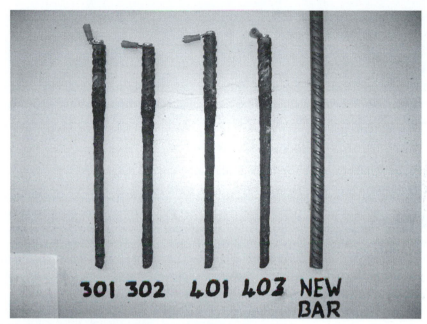

FIGURE 31.7 Steel Bars Extracted from Corrosion Samples in Figure 31.6

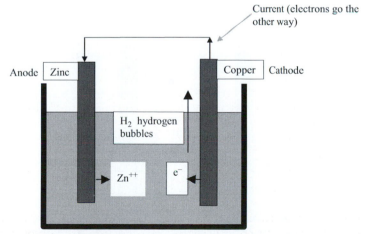

FIGURE 31.8 Corrosion of Dissimilar Metals

to fix down lead roofing, the copper will be protected but the lead will corrode around it. If aluminium components are connected to a steel structure, they will corrode.

31.2.5 ACIDS

It was noted in Section 7.5 that acids contain free positive hydrogen ions (measured as a low pH, typically in the range 1–6). Provided the metal has a potential below that of hydrogen, the hydrogen ions will combine with the electrons in the metal to release hydrogen gas (Fig. 31.9).

$$2H^+ + 2e^- \rightarrow H^2 \uparrow$$

The metal ions will then combine with the acid in solution, and the process will continue until either the metal or the acid is exhausted.

31.2.6 OXYGEN

Oxidation in air is a very slow process for most metals, and is not of great significance at normal temperatures. Oxygen from the air reacts directly with the metal, without substantially disrupting its structure. In general, the oxidised layer is impervious, so the process stops. Apart from a loss of "shine," no damage is done.

If oxygen is present in the water, it will react at the cathode (Fig. 31.10):

$$2H_2O + O_2 + 4e^- \rightarrow 4(OH)^- \tag{31.9}$$

This uses up electrons at the cathode (increasing its potential), and provides hydroxyl ions to react with the iron ions in solution, and thus greatly accelerates the corrosion. If there is a good supply of

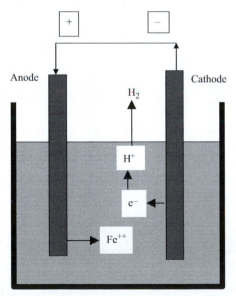

FIGURE 31.9 Acid Attack on Metal

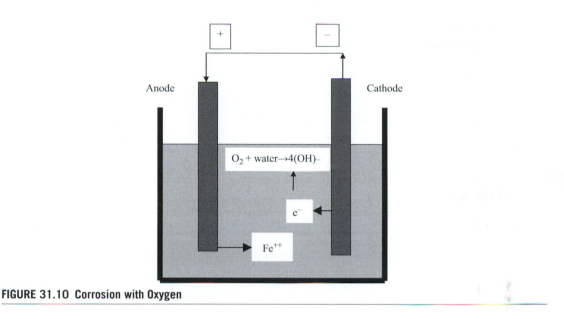

FIGURE 31.10 Corrosion with Oxygen

oxygen, the final product is ferric hydroxide $Fe(OH)_3$; this is common "red rust." If the air supply is limited, however, the product is Fe_3O_4 that is "black rust" (see Section 25.3.1 and Fig. 31.6).

31.3 THE EFFECT OF pH AND POTENTIAL

As discussed in Sections 31.2.3 and 31.2.5, the corrosion rate depends on both the pH and the potential of the steel relative to the solution. The combined effect is shown in Fig. 31.11 which is known as a Pourbaix diagram.

The pH of normal water is around 7 (neutral pH), so it may be seen that at potentials above −0.44 corrosion is likely to occur. The pH of concrete is around 12.5, so it is normally protected by a passive layer of ferric oxide. Chlorides (see Section 25.3.3) will react with this layer, and break it down, and carbonation (Section 25.3.2) will reduce the pH, so both these processes can move steel into the Fe^{++} corrosion region.

For laboratory work (Figures 31.5–31.7), accelerated corrosion may be caused by raising the potential of a steel sample in a salt solution. However, it may be seen from Fig. 31.11 that only low voltages, typically 0.1 V, should be used, or different corrosion products such as the Fe^{+++} will be formed, and the results will not be relevant to real structures. These experiments should not be confused with the accelerated transport tests described in Section 22.8.3, where high voltages are correctly used for chloride electromigration.

31.4 MEASURING CORROSION RATES WITH LINEAR POLARISATION
31.4.1 LABORATORY MEASUREMENT APPARATUS

This technique is known as linear polarisation resistance measurement, and is carried out with a potentiostat which is basically, a power supply that can apply a voltage relative to a potential on a reference

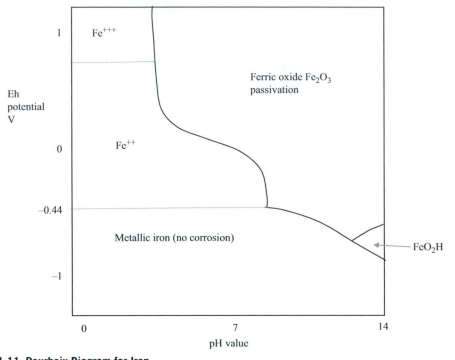

FIGURE 31.11 Pourbaix Diagram for Iron

electrode, without putting a current through it (as it would damage it). Thus, the potentiostat applies voltages to the secondary electrode, while controlling them with measurements from the reference electrode. Fig. 31.12 shows the arrangement for a test of a steel sample, cast into concrete.

The test is carried out by measuring the rest potential, and applying a voltage slightly above or below it, and measuring the current.

31.4.2 CALCULATION OF CURRENT

A schematic equivalent circuit for the steel in solution is given in Fig. 31.13. This shows the processes taking place in Fig. 31.9 or Fig. 31.10, if they were connected to a potentiostat to provide the external current I_x.

The voltage V will be the same across the anode and the cathode because they are connected as shown. The cathode voltage may be expressed similarly to the anode voltage in equation (31.4), as:

$$V = V_{c0} - B_2 \mathrm{Log}\left(\frac{I_c}{I_{c0}}\right) \tag{31.10}$$

If there is no external applied voltage, the voltage is known as the rest potential E_o and the current flowing round the "loop" is the corrosion current I_{corr} that will occur when the potentiostat is disconnected. This is the current that is of interest to engineers because it is a measure of the rate at which iron is being lost from the steel at the anode.

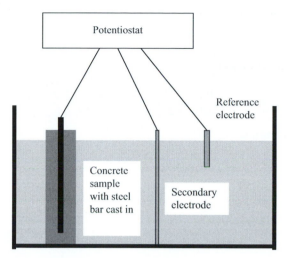

FIGURE 31.12 Diagram of Corrosion Experiment

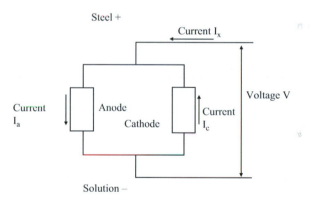

FIGURE 31.13 Schematic Equivalent Circuit for Steel in Solution

This rest potential is the voltage across both the anode, and the cathode. Thus:

$$E_0 = V_{a0} + B_1 \text{Log}\left(\frac{I_{corr}}{I_{a0}}\right) = V_{c0} - B_2 \text{Log}\left(\frac{I_{corr}}{I_{c0}}\right) \tag{31.11}$$

Thus, subtracting from equations (31.4) and (31.10)

$$V - E_0 = B_1 \text{Log}\left(\frac{I_a}{I_{corr}}\right) = -B_2 \text{Log}\left(\frac{I_c}{I_{corr}}\right) \tag{31.12}$$

(see Chapter 1 for the subtraction of logarithms)

$$\text{but when } x \text{ is close to } 1 : x - 1 \approx \text{Ln}(x) \tag{31.13}$$

$$\text{Thus}: \frac{(x-1)}{\text{Ln}(10)} \approx \text{Log}(x) \tag{31.14}$$

Thus, when I_a and I_c are close to I_{corr}

$$V - E_0 = \frac{B_1}{\text{Ln}(10)} \times \left(\frac{I_a}{I_{corr}} - 1 \right) = -\frac{B_2}{\text{Ln}(10)} \times \left(\frac{I_c}{I_{corr}} - 1 \right) \tag{31.15}$$

With the following definitions:

$$\text{Constant } B = \frac{B_1 B_2}{(B_1 + B_2)\text{Ln}(10)} \tag{31.16}$$

and

$$\text{Polarisation resistance } R_p = \frac{B}{I_{corr}} \tag{31.17}$$

Equation (31.15) reduces to:

$$\text{External current } I_x = I_a - I_c = \frac{V - E_0}{R_p} \tag{31.18}$$

This is known as the Stern–Geary equation, and the constants B, B_1 and B_2 are known as the Tafel constants. The polarisation resistance is called a resistance because it is the ratio of voltage to current (see Section 6.6). The aim of the test is to measure it, so that the corrosion current I_{corr} can be calculated from equation (3.17). The constant B does not vary significantly, and has been found to have a value of around 26 mV. When the current is known, the rate of mass loss can be calculated from the known mass and charge of an Fe^{++} ion (see tutorial question 1).

The procedure is:

1. Measure the rest potential E_0 between the corroding metal and the solution.
2. Connect the potentiostat and apply a voltage V a few mV above or below the rest potential.
3. Measure the resulting current.
4. Use equation (31.18) to calculate R_p and then equation (31.17) to calculate the corrosion current.

The combination of the two logarithmic relationships to form a linear sum is shown in Fig. 31.14. This effect is why the test is called "linear" polarisation. Figure 3.14a shows the anode current, the cathode current, and the difference between them, that is the external current provided by the potentiostat. It also shows a straight line from equation (31.18). The gradient of this line is $1/R_p$ and this is used to calculate the corrosion current. Figure 3.14b shows the situation if the anode current (and thus the corrosion) is increased. The gradient is greater, and thus R_p is lower, indicating a higher corrosion current.

31.4.3 REST POTENTIAL MEASUREMENT

Looking again at equation (31.11)

$$E_0 = V_{c0} - B_2 \text{Log}\left(\frac{I_{corr}}{I_{c0}} \right) \tag{31.19}$$

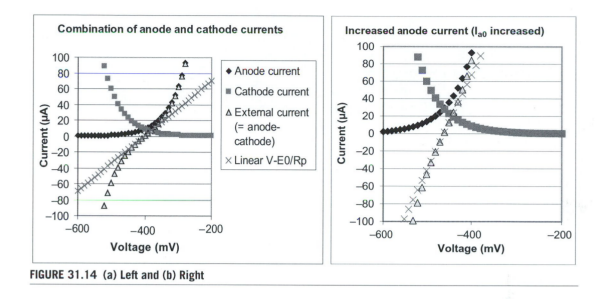

FIGURE 31.14 (a) Left and (b) Right

It may be seen that when comparing systems with similar cathode conditions (i.e., the same I_{c0} and V_{c0}) as the rest potential E_o increases the log of the corrosion current I_{corr} decreases. This is the basis of potential mapping described in Section 27.4.5. The rest potential is measured, and a high potential indicates a low corrosion rate.

31.5 CORROSION OF STEEL IN CONCRETE

31.5.1 THE CORROSION CIRCUIT

The discussion in Section 31.4 is for steel in a solution. The corrosion circuit for steel in concrete is shown in Fig. 31.15.

The current flows through the pore solution in the concrete from the anode to the cathode, and back through the steel. At the anode, the iron ions are lost from the surface of the steel, causing the damage. At the cathode, the electrons flow into the pore solution where hydroxyl ions are formed.

There are two main differences in the equivalent circuit in concrete, compared to corrosion in a solution:

1. The circuit must pass through the concrete that has a resistance.
2. The steel/concrete interface has a capacitance. This is known as the "double layer capacitance," and is caused by charge build up at the interface.

The equivalent circuit is shown in Fig. 31.16. The double layer capacitance carries no direct current (see Section 6.7) so, in normal conditions, it will have no effect. However, if the current is suddenly increased to take a reading, it will carry some initial current.

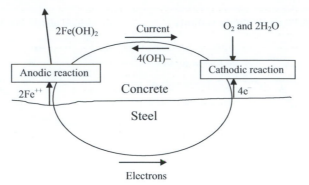

FIGURE 31.15 Corrosion of Steel in Concrete

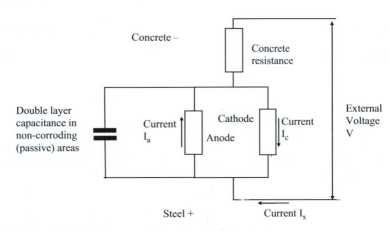

FIGURE 31.16 Equivalent Circuit Diagram of Steel Corrosion in Concrete

31.5.2 LINEAR POLARISATION MEASUREMENTS IN CONCRETE

The current through a capacitor depends on the rate of change of the voltage across it. The characteristics of this circuit are therefore as follows:

1. If a voltage different from E_0 is applied to it, there will be a high initial current through the capacitor, but this will decay to zero. Thus, in order to make a linear polarisation resistance measurement, it is necessary either to:
   a. wait about 30 s after applying the voltage. This has the disadvantage of causing possible changes to the corrosion process or,
   b. apply a very slowly changing voltage or,
   c. apply a pulse of voltage, and make a measurement when it is switched off.
2. When measuring the polarisation resistance, the concrete resistance will also be measured. Fortunately, the capacitance has a very low resistance to alternating current, so this may be used (50–100 Hz) to measure the concrete resistance, and it may then be subtracted.

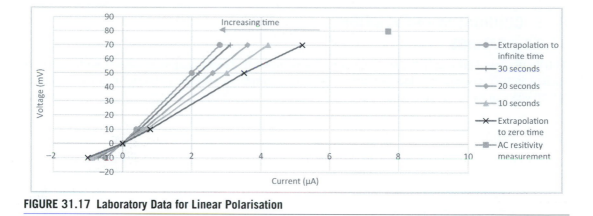

FIGURE 31.17 Laboratory Data for Linear Polarisation

Figure 31.17 shows some typical laboratory data for linear polarisation. The different lines are for different time delays between applying the voltage, and taking the reading. Typically, a delay of 30 s is sufficient.

31.5.3 *IN SITU* MEASUREMENT OF LINEAR POLARISATION

The difficulty with taking linear polarisation readings on an existing structure is that the reinforcing steel in a structure is generally connected together, and the linear polarisation test needs to be carried out on a small sample of known size, so the mass loss per unit can be calculated. On some major structures, special lengths of steel have been installed that are electrically isolated from the main reinforcement. However, on most structures, this has not been done. Some good results have been obtained using a "guard ring." as shown in Fig. 31.18. The voltage follower is an electronic device that simulates the effect of isolating the short length of bar under the test instrument from the rest of the reinforcement. Fig. 31.18 also shows how the arrangement in Fig. 31.12 can be adapted for use on structures.

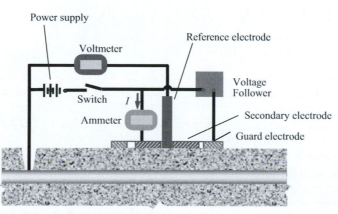

FIGURE 31.18 Linear Polarisation with Guard Electrode

31.6 CORROSION PREVENTION

31.6.1 COATINGS

This is the standard method. Steel sections may be painted. Steel cladding is coated with materials such as polyvinylidene fluoride, polyester with acrylic beads, or silicone polyester that have a life of 30–40 years in most environments.

31.6.2 WEATHERING STEELS

Carbon steel with a 0.2% copper content forms a very stable oxide layer (in the absence of chlorides). It is, therefore, very durable, but equally ugly.

31.6.3 STAINLESS STEELS

These are alloys of steel with some chromium, and some other elements. Its use is increasing (Fig. 31.19). Most stainless steels corrode to some extent, and the weathering properties should be checked (Fig. 31.20).

31.6.4 CATHODIC PROTECTION WITH APPLIED POTENTIAL

This method makes the metal cathodic (negative) relative to the solution and, thus, stops the anodic reaction, and can be applied for considerable lengths of time. In some regions, it is now being applied to new structures, and used continuously. On some structures, provision has been made to install cathodic protection by providing connections to the reinforcement, etc. but it has been left to be actually

FIGURE 31.19 This Bridge in Singapore has been Constructed Entirely out of Stainless Steel

FIGURE 31.20 A Sample Panel of Stainless Steel Cladding Panels for use on the Car Park Behind that is Under Construction

The sample panel enables the architect to see how the proposed panels weather in the local climate.

installed if, or when, needed. Experience has shown that this is not a good idea because, when required, the connections have often been found to be incomplete or damaged. Figures 31.21 and 31.22 show a cathodic protection system. The power requirements of a system like this are minimal, and it may be connected to the street lighting circuit.

31.6.5 CATHODIC POTENTIAL WITH SACRIFICIAL ANODES

These are lumps of magnesium or zinc, which corrode in preference to the substrate that they are protecting. They are typically made from zinc, for use in salt water, and magnesium, for fresh water, when they are fixed to steel structures. They need to be renewed every few years, but the structure itself is protected. Small anodes may be fixed to reinforcement in concrete, during a patch repair. Zinc mesh has also been used, set into a render on the surface of concrete.

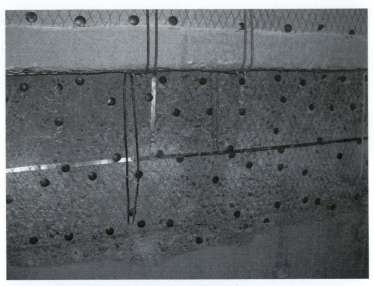

FIGURE 31.21 Titanium Mesh Anode Installed for Cathodic Protection

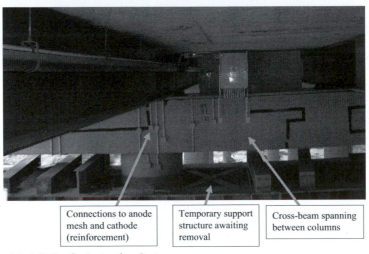

| Connections to anode mesh and cathode (reinforcement) | Temporary support structure awaiting removal | Cross-beam spanning between columns |

FIGURE 31.22 Completed Cathodic Protection System

31.7 CONCLUSIONS

- Metals will lose ions to solutions, but this process will normally stop when they reach their rest potential.
- The corrosion will only continue if the electrons in the sample can be dissipated.
- A positive applied voltage will corrode a metal, and a negative potential will protect it.

- Corrosion may be detected with rest potential measurements, but linear polarisation is needed to measure corrosion rates.
- When measuring corrosion rates in concrete, account must be taken of the concrete resistance, and the double layer capacitance.
- Cathodic protection may be applied with a potential from a power supply or a sacrificial anode.

TUTORIAL QUESTIONS

1. A steel sample is cast into a concrete structure, and is kept electrically insulated from the reinforcing cage. During a survey of the structure, the rest potential of the sample is found to be -400 mV, relative to a reference electrode on the concrete surface. By applying a voltage relative to the reinforcing cage, the following observations are made:

Voltage Relative to Reference Electrode (mV)	Current Through Sample and Reinforcing Cage (μA)
-380	30
-390	15
-400	0
-410	-15
-420	-30

  a. Assuming $B = 26$ mV, calculate the corrosion current in the bar.
  b. Assuming the mass and charge of a Fe^{++} ion are 9.3×10^{-26} kg and 3.2×10^{-19} C, calculate the rate of loss of mass from the bar.
  c. On a subsequent occasion, the rest potential of the bar is observed to have dropped to -600 mV. Assuming the cathode conditions have not changed, calculate the new rate of mass loss (assume $B_2 = 0.12$ V).

Solution:

  a. From the table: when $I_x = 30$ μA, $V-E_0 = 20$ mV
     Thus, $R_p = 20 \times 10^{-3}/20 \times 10^{-6} = 666$ Ω (31.18)
     Thus, $I_{corr} = 26 \times 10^{-3}/666 = 39$ μA   (31.17)
  b. 1 A = 1 C/s thus the rate = 39×10^{-6} C/s
     $= 39 \times 10^{-6} \times 9.3 \times 10^{-26}/3.2 \times 10^{-19} = 1.13 \times 10^{-11}$ kg/s
  c. Using the second part of equation (31.11) referring to the cathode:
     For the initial condition when $E_0 = -0.4$ V
     $-0.4 = V_{c0} - 0.12 \log (39/I_{c0})$
     When E_0 has dropped to -0.6 V
     $-0.6 = V_{c0} - 0.12 \log(I_{corr}/I_{c0})$
     thus, solving by subtraction:
     $0.2/0.12 = \log(I_{corr}/39)$

thus

$I_{corr} = 1816 \ \mu A$

Calculating the mass loss from the constants in (b)

Mass loss $= 1816 \times 10^{-6} \times 9.3 \times 10^{-26}/3.2 \times 10^{-19} = 5 \times 10^{-10}$ kg/s

2. a. The reinforcement in a concrete structure is connected to an electrical power supply that can maintain it at different potentials relative to the concrete surface. Describe the processes that take place when it is held at different voltages, both at the potential that is present without the power supply connected, and above and below it.

b. The following observations are made:

Voltage of Steel Relative to Concrete Surface	Current Through Steel
−0.4	0
−0.2	100 μA
0	1 mA

By assuming that these currents arise only from anodic processes, calculate the constants V_{ao} and I_{ao} and explain their significance.

Solution:

a. The voltage without the power supply is the rest potential, and, at this potential, the power supply will have no effect. Two processes will be producing equal currents:
  - The anodic process that involves the loss of positive metal ions from the steel.
  - The cathodic process that involves the loss of electrons from the steel, that then combine with oxygen and water, to form hydroxyl ions.

  The current flowing is a measure of the corrosion rate.

  When the voltage is increased, the cathodic process will be replaced by the power supply, and will stop. Depending on the pH, a stable oxide layer may form, and cause passivation. If not, ferrous or ferric ions will be lost from the steel, with consequential loss of section.

  If the voltage is decreased, the anodic process will be stopped, and the steel will be protected.

  The voltage will cause electromigration, as well as directly affecting the corrosion. A negative voltage will move negative ions (e.g., chlorides) away from the steel.

b. Since it is assumed that all of the current is anodic, the zero current must come from the flow of Fe^{++} back into the steel, at the exchange voltage.

thus $V_{a0} = -0.4$ V

At higher voltages, the flow in will be negligible, thus (using the data from the table):

$$-0.2 = -0.4 + B_1 \ (\log 10^{-4} - \log I_{a0}) \tag{31.4}$$

and

$$0 = -0.4 + B_1 \ (\log 10^{-3} - \log I_{a0}) \tag{31.4}$$

solving these gives $B_1 = 0.2$ and $I_{a0} = 10^{-5}$ A

V_{a0} is the voltage at which the flow of ions in and out of the metal is balanced, and I_{a0} is the flow at that voltage.

3. The Tafel constants for a corroding steel sample in concrete are $B_1 = B_2 = 0.15$ V
   a. If the polarisation resistance is measured as 1500 Ω, what is the corrosion current?
   b. If the rest potential subsequently falls from -300 to -500 mV, and the cathode conditions do not change, what is the new corrosion current?

Solution:

a. $$B = \frac{0.15 \times 0.15}{(0.15 + 0.15) \times 2.303} = 32 \, \text{mV} \tag{31.16}$$

$$I_{corr} = \frac{B}{1500} = 21 \, \mu A \tag{31.17}$$

b. In the original condition when $E_0 = -0.3$ V
   $$-0.3 = V_{c0} - 0.15 \log (21/I_{c0}) \tag{31.11}$$
   When E_0 drops to -0.5 V
   $$-0.5 = V_{c0} - 0.15 \log(I_{corr}/I_{c0}) \tag{31.11}$$
   Thus,
   $$0.2/0.15 = \log(I_{corr}/21)$$
   Thus,
   $$I_{corr} = 449 \, \mu A$$

NOTATION

B Tafel constant (V)
B_1 Tafel constant (V)
B_2 Tafel constant (V)
B_1' $B_1/\ln(10)$ (V)
E_0 Rest potential for circuit (V)
I_a Anode current (A)
I_{a-} Anode exchange current (A)
I_{a0} Anode current at anode rest potential (A)
I_{corr} Corrosion current (A)
I_c Cathode current (A)
I_{c0} Cathode current at cathode rest potential (A)
I_x External current (A)
R_p Polarisation resistance (Ω)
V Voltage (V)
V_a Anode voltage (V)
V_{a0} Anode rest potential (V)
V_c Cathode voltage (V)
V_{c0} Cathode rest potential (V)

ALLOYS AND NONFERROUS METALS

CHAPTER OUTLINE

32.1 INTRODUCTION

Steel is by far the most commonly used metal in construction and it is discussed in Chapter 30, however, several other metals are used in significant quantities. As with iron, that is most frequently used alloyed with carbon to make steel, most other metals are rarely used in their pure form. These alloys generally have superior properties to pure metals.

32.2 ALLOYS
32.2.1 TYPES OF ALLOY

It was noted in Section 3.2.2 that metals are crystalline, the atoms that form them are arranged in regular arrays, and their properties are substantially dictated by this microscopic structure, and any imperfections in it.

When metals are mixed to form alloys, the outcome depends on whether the two are soluble in one another.

Figure 32.1 shows schematically how metal atoms are positioned in a regular grid, and this must accommodate a different element. It may be seen that this is only possible if the atoms are of similar sizes. There are three possibilities for the resulting alloy, depending on the solubility of one metal in the other:

- Completely soluble, that is, atoms of one will fit exactly into the structure of the other, without disturbing it (e.g., copper and nickel).
- Partially soluble, that is, the atoms cannot form structures together, but crystals of each will mix (e.g., copper and zinc, i.e., brass).
- Insoluble, for example, molten iron will float on molten lead, they will not mix.

32.2.2 METALS THAT ARE COMPLETELY SOLUBLE

Figure 32.2 shows the effect of cooling a liquid mixture of two metals, X and Y, which are completely soluble. If an alloy of 45% metal X and 55% metal Y is cooled, solid will first appear at the temperature of points a with composition a_S. As the temperature falls, a solid with b_S percent of X will be present, so the remaining liquid will be richer in metal Y (point b_l). Eventually, at temperature c the remaining liquid solidifies, and the composition of the solid is c_S. The resulting crystals will thus be rich in metal X near the core, and rich in metal Y near the outer surface. Between temperatures a and c, the liquid and solid exist together in a "plastic" state. This property, in this temperature range, is, for example, used for working lead solder.

32.2.3 METALS THAT ARE PARTIALLY SOLUBLE

Figure 32.3 shows two partially soluble metals: A and B. In this situation, a cooling metal will either form a crystal structure of the size and type of pure metal A (α structure), or one similar to metal

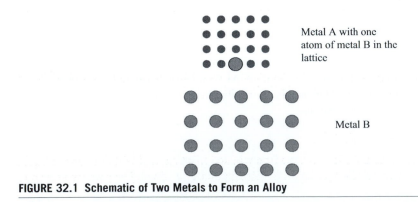

Metal A with one atom of metal B in the lattice

Metal B

FIGURE 32.1 Schematic of Two Metals to Form an Alloy

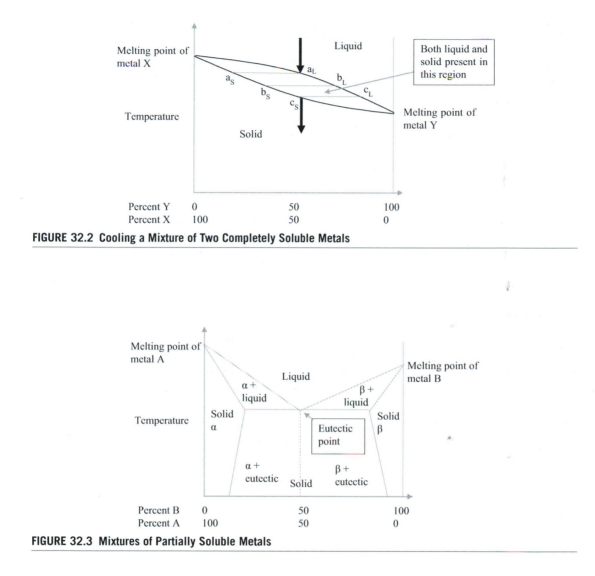

FIGURE 32.2 Cooling a Mixture of Two Completely Soluble Metals

FIGURE 32.3 Mixtures of Partially Soluble Metals

B (β structure), depending on the proportions of the alloy. As it cools further, it may form some "eutectic" that is a different mixed structure, with some similarities to α and β.

32.2.4 METALS THAT ARE INSOLUBLE

Molten iron will float on molten lead. As it cools, the iron will solidify first, and will float on the liquid lead because it has a lower density. Cooling below the melting point of lead will just give two separate solid layers of iron and lead.

32.3 COMPARISON OF NONFERROUS METALS

The four most common nonferrous metals used in construction are: lead, zinc, aluminium, and copper. Some properties are shown in, for comparison, in Table 32.1. These different properties are exploited for specific applications, as discussed in the following sections.

32.4 COPPER

32.4.1 APPLICATIONS FOR COPPER

It may be seen from Table 32.1 that copper has a high standard electrode potential, and is thus resistant to corrosion. It also has a good strength and modulus. This makes it very suitable for water pipes. The electrical resistivity is also low, making it suitable for electrical wires. The thermal conductivity is high, this means that copper pipes often need external insulation, but it makes them suitable for calorifiers, where heat is transferred through the wall of a coil of pipe into the surrounding tank.

32.4.2 GRADES OF COPPER

The three grades of copper are:

* Deoxidised copper, used for copper tube, suitable for welding.
* Fire refined tough pitch copper has higher strength, thermal, and electrical conductivity and resistance to corrosion. Used for roof coverings. These turn a pleasant green, as surface corrosion occurs. This corrosion product may stain adjacent materials.

Table 32.1 Properties of Common Metals

	Lead	Zinc	Aluminium	Copper	Iron
Standard electrode potential (V)	−0.12	−0.76	−1.7	+0.34	−0.44
Density kg/m^3 (lb/yd^3)	11,300 (19,100)	7,100 (12,000)	2,700 (4,560)	8,800 (14,900)	7,800 (13,200)
Specific heat J/kg°C (BTU/lb°F)	127 (0.030)	388 (0.093)	880 (0.210)	390 (0.093)	480 (0.115)
Thermal conductivity W/m°C (BTU/hft.°F)	35 (20)	116 (67)	200 (116)	400 (231)	84 (49)
Coefficient of thermal expansion μstrain/°C	29.5	Up to 40	24	17	11
Melting point °C (°F)	327 (620)	419 (786)	659 (1,218)	1,083 (1,981)	1,537 (2,863)
Elastic modulus GPa (ksi)	16.2 (2,349)	90 (13,050)	70.5 (10,222)	130 (18,850)	210 (30,450)
Tensile strength MPa (psi)	18 (2,610) (short term)	37 (5,365)	45 (6,525)	210 (30,450)	540 (78,300)
Electrical resistivity (Ωm)	2.05×10^{-7}	5.9×10^{-8}	2.8×10^{-8}	1.7×10^{-8}	1.1×10^{-7}

For definitions of thermal properties see Chapter 4. For resistivity see Chapter 6. For electrode potential see Chapter 31.

- Electrolytic tough pitch high conductivity copper. Contains fewer impurities, has higher electrical conductivity, used for electrical conductors.

Copper is the most noble of the common construction metals, so it may be protected from corrosion by acidic water by, for example, sacraficial aluminium rods in copper hot water cylinders (see Section 31.2.4).

32.4.3 ALLOYS OF COPPER

Table 32.2 gives the names and basic properties of the main alloys of copper.

32.5 ZINC

Zinc is used for roofing applications. It is generally resistant to inland and marine atmospheres, but attacked by industrial pollution. However, its main application is for zinc plating (see Section 32.8).

32.6 ALUMINIUM

32.6.1 PRODUCTION OF ALUMINIUM

Aluminium production is energy intensive. About 17,000 kWh of electricity are required to produce 1 tonne of aluminium. This means that it has a relatively high cost. However, it is very suitable for recycling.

It is rarely used in its pure form. Aluminium alloys are typically made with magnesium, manganese, copper, and silicon, and may be up to 15 times stronger than pure aluminium. As with steel, heat treatment and cold working also increases the strength (see Section 30.3).

32.6.2 APPLICATIONS FOR ALUMINIUM

The very high electrode potential would be expected to make aluminium highly susceptible to corrosion, but, as soon as it is produced, it forms a stable oxide layer that protects it. The good strength to weigh ratio makes it highly suitable for light-weight structures, such as temporary buildings. It is used for window frames, but its high thermal conductivity often leads to condensation on the inside surface. The high electrical conductivity means that it is used for electrical wiring, but this can lead to corrosion problems.

Table 32.2 Alloys of Copper

Alloying Metal	Alloy Name	Properties
Zinc	Brass	Generally harder than copper
Tin	Bronze	Tough and has good resilience to abrasion and corrosion
Aluminium	Aluminium bronze	High strength and corrosion resistance. Used for pipe fittings and fasteners
Silicon	Silicon bronze	Good strength, ductility, and corrosion resistance in salt water. Used for door fittings, railings and hinges

32.6.3 ANODISED ALUMINIUM

The stable oxide layer, which protects aluminium from corrosion may be artificially enhanced by "anodising." In this process, the aluminium is connected to a circuit, as shown in Fig. 31.4, where it forms the anode. The solution is prepared so that, rather than removing metal from the anode, the process significantly increases the thickness of the oxide layer, and enhances corrosion resistance. It may also be used to make an attractively coloured finish, making anodised aluminium highly suitable for architectural applications, such as cladding and roofing.

The voltage required by various solutions may range from 1 V to 300 V dc, although most fall in the range of 15–21 V. Anodised coatings have a much lower thermal conductivity and coefficient of thermal expansion than aluminium, so the coating may crack at high temperatures. The coating also has a high melting point and electrical resistivity that can make welding more difficult.

32.7 LEAD

The high electrode potential of lead gives low corrosion for roofing applications (Fig. 32.4) (unless copper nails are used – see Section 31.2.4). The high density causes larger loadings for large areas, but also prevents uplift in wind. The low melting point gives easy welding, and the low modulus gives easy working and forming to shapes. The low strength makes it easy to cut to shape.

32.8 PLATING

32.8.1 THE PURPOSE OF PLATING

Plating is used to form a thin layer of one metal over the surface of another. It may be used to give enhanced corrosion resistance or appearance. It is used where the metal used for the plating is unsuitable for the body of the element, due to high cost or inadequate strength. Plating may be carried out by electroplating or hot dip.

FIGURE 32.4 A Lead Roof

32.8.2 ELECTROPLATING

Electroplating uses a circuit similar to that shown in Fig. 31.4. If an appropriate solution is used, the metal ions will be deposited on the cathode in an even layer that can be built up over several hours. Chromium is often used to give a highly reflective chrome plate. Copper and zinc plate are also common. Often the plating is built up of several layers of different metals, to give the desired finish. Tin, cadmium, and nickel electroplating is also used (Fig. 32.5).

32.8.3 HOT-DIP PLATING

Hot-dip plating is carried out by dipping the element to be plated into a tank of molten metal. It is most commonly used for hot-dip zinc plating, that is known as galvanising. Zinc is well suited for this, and gives good durability because, as well as forming a physical barrier, its low electrode potential will protect the parent metal (normally steel). Zinc also has a low melting point that reduces the energy demand.

Bright zinc plating is an electroplating process, and gives a far thinner, but more attractive, finish than the thicker, and thus more durable, dull grey of galvanizing.

32.9 CONCLUSIONS

* Alloys display different properties, depending on the solubility of the metals.
* Even if the metals are fully soluble, a typical casting will not have a uniform distribution of them.
* Partially soluble metals may solidify into structures resembling either of them, or the eutectic structure.

FIGURE 32.5 A Small Scale Electroplating Tank

The samples are suspended from hooks on one bar, and the anodes are suspended from the other bar.

- Copper, zinc, aluminium, and lead all have different properties that can be exploited in construction.
- Plating is an effective method to improve the appearance or durability of metals.

TUTORIAL QUESTION

1. Describe how the following processes are carried out, and describe a common application for each of them:
   a. Electroplating.
   b. Anodising.
   c. Cathodic protection.

Solution:

a. Electroplating. If two metals are immersed in a solution, and a direct electric current flows between them, the positive metal will dissolve, and the negative metal will be plated with it. Typical application: Bright zinc plated screws.

b. Anodising. Aluminium is highly reactive, but forms a stable protective layer of oxide. This layer may be artificially enhanced by "anodising." By using the different solutions for this attractive finish, various colours may be achieved. Typical application: window frames.

c. Cathodic protection. This method makes the metal cathodic (negative) relative to a solution, and thus stops the anodic reaction. It works either with a power supply, or with "sacrificial anodes," lumps of magnesium or zinc, that corrode in preference to the substrate they are protecting. Typical application: pipelines.

TIMBER

33

CHAPTER OUTLINE

33.1 INTRODUCTION

Timber is an excellent construction material. However, its mechanical properties are complex because they depend on the direction of loading, as noted in Chapter 3. This arises because trees grow differently at different times of year, producing softer wood early in the growing season, and harder wood later. This creates the annual growth rings that may be seen on cut samples. The strength within a ring is greater than the bond between rings. The strength parallel to the grain is even higher.

Timber is either hardwood or softwood. Hardwood is from broad-leafed trees, many of which are deciduous (i.e., they lose their leaves in winter). Softwood is from coniferous trees that are generally evergreen. Hardwoods are generally stronger and denser then softwoods. Balsa (hardwood) and yew (softwood) are obvious exceptions. Softwood trees grow far faster than hardwood. Typical growing times are 20–30 years, while hardwood trees may take over 100 years to reach suitable sizes for timber production. Almost all timber that is used for structural purposes in construction is softwood.

Trees clearly benefit the environment because they take carbon dioxide from the atmosphere and use the carbon to make wood.

33.2 THE ENVIRONMENTAL IMPACT OF FORESTRY

33.2.1 FORESTS THAT ARE NATURAL BUT ARE NOT REPLANTED

Widespread deforestation is still in progress in many tropical regions, and is carried out both to produce hardwood, and to clear the land for farming. Cutting these forests has a major environmental impact, especially on the greenhouse effect. Most hardwood for construction comes from tropical rainforests. Hardwoods from other regions are generally expensive, and used for furniture and similar applications.

There is considerable resistance to the use of tropical hardwoods because it is seen as contributing to the destruction of the rainforests. There are many conflicting arguments. A large proportion (over 50%) of tropical hardwood that is cut is burnt to clear the land, and not used for timber, so using more of it may not increase the number of trees cut down. Timber sales also provide vital income to many poor areas. Many countries are attempting to stop the destruction of the rainforests, but illegal logging is common. Buying timber from sustainable forestry may help fund efforts to prevent illegal logging.

The range of hardwood species available and their costs are changing constantly, as areas are "logged out" of a given species, and other ones are marketed. True Mahogany is now rare. Lauan from Southeast Asia, and other similar red hardwood species, are often substituted for it.

33.2.2 FORESTS THAT ARE NATURAL BUT ARE REPLANTED AFTER CUTTING

This is clearly preferable to forest clearance, but there will be loss of natural habitat, and diversity of species. There are particular problems when the planted trees are not native species to the area. These "exotic" species may be harmful to the local flora and fauna.

33.2.3 FORESTS THAT HAVE BEEN PLANTED AND ARE REPLANTED AFTER CUTTING

Most softwood is produced in this way. When these forests are harvested, there is little environmental impact, except for the effect of transport, etc. Trees take most carbon dioxide from the atmosphere when they are young, and some old trees actually emit methane that is a powerful greenhouse gas. If the trees are harvested, and the timber is used in a structure, the carbon dioxide is removed from the atmosphere and not returned (as it would be, if the timber was burnt or decayed naturally). The impact of softwood use on the environment/greenhouse effect is therefore probably beneficial.

33.2.4 REFORESTATION

There are areas that have not been forested for some time, but are now being planted with trees. This is being carried out on a large scale in many countries, including the United States and China. It is clearly generally very beneficial to the environment. However, by forming large connected areas of forest, it contributes to the rising numbers of forest fires. The choice and diversity of species is important to prevent disease, insect attack, or over-extraction of water (a known problem with eucalyptus). Timber production from reforested areas, followed by well managed planting, may be beneficial.

33.2.5 CERTIFICATION SCHEMES

There are a number of schemes in operation that certify the sustainability of various aspects of the production of timber, including the replanting procedures and the impact on local people and wildlife. Details of the processes used (and any reported abuses) of these schemes are available to download. Provided appropriate care is taken to confirm the source, timber can be a very sustainable resource with good, and probably increasing, availability.

33.3 PRODUCTION
33.3.1 CONVERSION

Round logs must be sawn to produce the square or rectangular sections, which are generally required in construction.

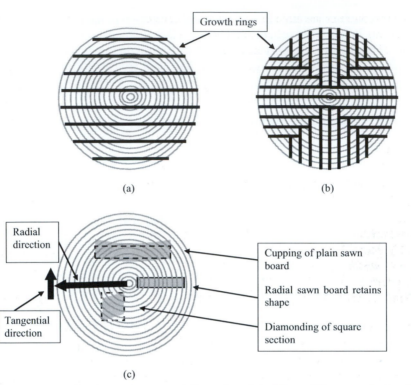

FIGURE 33.1 Common Log Sawing Arrangements, and Illustration of the Effect of Movement During Drying

(a) Through and through/live/plain cut, (b) quarter cut, and (c) showing the effect of movement during seasoning.

Figure 33.1 shows two typical sawing arrangements for this. The simple cutting arrangement in Fig. 33.1a is clearly the easiest to carry out. However, when the timber is subsequently dried, it will shrink and the tangential movement is about double the radial movement. This will mean that the plain sawn board from cutting method Fig. 33.1a will distort as shown in Fig. 33.1c, and thus the quarter cut method in Fig. 33.1b will produce better timber.

The planks must be cut carefully to keep them parallel to the grain. Sloping grain (slope > 1:10) will reduce strength.

It is also sometimes necessary to discard the heartwood at the very centre of the trunk because it is susceptible to cracking. In very fast grown timber, there may also be "resin pockets" that are voids within the timber that are full of resin. These sections must also be discarded.

33.3.2 SEASONING

When cut, timber has a high moisture content. Before use, most of this water must be removed. The reasons for this are:

1. When the moisture content of wood is reduced, it shrinks.

2. Moist timber is more susceptible to many types of decay.
3. Many types of finish will not adhere to moist timber.
4. The strength increases when wood is dried.

Some of the water is unbound within the microscopic structure, and some is bound to the structure. Removing the unbound water has little effect, but once it has all gone and the "fibre saturation point" (typically, around 30% moisture by weight) is reached, any further drying will cause shrinkage.

It is important to note that seasoning does not remove all of the moisture. It is intended to remove only sufficient water for the timber to be in hygral equilibrium with its environment when in use. This is, of course, only an approximation because there are seasonal and daily variations in temperature and humidity. Typically, at 60% RH, the moisture content should be 10–15%. The equilibrium moisture content at any given humidity will depend on the porosity and the microstructure.

The traditional method of seasoning is to stack the timber with spacers, so that the air can circulate around it, but it is protected from rain. In the United Kingdom, this will only bring the moisture down to about 20%. For dense hardwoods, it takes 1 year for every 25 mm of thickness.

For industrial timber production in temperate climates, the timber is normally dried in a carefully humidity and temperature controlled kiln. Kiln drying may be carried out on *green* timber, or on air-dried timber.

If seasoning is carried out incorrectly (normally too fast), the timber may deform by bending or twisting, and it may split, or the grain may partially collapse to give an uneven surface.

Seasoning is a partially reversible process. If timber is seasoned to a low-moisture content, for use in a heated building, and then inadequately protected on site, its moisture content will rise again.

33.3.3 GRADING

Timber is a natural product, and therefore variable. Apart from overall strength, which may be affected by climatic conditions while the tree was growing, timber will also contain defects such as knots, where branches have grown out of the trunk. Wood on the downwind side in areas of strong prevailing winds becomes strong in compression, but weak in tension. Wide growth rings generally indicate faster growth and lower strength. Timber must therefore be graded, that is, sorted, and the weak and defective pieces rejected. There are two ways of doing this:

Visual grading. This involves the use of a number of complex criteria covering the number of knots, etc. Each piece is individually inspected by a skilled operative, and either accepted or rejected.

Stress grading. This method uses a correlation that has been established between Elastic (Young's) modulus and strength. Timber is therefore passed between rollers (see Fig. 33.2), and the force required to bend it by a given amount is proportional to its strength. The machine is computer controlled, and a mark is applied to each piece to indicate its stress grade. Alternatively, X-rays may be used to establish the density of the timber, and this is correlated to strength.

Timber that is apparently not graded may be ungraded timber or, more probably, grade rejects. Thus, if a grade is not specified, timber will be of worse quality than the average obtained from the tree.

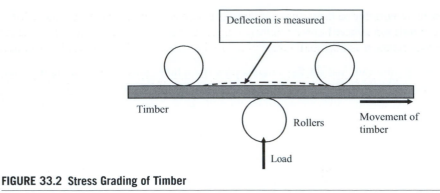

FIGURE 33.2 Stress Grading of Timber

33.4 ENGINEERED WOOD PRODUCTS

33.4.1 TYPES OF PRODUCT

Sawing tree trunks into planks produces a lot of waste, and cannot produce flat sheets or very large beams. There are, therefore, many processes that involve cutting up or pulping the timber, and then re-forming it into the required shapes. For some interior grade products, such as softboard, the natural resins in the wood are sufficient to hold the shape together, but for all engineering applications an adhesive is used, and the strength of the finished product depends significantly on the strength of the adhesive.

33.4.2 PLYWOOD

Plywood veneers are "peeled" from a complete trunk. This uses the strength of the wood along the grain, and in the tangential direction, but avoids the weak radial direction (Fig. 33.3). The veneers are then flattened, and glued together with alternate veneers having grain directions at right angles, and with the grain of the both outer veneers parallel to the longer edge of the sheet. This means that the bending strength is higher in this direction.

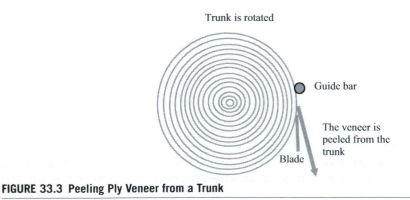

FIGURE 33.3 Peeling Ply Veneer from a Trunk

Marine, "weather, and boil proof", and exterior grade plywood have durable adhesive. Most other grades will fall apart when used outside. For permanent exterior work, always use marine grade in which every lamination is selected hardwood. The better grades have more, but thinner, laminations.

The two faces of plywood are often different, sometimes deliberately, for example, "Canadian exterior face one side" that is used for shuttering, and should be used with the fair face against the concrete.

33.4.3 OTHER BOARD PRODUCTS

There is a very wide range of board products available. The quality of the board depends on the quality of the wood used, and the quality of the adhesive. Boards with plastic or metal faces or even lead cores are available.

Particle boards (chipboards) are made by cutting the timber down to small "strands," and reassembling them with adhesive. Very efficient use of the tree may be made because even quite small branches may be processed in this way. In "parallel strand lumber," and "oriented strand board," the process makes all of the strands orient their grain in the same direction to optimise the strength.

Boards may also be assembled from many different combinations of strips and veneers of timber.

The other board products are not generally as strong or durable as plywood. However, the outer veneer can be a quality hardwood for decorative appearance, similar to a good plywood.

33.4.4 GLUED–LAMINATED TIMBER

"Glulam" is made by assembling timber sections, typically 25–50 mm thick, to form large beams. The relatively small timber sections can be bent, so this is an effective way of making curved beams. The strength of the finished beam is normally greater than a similar section of solid wood because defects are avoided, and optimum use is made of the orientation of the grain.

33.5 STRENGTH OF TIMBER

33.5.1 EFFECT OF MOISTURE

The strength of timber decreases with increasing moisture content, up to the fibre saturation point at which no further decrease occurs. This reduction will be typically around 40% for timber at normal moisture contents for structural use.

Timber shrinks significantly when it dries, and swells when it gets wet. This movement is highest in the tangential direction, but still significant in the radial direction (it is negligible in the longitudinal direction parallel to the grain).

33.5.2 ORTHOTROPIC BEHAVIOUR

Timber is *orthotropic*, meaning that it behaves differently under load, depending upon which of its three axes it is loaded on. Concrete and steel, by contrast, are generally *isotropic*, that is approximately the same in all directions.

Timber has three main axes:

L = longitudinal, that is, parallel to the grain,
T = tangential to the grain,
R = radial to the grain.

It has three compressive strengths, three tensile strengths, and three bending strengths, and associated elastic moduli for each of them.
It has six Poisson's ratios

v_{LT} = tangential strain/longitudinal strain
v_{RT} = tangential strain/radial strain
v_{RL} = longitudinal strain/radial strain
v_{TL} = longitudinal strain/tangential strain
v_{LR} = radial strain/longitudinal strain
v_{TR} = radial strain/tangential strain

Thus, if a piece of timber is subjected to stresses σ_L along the grain, and σ_R radial to the grain, and modulus E_L and E_R, the expansive strain tangential to the grain will be:

$$\varepsilon_T = \frac{\sigma_L v_{LT}}{E_L} + \frac{\sigma_R v_{RT}}{E_R} \tag{33.1}$$

33.5.3 FLEXURAL STRENGTH

The most common test for timber is to measure its flexural strength in the longitudinal direction. In Section 22.6.2, the flexural strength test for concrete was described, and it was noted that this is a "4-point" test with the load at two points on the top. For timber, a simpler "3-point" test may be used with just one load point on the top (Fig. 33.4).

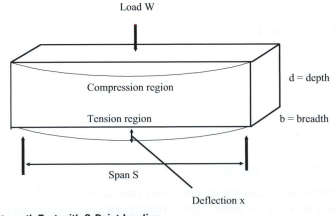

FIGURE 33.4 Flexural Strength Test with 3-Point Loading

The bending strength is given by equation (33.2), and the modulus by equation (33.3).

$$\text{Strength} = \frac{3WS}{2bd^2} \tag{33.2}$$

where S = span, W = load at failure, b = breadth, and d = depth of specimen.

$$E = \frac{S^3}{4bd^3} \times \frac{W}{x} \tag{33.3}$$

where E = modulus of elasticity, x = deflection.

33.5.4 CREEP

The creep of timber depends significantly on the moisture content. It is also very significantly increased by cyclic changes of moisture content caused by, for example, cycles of heating in a building. Tests have shown higher creep in timber subject to cycles of high humidity, while subject to 12% of its failure stress, than at a constant high humidity at 37% of its failure stress.

33.5.5 TIMBER STRENGTH CLASSES FOR THE EUROCODES

The Eurocode classes show the bending strength, for example, a timber with grade C14 has a strength of 14 MPa in bending.

Some strength classes are:

C14–C40 for coniferous wood (softwood),
D30–D70 for deciduous (hardwood), and
GL20–GL36 for glulam made with C35 laminations.

33.5.6 TIMBER STRENGTH CLASSES IN THE UNITED STATES

Using American Lumber Standards Committee rules, machine graded softwood lumber in the United States is graded by bending strength and modulus of elasticity in bending. Both of these properties may be determined by a single stress-grading test.

Thus, a grade designation 2250f-1.7E has a bending strength of 2250 psi, and a modulus of 1.7×10^6 psi. Tensile and compressive strengths parallel to the grain may also be specified.

33.6 JOINTING TIMBER
33.6.1 NAILS AND SCREWS

The permissible loads on nails and screws are given in the design codes. There are many types of nail (e.g., ringnails) that give better performance than plain round nails. Screws are stronger than nails, and with good quality power tools, they take little longer to install.

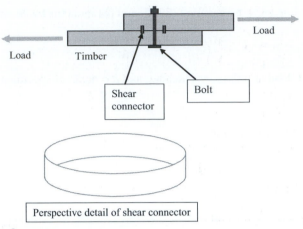

FIGURE 33.5 Use of Shear Connector

33.6.2 GLUED JOINTS

These are most effective for multiple laminates. Note that, for fixing plywood, the shear strength will be limited by the bond between the laminates of the ply.

33.6.3 BOLTED JOINTS

Bolted joints may be made far more effective by shear connectors that spread the load to prevent local failure of the timber around the bolt (Fig. 33.5).

33.6.4 METAL PLATE FASTENERS

These are metal plates with a large number of "nails" on the surface that are pressed into the timber. Plates of this type (e.g., "gang nail plates") are by far the most common type of joint for roof trusses. They have the advantage of being cheap, easily fixed by machine, and not requiring overlap of the timber members. They are frequently used in truss rafters that are prefabricated trusses used in house roof structures, etc. (see Fig. 33.6).

33.6.5 PLYWOOD GUSSET JOINT

Metal plate connectors require machines to install them, and cannot generally be installed on site. For site applications, a plywood gusset may be used on site, in place of a plate fastener for joints, such as those shown in Fig. 33.6. The plywood gusset would normally be larger than the metal plate, and fixed with nails or screws.

FIGURE 33.6 Typical Application of Metal Plate Fasteners

33.7 DURABILITY OF TIMBER

33.7.1 MECHANICAL DAMAGE

Mechanical damage is usually readily visible, but cases have been reported where, for example, scaffold boards dropped on end from a height have not shown signs of damage, but have subsequently failed due to buckling of the grain (see Fig. 3.15).

33.7.2 WEATHERING

Weathering is the general effect of external exposure. Often, it only involves loss of colour.

33.7.3 DRY ROT

Dry rot is the term used to describe a highly damaging fungus that grows on timber. It actually requires some moisture, and eventually dies at moisture contents below 20% RH, and it is inactive below 20°C (68°F). It can spread very quickly, and the tendrils can grow across brickwork to infect other timber, and is exceptionally difficult to eradicate. However, it only grows in very still unventilated air.

33.7.4 WET ROT

Wet rot is another fungus, and is often difficult to distinguish from dry rot. It can be stopped by rapid drying.

33.7.5 MARINE BORERS

These are usually found in warm salt water. They are becoming an increasing problem in marine structures, as sea temperatures rise and they spread to new areas.

33.7.6 WOOD DESTROYING INSECTS (e.g., WOODWORM)

There are a number of different wood boring insects that attack timber. They lay eggs on the surface or in cracks, and the larvae tunnel through the wood, usually sapwood. The visible holes are exit holes. If a building is dried and heated, they often die, and signs of the holes in timber often do not indicate active infestation. This may be checked by looking for recent deposits of the fine powder that comes out of the holes.

33.7.7 CHEMICAL DEGRADATION

If timber gets wet, the pH of the water will have the opposite effect to that with steel. Timber is preserved in an acidic environment, but alkaline solutions are used to break it down to make paper pulp.

33.8 PRESERVATION OF TIMBER

33.8.1 NATURALLY DURABLE TIMBERS

Knowing whether timber will be durable is generally a question of identifying the species. There is a slight tendency for the darker timbers to be more durable, but this is certainly not universal. Unlike concrete, strength and hardness do not indicate durability. Beech and maple are strong but not durable, cedar is soft but durable. The range of durabilities is very large; choosing the correct timber makes a substantial difference, for example, from 2 years to 200 years to failure. Some species are naturally durable, but beware of "exotics," such as European softwoods grown in New Zealand. Different parts of the log may be more durable than others. Most softwood comes from immature trees with a high proportion of sapwood that is never durable, unless treated with preservative.

33.8.2 TIMBER PRESERVATIVES

Wood preservatives, if correctly applied, can make timber durable. There are large numbers of different preservatives. Different preservatives are needed for different types of degradation.

In order of increasing effectiveness, application methods are:

- Brush.
- Spray.
- Dipping.
- Steeping (long-term immersion).
- Hot and cold open tank.
- Pressure impregnation.

Many preservatives, especially those for brush or spray application, were solvent-based and have been replaced with water-based ones that are generally less effective, in order to comply with environmental regulations regarding solvent release to the atmosphere. Some preservatives, especially those used to eradicate wood destroying insects, may be highly toxic and special low toxicity fluids should be used in schools, and other areas where young children use buildings.

FIGURE 33.7 Metal End Connector to Keep Timber Frame Dry (Note the Polythene Wrapping Used to Keep the Timber Dry During Construction)

If the timber is subsequently cut, it is almost always necessary to treat the cut surfaces, and the durability will only be as good as this treatment.

33.8.3 KEEPING TIMBER DRY

If timber remains dry, it will not decay (even dry rot actually requires moisture). Thus, the best way to preserve it is to keep it dry. It was noted in Section 33.5.1 that timber shrinks significantly when it dries, and swells when it gets wet. This will damage paints and fillers used to protect it, and cause further water damage.

An impermeable surface coating is difficult to achieve, but may keep the timber at optimum moisture content for strength and durability. A coating may peel off due to moisture build-up beneath it. A porous (microporous) coating is easier to maintain. Timber will weather under a varnish, due to the effect of ultraviolet light on it, so paint is far more durable.

Timber will not generally crack after seasoning, unless it is decaying or mechanically overloaded (possibly by restraint during shrinkage). Continual movement due to moisture and temperature changes makes filling cracks difficult. Detailing of sealant joints is discussed in Chapter 39.

Good detailing is needed to keep timber dry (Fig. 33.7).

33.8.4 VENTILATION OF TIMBER ELEMENTS

In addition to keeping them dry, timber structures should be ventilated. Figure 33.8 shows typical ventilation of a brick-built house. The under-floor ventilation is through air bricks (this may also be required to prevent the build-up of radon gas. See Section 11.3). The eaves vents through the soffit boards are a common detail, but care should be taken not to block them with insulation. Ridge vents are included in most dry-hip roofing systems (see Fig. 34.6). Adequate ventilation of the timber should be detailed whenever a new structure, or any works on an existing structure, are designed. Major problems have been encountered with dry rot when refurbishment works have blocked the ventilation to roof structures.

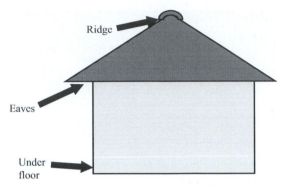

Ridge

Eaves

Under
floor

FIGURE 33.8 Ventilation of a Typical Structure with Wood Floors and Roof Rafters

33.9 BAMBOO

Bamboo is technically a grass, but it is effectively a timber for construction purposes. Bamboo grows in a number of countries with diameters of 100 mm (4 in.) and above. It grows very fast, reaching heights of 30 m (100 ft.) or more in just a few months (Fig. 33.9). Once it has grown, the diameter of the stem does not increase, but the strength improves to give an optimum harvesting age of 4–5 years. Due to its abundance and fast growing cycle, it offers great opportunities for use in structures.

There are various engineered products available from bamboo, including boards and square sections, but the original circular sections have excellent structural properties. The weakness of bamboo structures is at the joints. These are often made with bolts that are simply drilled through the stems, and lead to localised failure (Figs. 33.10 and 33.11). Filling the stem with cement mortar in the region of

FIGURE 33.9 Bamboo Culm Growing in Colombia

Within a few months, this will be up to 30 m tall.

FIGURE 33.10 Bamboo Roof

The culms have just been roughly cut to shape where they meet, and several already have metal clips around them because they have split.

FIGURE 33.11 Tidy Metal End Connectors on Bamboo

FIGURE 33.12 Bending Test on Bamboo

the joint is a commonly used remedy, but it is not recommended because it traps damp, and creates an alkaline environment that encourages decay.

Figure 33.12 shows a bending test on bamboo.

33.10 CONCLUSIONS – TIMBER CONSTRUCTION

33.10.1 SPECIFICATION

- Choose your species carefully. Pale hardwoods are often less durable than softwoods. Beware of exotic species, such as European species grown in New Zealand.
- Always specify a stress graded strength class that has been tested for strength. If not, you will get stress grade rejects.
- Ensure that your board products are adequate for the exposure. Note that marine plywood is more durable than exterior grade.
- Check that your source is sustainable.

33.10.2 DESIGN

- Timber is "anisotropic," that is, the strength and modulus is different on each of the three axes.
- Moisture cycles caused by central heating will cause high creep.
- Rot (even "dry rot") cannot occur if timber is completely dry.
- Detailing should ensure that:
  - Joints are protected from water penetration as they open, when the timber moves.
  - End grain is not exposed.
  - Contact with the ground is minimised.
  - Cutting after preservative treatment is minimised.
  - Rainwater is easily shed from all surfaces.
  - Joints with masonry, concrete, or steel are well sealed.
  - For internal timber, ventilation is essential, for example, ridge and eaves vents in roofs, where a felt is used, and air bricks to under-floor voids.

33.10.3 **CONSTRUCTION**

- Check all timber as it is being unloaded:
  - Are the stamps (e.g., for marine ply) correct?
  - Does it look poor (e.g., full of knots)? As with all materials – if they look wrong, they probably are wrong. Instances of incorrectly certified timber have occurred.
- Timber is seasoned to a moisture content. Try to get wood that has been seasoned to the moisture content it will have in use. If it is for interior use, do not let it get wet again (prime it as soon as possible).

TUTORIAL QUESTIONS

1. The figure below shows the load versus displacement output for a timber sample 21 mm wide by 19 mm deep, over a span of 250 mm. The equation on the graph is for a straight line fit to the region shown. Calculate the bending strength and modulus.

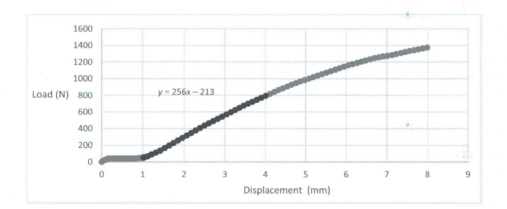

Solution:

Failure load = 1,380 N (from inspection of graph)

Strength = $3 \times 1{,}380 \times 0.25/(2 \times 0.021 \times 0.019^2)$ = 68.26 MPa   (33.2)

Gradient = W/x = 256 N/mm = 256,000 N/m   (from equation on graph)

Modulus = $256{,}000 \times 0.25^3/(4 \times 0.021 \times 0.019^3)$ = 6,942 MPa   (33.3)

2. The figure below shows the load versus displacement output for a timber sample ¾ in. square, and 0.83 in. long, tested in shear with the load on the ¾ in square ends. Calculate the failure shear stress and the shear modulus.

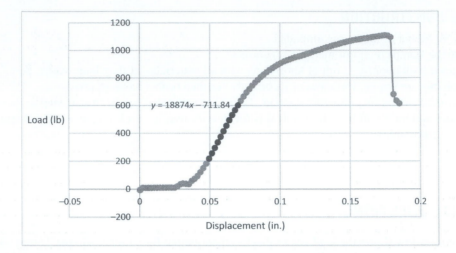

Solution:

Area resisting shear $= 0.75 \times 0.83 = 0.622$ in.^{2}

Load at failure $= 1,100$ lb

Stress $= 1,100/0.622 = 1,768$ psi   (2.4)

Gradient of graph $= 18,874$ lb/in.

Modulus $= 18,874 \times 0.75/0.622 = 22,758$ psi   (2.6) and (2.8)

NOTATION

b Breadth (m)
d Depth (m)
E Young's modulus (Pa)
S Span (m)
W Load (N)
x Deflection (m)
ε Strain
v Poisson's ratio
σ Stress (Pa)

NOTATION SUBSCRIPTS

L   Longitudinal, i.e., parallel to the grain
T   Tangential to the grain
R   Radial to the grain

MASONRY

CHAPTER OUTLINE

34.1 INTRODUCTION

The term "masonry" refers to construction with "masonry units," such as bricks, blocks, or stone bonded with mortar. The production and use of mortar is described in Section 28.2. Masonry is not generally used as the main load bearing element in large new structures, and its primary function is to provide an attractive, durable, and weatherproof exterior skin, or internal partitioning, in buildings. However, many older masonry structures exist, and these frequently require structural analysis and maintenance (Fig. 34.1).

34.2 CLAY BRICKS
34.2.1 MANUFACTURE

There are many types of clay, and each will give particular characteristics to the bricks. Some clays will need water added to them to be workable.

Clay is formed into shapes by various processes, such as:

- Semidry process. Uses clay with moisture content of approximately 10% that is ground and pressed into moulds.
- Stiff plastic process. Clay with about 15% moisture is extruded. This is often used for engineering bricks.
- Wire cut process. Clay with about 20% moisture is extruded, and cut to thickness with tensioned wires.
- Soft-mud moulding. Uses clay with up to 30% moisture. The moulds are sanded to prevent sticking. This process is used for handmade bricks.

The bricks are then fired in a kiln at 900–1200°C (1650–2200°F). The stages are:

- 0–100°C (32–212°F) – free water is lost
- 100–200°C (212–392°F) – weakly bound water is lost
- Up to 1200°C (2200°F) – vitrification

FIGURE 34.1 Students Below the Pontcysyllte Aqueduct, Built in 1805 to Carry the Llangollen Canal Over the River Dee Valley in Wales

The piers are masonry, and the trough is cast iron.

In some types of brick, there is a substantial organic content in the clay, and this burns, and saves a considerable amount of energy, and gives a non-uniform, visually attractive finish.

34.2.2 STANDARD SIZES

The length of a brick is twice the width plus one joint, or three times the height plus two joints. This enables upright bricks, and header bricks to be built into a wall without cutting them (Fig. 34.2).

FIGURE 34.2 Length/Width/Height Ratios for Bricks

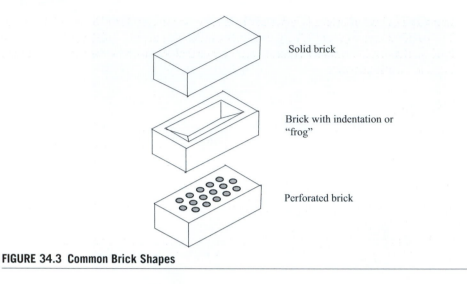

Solid brick

Brick with indentation or "frog"

Perforated brick

FIGURE 34.3 Common Brick Shapes

Metric bricks are $215 \times 102.5 \times 65$ mm. The mortar joint in metric brickwork is 10 mm. Metric brickwork is built with four courses in 300 mm vertically, and four bricks in 900 mm horizontally.

Imperial bricks are $7.625 \times 3.625 \times 2.25$ in. Thus, United States imperial brickwork has three courses to 8 in. vertically, and one brick to 8 in. horizontally.

The size of bricks varies in different countries, and many special sizes and shapes are also available.

34.2.3 SHAPES

Shapes of bricks are shown in Fig. 34.3.

The reasons for providing the indentations (frogs) and perforations are as follows:

- They assist in forming a strong bond between the bricks.
- They reduce the effective thickness of the brick, and hence the firing time.
- They reduce the material content, and hence the cost and weight.

Most specifications call for bricks to be laid "frog up." Laying "frog down" may be faster, will use less mortar, and will give a weaker structure.

34.2.4 TYPES

The three main types of brick are:

- Commons. General purpose bricks.
- Facings. With some or all faces good for appearance.
- Engineering bricks. With high strength, and often low permeability.

The compressive strength of clay bricks can vary over a very wide range, from 4 MPa to 180 MPa (0.6–26 ksi). Facings are normally in the range 12–120 MPa (1.7–17 ksi), and engineering bricks are generally in the range 50–70 MPa (7–10 ksi).

Note that the strength required of bricks is generally low, for example, a 50 m (165 ft.) high wall of bricks, with a density of 2000 kg/m^3 (3400 lb/yd^3), will only produce a stress of 1 MPa (145 psi) at the base. The strength of brickwork is normally controlled by other factors, such as the mortar strength, the bond strength, and the workmanship.

34.2.5 APPEARANCE

This is important in facing bricks, and the correct colour brick can command a very high price.

Red is the typical colour, and comes from iron oxide. Red–purple multicoloured stocks are made from "soft mud" incorporating organic material. The word "stock" is used for bricks from one region with a specific appearance.

White bricks are made from fireclay and chalky clays.

Yellow bricks are made from brickearth and chalk. Many traditional buildings in London were built with these.

Blue bricks are common as engineering bricks. The colour comes from iron compounds.

Other colours and finishes, such as sand facing, are often applied to the face only. If a faced brick is cut or damaged, the finish is lost.

34.2.6 TESTING FOR STRENGTH

Bricks may be tested for strength, using methods similar to those for concrete. When testing strength of bricks:

- The treatment of frogs should simulate end use, that is, they should be filled with mortar.
- Strength varies with moisture content. Bricks should be saturated for testing.
- A minimum of 10 randomly selected bricks should be tested.
- Strength generally correlates with density (if correct firing has been used).

34.2.7 WATER ADSORPTION

Adsorption of water by bricks is an indicator of durability, but not a very reliable predictor of it. Adsorption varies from 3% to 30%, measured on a 5-h test with the bricks immersed in boiling water. Some adsorption is desirable because good suction properties are essential for good adhesion of mortar and rendering.

34.2.8 EFFLORESCENCE

This occurs when salt is brought to the surface by water, and deposited by evaporation (Fig. 34.4). The salts are generally magnesium or sodium salts or, less frequently, calcium, or potassium.

- The problem is mainly visual, and will normally disappear with time, as the salt washes out. It may be stopped by keeping the brickwork dry or exhausted by regular washing.
- Crypto-efflorescence occurs below the surface, in underfired bricks, and causes crumbling.
- The source of the salt may be external (soil or water), but is usually internal.
- Efflorescence from concrete or mortar is different, and is caused by leaching of lime. A wall may have both salt efflorescence from the bricks, and lime efflorescence from the mortar.

FIGURE 34.4 Severe Efflorescence on New Brickwork

- If the mortar is not sulphate resisting, it may be attacked by sulphates from the efflorescence from the bricks.
- Treatment with a water repellent, such as a silane, will help.

34.2.9 CHEMICAL RESISTANCE

Bricks are generally resistant to chemicals with which they come into contact. Some more porous bricks decay in polluted atmospheres. Engineering bricks may be made substantially acid resisting (but note that cement mortar is not resistant to acids).

34.2.10 FROST RESISTANCE

Some bricks are suitable for internal use only. They must be protected from frost on site, in winter, and protected from damp in the structure. Ordinary grades of bricks are suitable for most walls where they are protected by a roof. Special or engineering grade bricks are required for parapets, exposed external walls, and paving (Fig. 34.5). The laboratory tests for frost resistance have a poor reputation for reliability; evidence of successful applications is considered better.

34.2.11 MOISTURE MOVEMENT

Bricks leave the kiln completely dry, and start to take up moisture. This causes expansion of 0.1 or 0.2%. It is therefore recommended that bricks should not be used for 7 days after firing. This may cause delays because brickworks often hold virtually no stockpiles and deliver bricks very quickly after manufacture.

34.3 CALCIUM SILICATE BRICKS

To make calcium silicate bricks, silica sand is mixed with high calcium lime at a sand–lime ratio of 10 or 20. The mix is then compressed into moulds and "autoclaved" at about 170°C (340°F) for several hours. Some gel, similar to the calcium silicate hydrate gel of the type that is formed by cement, is formed, and this bonds the sand particles together. The main properties of these bricks are:

- Good regularity, smooth faces, and sharp corners.
- Low salt content – little efflorescence.
- Fairly high moisture movement.
- Wide variety of strengths.
- Durability similar to concrete. May deteriorate in polluted sulphur-containing environments.

34.4 CONCRETE BRICKS

These are popular in areas where there is no clay to make clay bricks. Their appearance is similar to other bricks (the colour is provided with pigments). Their manufacture and properties are similar to other concrete elements. On site, they are more difficult to cut and less pleasant to handle than clay or calcium silicate bricks (Fig. 34.5).

34.5 CONCRETE BLOCKS

34.5.1 BLOCK SIZES

Concrete blocks typically occupy the volume of about six bricks, and come in a wide variety of shapes, and sizes. The large sizes can be laid without problems because they have a far lower density than most bricks.

FIGURE 34.5 Brick Paving with Concrete Bricks

FIGURE 34.6 Production of Aggregate Concrete Blocks

These are dense blocks made with waste quarry fines as aggregate. The blocks are made with semidry concrete, and compacted with intense vibration and pressure, and immediately removed from the moulds.

34.5.2 AERATED CONCRETE BLOCKS

To make these, a binder is mixed with aluminium powder that gives off hydrogen gas in contact with water that forms a mass of small bubbles in the mix that is subsequently autoclaved to give a lightweight product. The binder is made with combinations of OPC/PFA/GGBS/lime, and silica sand (that reacts at autoclave temperatures). These blocks are always solid, but cannot be used externally.

34.5.3 AGGREGATE CONCRETE BLOCKS

These are normally made with no-fines concrete, with lightweight aggregate (e.g., sintered PFA). They may be cast into all shapes, often with hollow cores that may be filled with expanded polystyrene for insulation. They come in many grades, varying from strong, dense durable grades to very light blocks, for internal partitions (Fig. 34.6).

34.6 NATURAL STONES
34.6.1 APPLICATIONS OF DRESSED STONE

These are used generally for decorative facades in new work or restorations. The common types are:

34.6.2 GRANITE

This is very expensive to produce because the parent rock has no bedding planes to assist with cutting. It is very durable, even in quite thin panels; polishes well and is almost self-cleaning.

34.6.3 **SANDSTONE**

This is relatively easily worked into ornamental shapes. It will not polish, and is not durable in polluted atmospheres. It turns black in polluted air, but may be cleaned if the air quality improves. There is a wide range of porosities, and stone with different porosity should not be mixed because the less porous will shed water onto the more porous.

34.6.4 **LIMESTONE**

Very many grades of limestone are available (e.g., Portland stone). Some grades are durable, but others are not. It may be polished to look similar to some granite.

34.6.5 **MARBLE**

This is generally expensive and used for interior applications.

34.6.6 **RECONSTITUTED STONE**

This is concrete made with cement and ground stone.

34.7 **ROOFING TILES**

Tiles are made from clay or concrete. The important properties are frost resistance and appearance. Frost resistance of clay tiles is measured by water adsorption. In older roof structures, the tiles are flat and are only held in place with nails, and the durability is limited by the life of the nails (Fig 34.7).

These ridge tiles are fixed with metal clips and have metal vents below them. This is a dry-hip system

These hip tiles are bedded in mortar

FIGURE 34.7 Clay Tiles

34.8 SLATES

These are expensive, but generally durable when used for roofing. Attention should be paid to the durability of the nails required to hold them. Resin-bound synthetic replacements are available. Asbestos was used for artificial slate, but asbestos products are no longer sold due to the health hazards. Care should be taken to identify these on older structures.

34.9 MASONRY CONSTRUCTION DETAILING
34.9.1 THE IMPORTANCE OF MASONRY DETAILING

Masonry is frequently specified because of its attractive appearance. This is only achieved with good detailing, and it is the responsibility of the designer to detail where every individual masonry unit will be positioned. Simply indicating an area to be built up with masonry is not sufficient. All windows and doors should, if possible, be located with a number of complete bricks between them, see Fig. 34.8 (there is often a facility in design software to check this).

34.9.2 BRICKWORK BOND

Some different brickwork bonds are shown in Fig. 34.9. The bond should be specified for all brickwork.

FIGURE 34.8 Well Planned Brickwork

The windows have exactly five bricks between them and are 12 courses high.

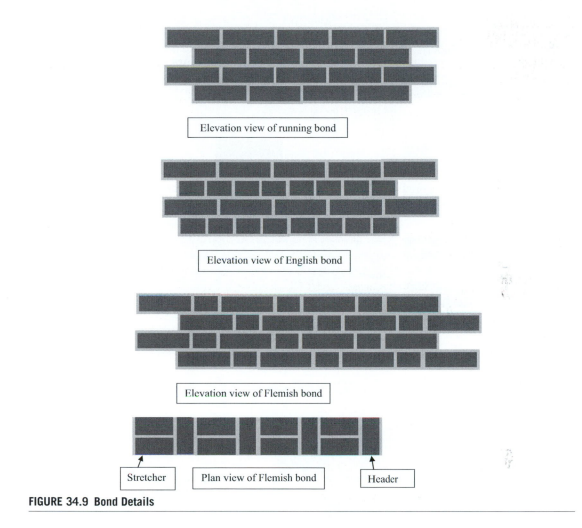

Elevation view of running bond

Elevation view of English bond

Elevation view of Flemish bond

Stretcher　　Plan view of Flemish bond　　Header

FIGURE 34.9 Bond Details

34.9.3 POINTING DETAILS

Figure 34.10 shows pointing details. A detail should always be specified. A simple detail should be used, unless the work can be viewed from very close up.

34.9.4 PROTECTION FROM DAMP

Even if the masonry units are of a suitable grade for exterior use, masonry walls should be designed to minimise the amount of water on them.

Figure 34.11 shows a typical detail for a window sill. As with other materials, such as concrete and timber, the detail should always minimise the amount of water running down the face.

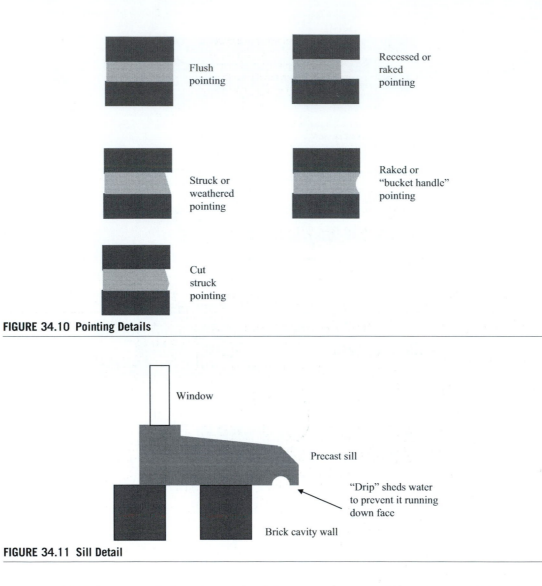

FIGURE 34.10 Pointing Details

Flush pointing

Recessed or raked pointing

Struck or weathered pointing

Raked or "bucket handle" pointing

Cut struck pointing

Window

Precast sill

"Drip" sheds water to prevent it running down face

Brick cavity wall

FIGURE 34.11 Sill Detail

Figure 34.12 shows a typical brick wall detail. Low porosity (normally engineering) bricks should be used below dpc. In older structures, the dpc itself was formed with courses of engineering bricks laid in a cement mortar. Any rainwater that penetrates the outer skin will run down the inner face, and out through the weep holes (Fig. 34.13).

34.9.5 FEATURES

If a large panel of brickwork is built in a high visibility location, every minor blemish will show up. This problem may be avoided by forming features in the brickwork, such as patterns formed with bricks of different colours, or headers that project from the wall. These will draw the eye from the imperfections (Fig. 34.14).

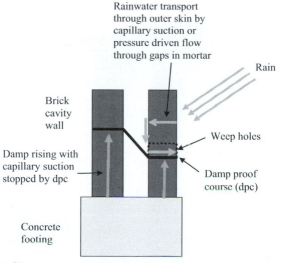

FIGURE 34.12 Brick Wall Detail

FIGURE 34.13 Weep Hole with Plastic Fitting in it to Keep it Clear

34.9.6 THERMAL PERFORMANCE

The thermal conductivity of a cavity wall was discussed in Section 4.5. The cavity was traditionally left empty, but it is now normally filled with insulation. Specifying a greater width of cavity will clearly improve the performance of the wall. If buildings with solid walls are being refurbished, additional insulation may be added to the inside of the walls (but this makes the rooms smaller), or by overcladding the outside (but this may spoil the appearance).

FIGURE 34.14

A simple pattern feature built into brick wall diverts the eye from the bricked up opening (bottom left) and colour change (top courses).

34.9.7 WALL TIES

The structural performance of a cavity wall depends on adequate ties being provided between the two skins, in order to prevent buckling. These are normally metal, and are laid into the bed joints at approximately one per square metre of wall. They are designed to make water drip off, in order to prevent it running across them (Fig. 34.15). For high thermal performance, plastic wall ties with low thermal conductivity may be used.

It was noted in Section 28.2.5 that the mortar is expected to be resistant to carbonation, so it maintains an alkaline environment to prevent corrosion of wall ties. They are normally galvanised, but if they do fail, they expand and push the wall apart. They must then be cored out and new "friction grip" ties installed in holes drilled in from the outside.

34.10 MASONRY CONSTRUCTION SUPERVISION

34.10.1 SAMPLE PANEL

At the start of the work, a sample panel should be built. This may show up problems with colour, type of pointing, efflorescence, etc. In the event of any dispute over the quality of the work, this may be used as a reference for the required standard (Fig. 34.16).

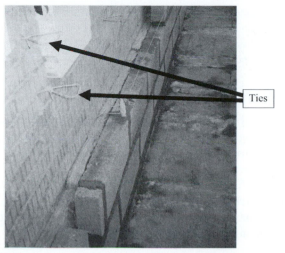

FIGURE 34.15 Cavity Wall Ties

The wall is built with a brick outer skin, block inner skin, and insulation between.

FIGURE 34.16 Sample Panel Built to Include Proposed Features of Finished Work

34.10.2 LINE AND LEVEL

Consider the consequences if you are responsible for the construction of a long wall, with one gang of bricklayers working from each end, and one gang is building it up in 105 courses, and the other has 106 courses. Ideally, there will be an expansion joint in the middle, but even then it will look bad. Similarly, problems will occur if the window locations have been correctly designed to fit with whole bricks, and the bricklayers have laid the joints slightly tight, and fitted in an extra brick. To avoid these problems the engineer must ensure that every brick is laid to the correct line and level. Level may be achieved by accurately levelling the base and using a "gauge rod" – a piece of wood marked out with the height of each course. When you stand at the end of a wall, and sight down one of the bed joints, it should be perfectly straight and level.

34.10.3 COLOUR VARIATIONS

It was noted in Section 28.2.5 that achieving an attractive, uniform appearance for mortar requires accurate proportioning for mixing. If necessary, gauge boxes should be used. These are wooden boxes made to the required volume for each material (e.g., sand, cement) that are filled precisely for each batch. The water should also be accurately measured.

Bricks will vary in colour from batch to batch. This will cause unsightly lines in the brickwork, unless the bricks from each batch are mixed by working from two pallets at a time, and never finishing one load of bricks before the next is delivered. Thus the colour change will take place gradually, over several courses, rather than along a single line between two courses.

34.10.4 ACCESS AND SUPERVISION

Bricklayers cannot build good brickwork from a badly positioned scaffold. Laying bricks "overhand" (from the back of the wall) is possible, but difficult and slow. It is the responsibility of the engineer to ensure that the bricklayers have a well-positioned, safe scaffold, and also that the completed work is adequately protected from the effects of nearby concreting operations, etc. The completed work should be regularly inspected to ensure that the cavity is kept clean because mortar may form a bridge across it, and cause damp penetration to the inner skin.

34.11 CONCLUSIONS – MASONRY CONSTRUCTION

34.11.1 SPECIFICATION

- Try to find a building built with your chosen brick, and check it for efflorescence, etc.
- Use low porosity bricks below dpc, and in exposed areas.
- If in doubt, specify SRPC in the mortar.

34.11.2 DESIGN

- Try to avoid large plain panels. Always put in a feature to divert the eye from colour variations.
- Avoid unnecessary exposure to frost. For example, put tiles to protect the top of a wall.
- Work out where each brick goes, and try to make each element a complete number of bricks.
- A struck or recessed joint may be more expensive than a bucket handle joint, and the difference may not be visible (depending on viewing distance).

34.11.3 CONSTRUCTION

- Try to build a sample panel in an exposed position, well before the start of construction.

- Always plan out the brickwork, and set out the height of the courses and the horizontal positions. Ensure that a gauge rod is in use, and check that it is accurate.
- Minimise colour variation by ensuring that the mortar is batched accurately, and the bricks are mixed, if they have variable colour.
- Always open one pallet of bricks before the previous one is finished, in order to avoid visible changes of colour when the new one is started.
- Good, durable masonry is a product of good management, such as access, quality control, protection of completed work, etc.
- Masonry normally fails in buckling. In cavity walls, the cavity ties are vital to the structure.
- In cavity walls, ensure that the cavity is kept clean and tray dpcs are not damaged, or water penetration will result.

PLASTICS

35

CHAPTER OUTLINE

35.1 INTRODUCTION

It is the wide variety of forming processes that has caused the great popularity of plastics. A typical example is injection moulding. In this process, granules or powder are injected into a mould where they set to give an accurately formed shape. The mould is expensive, but the production costs are low. The most common raw material for plastic is oil, but the quantities used are low, relative to those used for fuels – so, increases in the price of oil have not reduced the use of plastic.

Plastics are divided into two types:

- Thermoplastics always soften when heated.
- Thermosetting plastics polymerise and set when mixed with a "hardener," and will not soften when heated. Setting is accelerated by catalysts, heat, pressure, or even γ radiation.

35.2 TERMINOLOGY

- Polymer: a material formed of large molecules that are built up (polymerised) from a large number of small molecules (monomers). The usual (but not only) example is the organic polymers that include plastics, and natural polymers such as rubber.
- Organic materials: these are materials originating from living organisms. All materials containing carbon are defined as organic in science, but the term has now been adopted for naturally grown food.
- Plastics: this term is used for a range of organic materials. The term "plastic state" of concrete is applied before it sets, and relates to the generally low modulus of elasticity and high creep of plastics.

35.3 MIXING AND PLACEMENT

For specialised applications, thermosetting plastics may be mixed on site. Two common types are epoxies and polyesters. Both of these are made by mixing a resin and a hardener. For epoxies, the resin and hardener are supplied separately, and must be mixed in the correct proportions. If the proportions are wrong, not all of the material will react, and the strength will be reduced. For polyesters, the resin and hardener are supplied premixed, but they will not react for several years. To make them set, a small amount of catalyst is added. Adding higher doses of catalyst gives a faster set. Polyesters may be recognised by the characteristic "fibreglass" smell.

The following precautions should be taken:

- The components are toxic – always wear gloves. The organic catalysts used with polyesters are particularly carcinogenic. The dust arising from cutting/abrading the hardened resin is carcinogenic. The odourless vapour that evolves during curing of epoxy is toxic, so it should always be used in a well-ventilated area.
- To achieve a low permeability, the resins must be cured correctly, in very dry conditions (the opposite of the conditions required for curing concrete). A white "bloom" on the surface indicates the presence of moisture during curing.
- The setting reactions are exothermic (especially polyesters). If use is delayed after mixing, the set may be delayed by placing the material in a shallow metal container so the heat can dissipate.

35.4 PROPERTIES OF PLASTICS
35.4.1 STRENGTH AND MODULUS

For most plastics, strengths (and, in particular, strength to weight ratios) are high for short term loading, but creep is high and the modulus of elasticity is low. Some polymer fibres, such as aramid (aromatic polyamide), have a very high tensile strength.

35.4.2 DENSITY

The density of plastics is generally in the range 900–2200 kg/m^3. This means that some plastics will float on the surface of water, and others will float at different depths below it, or sink to the bottom. Great care should be taken to avoid polluting the sea with waste plastic because this range of densities means that it will be present throughout the ecosystem, and present a hazard to all marine life.

35.4.3 THERMAL PROPERTIES

The thermal conductivity of plastics is similar to wood, but the thermal capacity is higher. The coefficient of thermal expansion is often high.

35.4.4 RESISTIVITY

Generally, plastics are insulators.

35.4.5 PERMEABILITY

Many plastics are more permeable than they look. Polythene sheet actually permits quite high moisture transmission. Many plastics (e.g., polytetrafluoroethylene (PTFE) that is commonly used in lab apparatus) are highly permeable to gases.

35.5 MODES OF FAILURE (DURABILITY)

35.5.1 BIOLOGICAL

Being organic, many plastics are nutritious to some forms of animal/insect/fungus, etc. Biocides may be added during manufacture.

35.5.2 OXIDATION

Oxidation causes embrittlement and loss of strength. It is generally slow, in the absence of heat or sunlight. Antioxidants may be used.

35.5.3 SUNLIGHT

Most plastics are damaged by long-term exposure to ultraviolet light. The process is known as degradation, or photoembrittlement. It may be reduced by adding an ultraviolet absorber, for example, carbon black, or silica fume. Pigments in plastics often fade in sunlight, giving loss of colour.

35.5.4 WATER

In permeable plastics, there may be loss of some components through leaching. Osmotic pressure from moisture ingress may also cause surface spalling, known as osmosis.

35.5.5 LEACHING

There has been a specific problem with polyvinylchloride (PVC) leaching plasticiser when in contact with insulating materials, such as expanded polystyrene. Electrical wiring in loft spaces has been found to have premature embrittlement of the PVC insulation.

35.6 TYPICAL APPLICATIONS IN CONSTRUCTION
35.6.1 PLASTICS FOR GLAZING

Clear plastics may be used in windows, in place of glass. They have a far higher coefficient of thermal expansion than glass, and allowance must be made for this with a suitable flexible sealing system. They also have far poorer resistance to scratching, so in the long term the clarity may become poor. The cheaper material is acrylic (e.g., "Perspex"), but this has poor impact resistance. Polycarbonates are more expensive, but have excellent impact resistance, so they are "vandal-proof," and are used in bullet-proof laminates.

35.6.2 POLYTHENE (POLYETHYLENE)

Polythene is used for many applications, and is available in high-density polyethylene (HDPE) form that is less permeable. HDPE is used as a damp-proof course (dpc) membrane, for waste containment in landfills, and also as a protection membrane for reinforced concrete, in contact with highly saline or sulphate-bearing groundwater.

Polythene sheet is available in different thicknesses. A 1000-gauge polythene is 0.01 in. (0.254 mm) thick. It is typically used as a curing membrane above and below concrete slabs.

35.6.3 POLYMER GROUTS AND CONCRETE

These are made in the same way as cementitious grouts and concrete, but with the cement matrix replaced with a polymer, such as epoxy or polyester.

Polymer grouts are used for similar applications to cementitious grouts (see Section 28.4), but are more expensive and have far greater bond strength, and are normally used in applications where far lower quantities are required.

Figure 35.1 shows a typical polymer grout fixing to concrete. This is used when the fixing is required to an existing concrete element, unlike the detail shown in Fig. 28.2 that is for a fixing cast in at the time of construction. The polymer grout is used because it has far superior bond and shear strength to the cementitious grout, that can only be used when a plate is cast in to anchor the bottom of the bolt. The additional cost and time required for the polymer material is thus justified.

Another typical application for polymer concrete is in thin sections, for patch repairs or bridge overlays.

35.6.4 POLYMERS IN CONCRETE

Organic polymers are used for many purposes in concrete. The most common ones are admixtures such as plasticisers.

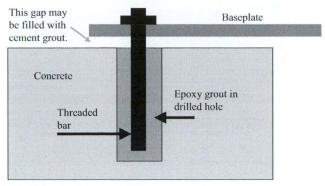

FIGURE 35.1 Polymer Grout Fixing to Concrete

Polymer impregnated concrete is made by vacuum impregnating a monomer into hardened concrete that is polymerised inside concrete, with heat or γ radiation. It is used in factory produced precast units.

Polymer modified concrete is made by adding catalysed polymer into ordinary concrete at the mixer, and polymerises *in situ*. It is used for concrete repairs with thicknesses of 50–100 mm, overlays for bridge decks, etc.

Polymer reinforcement is always a composite, and is discussed in Chapter 38. Steel reinforcement may be epoxy coated (see Section 26.2.6).

35.6.5 GEOTEXTILES

There are large numbers of different types of geotextiles. They may broadly be divided into woven materials, made with a polymer thread, and nonwoven that are either moulded to form a grid, or made up of random fibres, with some form of adhesion between them. In general, the woven types are more expensive.

Figure 35.2 shows a geotextile filter being used to keep the topsoil out of the granular material, above a permeable drain. Since the filter will not be subject to high stresses, a nonwoven material

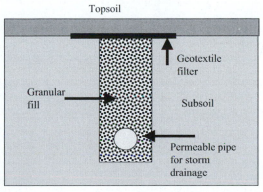

FIGURE 35.2 Geotextile Filter

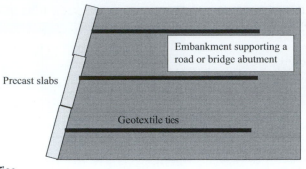

FIGURE 35.3 Geotextile Ties

is adequate for this application. For short-term applications, a natural fibre might be used, but the additional cost of a polymer is justified to give increased durability.

Figure 35.3 shows geotextile ties fixing precast slabs to the side of an embankment. These must carry a tensile load, so a woven material would be suitable. The alternative is steel ties that would not suffer from creep, but they would be at risk of corrosion.

Figure 35.4 shows geotextile use to stabilise a driveway. This application is common in both permanent, and temporary works.

35.6.6 PLASTIC PIPES

These are almost universal for above ground waste, and are increasingly used for below ground, and supply applications. They are resistant to chemicals, and good at accommodating movement during service. Water suppliers have very strict controls on the types of polymer that may be used for potable

FIGURE 35.4 Typical Use of Geotextile to Stabilise a Driveway Before Placement of Block Paving

water supply because the polymerisation is not normally completed during manufacture, and residual monomers can be leached out, and these may be toxic.

35.6.7 FORMED PRODUCTS

Plastics will be used throughout the fitting out of a building, in windows and doors, electrical fittings, and numerous other applications. They are chosen for interior application because of their low cost of manufacture, and for external applications for their durability.

35.7 CONCLUSIONS

- Thermoplastics soften when heated. Thermosetting plastics do not.
- The term organic materials technically means all materials containing carbon, but has been adopted to mean naturally sourced materials.
- Plastics often have a deceptively high permeability.
- Plastics degrade in the presence of oxygen and sunlight.
- Polycarbonate is stronger than acrylic, for glazing applications.
- Polymer grouts have high adhesion and shear strength.
- Geotextiles have a wide variety of applications in construction.
- Plastic for potable water supply pipes should have low leaching properties.

GLASS

36

CHAPTER OUTLINE

36.1 INTRODUCTION

36.1.1 APPLICATIONS OF GLASS IN CONSTRUCTION

Glass is used for construction in three different forms:

Glass for glazing. This must have sufficient strength to resist wind loads, but is not normally expected to carry structural loads.

Glass fibres. These are drawn from molten glass as very fine fibers. These are used for reinforcing composites.

Glass wool. This is random masses of fine fibres, and is used for insulation.

36.1.2 STRENGTH

The strength of a glass unit is largely determined by the effect of surface imperfections. This is the reason why glass fibres are so strong; they are thin, so they have few imperfections. Glass fibres

413

can have strengths up to 3000 MPa (435 ksi), new glass for glazing is about 200 MPa (30 ksi), but older glass will go down to about 20 MPa (3 ksi). For this reason, old glass is more difficult to cut, and more likely to break. In practice, the strength is not used to calculate loadings, and there are tables based on past experience that are published to give the thickness necessary for given window sizes.

36.1.3 RAW MATERIALS

Glass is basically silica (silicon oxide) that may be obtained from silica sand. In its pure form, it has a melting point of 1700°C (3100°F) that is too high for economic manufacture, so compounds such as sodium carbonate are added to reduce it. This makes soda glass that is water soluble, so calcium carbonate is added to stabilise it.

A typical composition of glass for glazing is:

SiO_2	Silica	75%
Na_2CO_3	Sodium carbonate	15%
$CaCO_3$	Calcium carbonate	10%

Other compounds are also added, for example, manganese dioxide, to remove the green colour caused by iron in the sand.

36.2 GLASS FOR GLAZING
36.2.1 MANUFACTURE

Three basic types of glass are manufactured:

- Plain flat glass: most plain glass is float glass. In this process, liquid glass is cooled to give a viscosity sufficiently high for forming, and it is then drawn across the surface of molten tin. This method may be used to produce very flat glass in large quantities.
- Textured, patterned and wired glass: the rolled glass process is used for the manufacture of these types. The glass is drawn in a horizontal ribbon on rollers. The texture or pattern may be imprinted onto it, and wire mesh may be cast into it. However, if flat glass is required from this process, it must be ground.
- Laminated glass: this is made with two or more sheets of glass (normally float glass) that are bonded together with layers of plastic between them. The plastic is flexible, so an impact load will not normally cause the whole composite to break through.

36.2.2 CUTTING

Glass may be cut to size by creating a surface imperfection, and then causing it to propagate through the solid. In practice, this involves scratching the surface with tungsten, or preferably diamond, tapping along the cut to spread the fracture, and then bending the sheet to break it. Laminated glass with two laminates may be cut by scoring both sides, in order to get fracture in both of them.

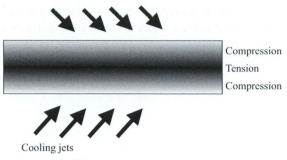

Compression

Tension

Compression

Cooling jets

FIGURE 36.1 The Toughening Process for Glass

36.2.3 TOUGHENING AND ANNEALING

Rolled glass must be annealed to relieve stresses. In this process, the glass is heated until it is sufficiently plastic for the internal stresses to be relieved, and then cooled slowly and uniformly, so no new stresses are developed.

Toughening involves heating glass uniformly, and then cooling the outer layers with air jets so that they then contract solidify. As the inner layers subsequently cool, the tendency to contract places them in tension, and the outer layers in compression (Fig. 36.1).

The properties of toughened glass are as follows:

- It is stronger in bending than plain glass because bending stresses one of the outer layers in tension, and it is initially in compression.
- If the material is broken, the stress distribution becomes unbalanced, and it shatters into small and relatively harmless fragments.
- For the same level of safety, it is cheaper and lighter than laminated glass.
- It cannot be cut, so it must be ordered to size. This has the advantage, however, that the stamp, which is put on every sheet, will always be there for checking.
- There are restrictions on dimensions of holes, near the edge of a sheet.

36.2.4 DURABILITY

Glass is intrinsically very durable. However, it is not resistant to alkalis (e.g., from runoff from concrete), and these may cause some surface damage, and resulting loss in light transmission.

36.2.5 INSULATION

Glass is an intrinsically poor insulator to both heat and sound. Double, or even triple glazing will improve both.

36.2.6 SOLAR HEAT GAIN

All objects radiate heat as electromagnetic radiation, all of the time, when they are at temperatures above absolute zero (see Section 4.9). Glass transmits wavelengths close to visible light (such as sunlight) well, but is a much poorer transmitter of longer infrared wavelengths given off by warm objects. Thus, sunlight will transmit energy into a room through glass, but the energy given off by the objects in the room will be prevented from escaping. This is the greenhouse effect, and causes excessive heating of some buildings.

The two main methods used to limit solar heat gain are:

- Tinted glass. This absorbs heat and re-emits some of it outwards. Care should be taken that the sheet of glass heats uniformly; if the perimeter remains cool, it may fail in tension. There are special glasses with low thermal expansion, but they must be very clearly labeled so that if they are broken, the replacement has the same properties.
- Partially reflecting glass. This has a thin film on the inner surface, which may be added during manufacture, or can be put on *in situ* in older buildings. The result is to give a finish that appears to reflect from one side only. However, no glass gives one-way vision under all circumstances, this will depend on the lighting on the "viewing" side being less intense than the other side.

The transmission properties of polycarbonates and acrylics (see Section 35.6.1) are very different from glass. Transmission is generally higher, and far more uniform, so buildings glazed with these materials do not have the greenhouse effect. It is interesting to note that greenhouses glazed with acrylics get almost as hot as glass ones, thus showing that the greenhouse effect is actually not very important in greenhouses.

Low-emissivity glass is deliberately designed to reflect the longer wavelengths. This is used in cold climates to retain the heat in buildings.

36.3 GLASS FIBRES

36.3.1 APPLICATIONS FOR GLASS FIBRES

Glass fibres are drawn from melted glass, and are used to reinforce concrete or polymers. "Fibreglass" is the term used for a polymer reinforced with glass. Glass reinforced polyester (GRP) is the most commonly used polymer composite (see Section 38.5).

36.3.2 TYPES OF GLASS FIBRE

Loose chopped strand is bundles of fibres typically 25–50 mm (1–2 in.) long. They may be mixed into concrete to give crack resistance. Glass reinforced cement panels can be less than 20 mm thick, so a fine mortar is used.

Chopped strand mat is bundles of fibres typically 50–100 mm (2–4 in.) long that are stuck together randomly with an adhesive to form a mat. This is then used as reinforcement in fibreglass, or glass reinforced cement panels.

Rovings (yarn) is bundles of fibre formed into a single long strand, supplied on a reel. This is used as reinforcement in non-ferrous reinforcing bars for concrete (see Section 26.2.6). It may also be used in fibreglass when a spray process is applied. Both liquid polymer and glass roving are fed to a spray gun, where a chopper cuts the roving into short lengths, before it is mixed with the resin at the nozzle, and sprayed onto the mould.

Woven roving is a fabric used to reinforce high quality fibreglass. It may be made with different quantities of fibres in different directions, to give higher strengths on a chosen axis.

36.3.3 SAFETY

The fine fibres of glass are hazardous if they are breathed in, so masks must be used. They will also irritate the skin, so gloves should be used.

36.3.4 DURABILITY

Glass is not durable in an alkaline environment, and the fine fibres will soon loose strength in an alkaline solution. If they are cast into a matrix, such as polycarbonate, this will protect them. However, if they are cast into concrete, they must be coated to provide protection, and even with this they will lose a large part of their strength with time.

36.4 GLASS WOOL

This is used for insulation. It has been noted before that glass is actually a poor insulator, but the fibres are very fine, and the material works by preventing the movement of air (convection). It is used in wall cavities, roofs, and floors.

36.5 CONCLUSIONS

- The strength of glass is determined by surface imperfections.
- Sheet-glass is cut by scoring the surface with tungsten or diamond.
- Glass is toughened by rapidly cooling the surface.
- Solar heat gain is caused by the unequal transmission of glass for different wavelengths or electromagnetic radiation.
- Glass fibres may be in the form of chopped strand or rovings.
- Woven rovings are used for high quality GRP.
- Glass fibre must be coated, if it is used in an alkaline environment, such as a cement matrix.

BITUMINOUS MATERIALS

37.1 INTRODUCTION

37.1.1 APPLICATIONS

Bituminous materials are used for road construction, roofing, waterproofing, and other applications. For the main application, which is road construction, the major concerns, as with concrete, are cost and durability.

37.1.2 DEFINITIONS

The terms used for these materials can be confusing, and tend to change with time and location. A web search is therefore likely to find a number of different terminologies in use.

- Binder. A material used to hold solid particles together, for example, bitumen or tar.
- Bitumen, asphalt. A heavy fraction from oil distillation. In North America, this material is commonly known as "asphalt cement," or "asphalt." Elsewhere, "asphalt" is the term used for a mixture of small stones, sand, filler, and bitumen, which is used as a road paving or roofing material. In view of this problem, the term "asphalt" is not used in this chapter.
- Tar. A viscous liquid obtained from distillation of coal or wood. It can be used as an alternative to bitumen, in many applications.
- Mastic asphalt. An adhering blend of bitumen and fine filler that is placed with trowelling.
- Asphalt mixture. A mixture of binder and fine and coarse aggregate.

37.1.3 SAFETY

When they are heated, binders will give off light solvent vapours that can be quite easily ignited to cause explosions. For this reason, vapours should be extracted from laboratories or, on site, an open flame may be used to burn them off as they form. The vapours are also carcinogenic, so good extraction from laboratories is essential.

If the binder itself catches fire, the flame will be spread by water, so a suitable fire extinguisher should be used.

37.1.4 PRODUCTION

Bitumen is made from crude oil by distillation. The crude oil is vaporised and then condensed in a distillation tower, with the lightest components (those with the lowest boiling points) condensing nearest the top. The main components are:

Top: Petrol (gasoline),
      Kerosene (paraffin),
      Diesel oil,
      Lubricating oil.
Base: Base bitumen.

By mixing base bitumen with lighter oil, different grade bitumens are produced. If increased demand for lighter oils means that more bitumen is being produced that is required, it can be "cracked" to break up the large molecules, and produce more light oils. This increases its value.

Tar is produced when gas is produced from coal. The coal is heated to high temperatures in the absence of air, giving off gas and crude coal tar, and leaving behind coke that is used in blastfurnaces. Crude tar may be distilled in the same way as crude oil, to produce different fractions.

37.2 BINDER PROPERTIES

37.2.1 VISCOSITY

All bituminous materials are viscous (see Section 8.2 for definition). The range of viscosities is, however, very large, and ranges from liquids to solids that have short-term elasticity. Increasing viscosity will increase cost because higher temperatures are required to reduce it sufficiently for road construction. The viscosity may be increased by "air blowing" that makes the bitumen react with oxygen.

37.2.2 SOFTENING POINT

This is the temperature at which the binder softens to a predetermined point.

37.2.3 ADHESION

Bituminous materials adhere to clean dry surfaces. Tars resist "stripping" from stone in the presence of water better than bitumens.

37.2.4 DURABILITY

With time, binders will oxidise, polymerise, and lose light oil components when exposed to air and heat. All of these processes tend to make them harder and, thus, more prone to cracking. They are virtually impermeable, if well compacted, and intrinsically resistant to plant growth. All binders are softened by high temperature and solvents. Softening with solvents occurs when a road accident results in fuel spillage. When this happens, the road surface often has to be replaced.

37.3 BINDER TESTING

37.3.1 PENETRATION TEST

The viscosity of binders is normally measured with a penetration test. A 1 mm (0.04 in.) diameter needle is loaded with a weight of 100 g (0.22 lb), and the distance it penetrates into a bitumen sample in 5 s is measured (at 25°C, 88°F). A bitumen is referred to as 70 pen if the penetration is 7 mm. Binders that can be tested in this way are referred to as "penetration grades."

37.3.2 ROTATIONAL VISCOMETER

This is similar to the device described in Section 22.2.8 for use with wet concrete. However, in this case it has a cylinder, which is rotated inside a second coaxial cylinder containing the bitumen that is sheared between them. The viscosity is calculated from equation (8.1).

37.3.3 SOFTENING POINT TEST

To measure the softening point, a small sample is melted, cast in brass ring, cooled, and then progressively reheated until a 10 mm (0.4 in.) diameter steel ball placed on top of it falls through.

37.4 BINDER MIXTURES

37.4.1 CUTBACKS

This term is used to describe mixtures of a binders and light volatile oils. They have low viscosity at low temperatures, until the volatile oil evaporates. They are the basis for cold rolled asphalt that does not require heating before use, and hardens in a few days after laying. There are significant environmental concerns about the release of the volatiles into the atmosphere.

37.4.2 EMULSIONS

An emulsion is a mix with water. When mixed with water, binders will generally settle out, so an emulsifier must be added to give a stable solution. Bitumen paints are made this way. The water evaporates, and the bitumen remains on the surface. Cold rolled materials based on emulsions do not suffer from the environmental problems caused by cutbacks because all that evaporates off is water, rather than the harmful solvents.

37.4.3 POLYMER MODIFIED BITUMENS

These are penetration grade bitumens with a small amount of added thermoplastic or rubber. These will generally make the properties more elastic and less viscous, so the durability and resistance to rutting, etc. is increased.

Rubber can be mixed with bitumen in the form of latex, sheet rubber, rubber powder, or ground tyre tread. The proportion of rubber is generally below 5%, and often below 0.5%. The resulting properties are:

- The viscosity and softening point are increased and penetration is decreased.
- The elasticity is increased.
- The sensitivity to temperature changes is decreased.
- The beneficial properties are lost with prolonged heating.

The increased elasticity and penetration resistance makes this material ideal for expansion joints and repairs in roads.

37.5 ASPHALT MIXTURES

37.5.1 CONSTITUENT MATERIALS

Binders are seldom used alone, and are usually mixed with aggregates. The four components of a typical mix are:

1. Coarse aggregate.
2. Fine aggregate.
3. Fine mineral filler, for example, limestone flour or cement.
4. Binder.

37.5.2 PROPERTIES

Adding aggregate to a binder has the following general effects:

- The cost is reduced.
- The strength is generally increased.
- If the mix has to be laid hot, the aggregate must also be heated (at considerable cost).
- It may fail due to loss of adhesion between the aggregate and the binder.

37.6 TESTING ASPHALT MIXTURES

37.6.1 PENETRATION TEST

This test is similar in principle to the penetration test for binders but on a larger scale. The steel pin is 6.35 mm (0.25 in.) diameter, and the load is 10 N/mm^2 (1.45 ksi).

37.6.2 COMPACTION TESTS

Samples may be compacted by falling weights or a gyratory compactor. The resulting volumes are represented schematically in Fig. 37.1. It may be seen that some of the binder is absorbed into the pores in the aggregate. This takes place over time, when the sample is heated. Laboratory samples must be "oven-aged" before compaction, to give time for this to happen.

In order to calculate the skeletal (net) volume and absolute density of the sample, it is necessary to take a separate part of the sample that has not been compacted, and break it up. It is then weighed dry

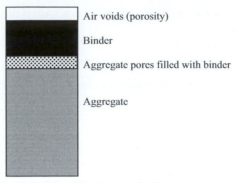

FIGURE 37.1 **Components of a Compacted Sample of Asphaltic Concrete**

and submerged, as described in Section 8.4. The process may be carried out in a vacuum chamber, so the sample is vacuum saturated to fill all the pores. This will give an absolute density using equation (8.4). This is also known as the theoretical maximum density because it is what would be achieved if compaction continued until no further voids remained.

After compaction, the outer dimensions of the sample may be measured to get the total volume, and it may be weighed to give the mass and, thus, the density. The voids in the total mix (VTM) is the porosity and is obtained from equation (8.5), using the absolute density obtained from the vacuum saturated sample.

Assuming the percentage of bitumen in the mix is known, this may be used to calculate the mass of aggregate. A sample of unused aggregate is weighed dry and submerged to give the specific gravity and, thus, the solid volume. The voids in the mineral aggregate (VMA) is the total sample volume minus the solid volume of aggregate, expressed as a percentage of the total sample volume.

Finally, the voids filled with binder (VFA) is the fraction of the VMA that has bitumen in it, that is, the part that is not porosity. This is obtained by subtracting the porosity (VTM) from the VMA, and expressing it as a percentage of the VMA.

These calculations are demonstrated in tutorial question 1.

A complication is the aggregate pores filled with binder shown in Fig. 37.1. When the aggregate is weighed dry and submerged, the water will not have penetrated the pores, so the VMA will not include the pores in the aggregate, and will be slightly low. The proportion of voids filled with binder, VFA, will also be low. If all of these pores (that could be measured by vacuum saturation) were included in VMA, and assumed to fill with binder during the oven-aging, this would increase VFA. However, this would now be an overestimate because, in practice, only about half the pores accessible to water are accessible to binder because it is much more viscous.

37.6.3 MARSHALL LOADING TEST

In this test, cylindrical specimens 63.5 mm (2.5 in.) long and 101.6 mm (4 in.) diameter that have been compacted in a falling weight compactor are heated to 60°C (140°F), and loaded in compression on their curved surfaces, at a constant rate of 50 mm/min (2 in./min) (Fig. 37.2). The maximum load (*stability*) and the deformation at maximum load (*flow*) is recorded.

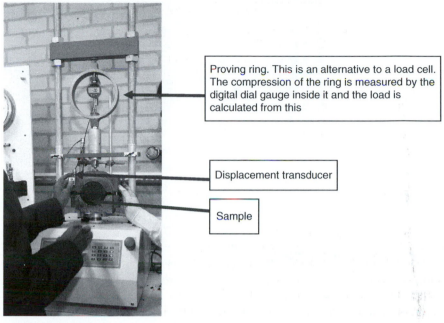

FIGURE 37.2 The Marshall Test

37.6.4 PERMEABILITY TESTS

The permeability of an asphaltic concrete used for road construction is important because, if water penetrates into the surface, it may cause stripping failure (loss of aggregate from the surface), frost damage, or cohesive failure leading to cracking. The permeability is measured as the flow under pressure, as shown schematically in Fig. 9.1.

37.6.5 DISSOLUTION OF BINDER

There are a number of tests that involve dissolution of the binder in solvents. The mix proportions and the nature of the binder may be determined.

37.7 MIX DESIGNS FOR ASPHALT MIXTURES

Mix design of asphalt mixtures is a compromise of cost, durability, flexibility, and skid resistance. Adding additional binder will decrease the void volume and increase flexibility and durability, but it will increase cost and reduce aggregate interlock, and binder extruded up to the surface by traffic loads will affect skid resistance. The mix design processes are therefore entirely guided by specifications that are based on experience.

Aggregate grading is discussed in Chapter 19. It is as important in asphalt mixtures, as it is in normal concrete made with cement.

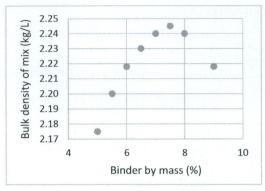

FIGURE 37.3 Bulk Density of Mix

The density in kilogrammes per litre is equal to the specific gravity.

37.7.1 THE MARSHALL METHOD

For this process, asphalt mixtures are made at several different binder contents, and tested using compaction and the Marshall loading test.

Figure 37.3 shows typical data for the bulk density of a number of mixes. At lower binder percentages, the density is reduced because the mix has poor workability, and is difficult to compact, leading to a high porosity. At higher binder percentages, the lower density of the bitumen (similar to the density of water) is significant.

Figure 37.4 shows the "aggregate density" – the aggregate volume divided by the total sample volume for the mixes.

The stability (maximum compressive load) from the Marshall loading test also rises to a maximum, and falls again as the binder percentage increases. The design binder content is obtained from an average of this maximum, and the maximum values of the two density plots. It is then checked against design minima for stability and flow, for different traffic loadings.

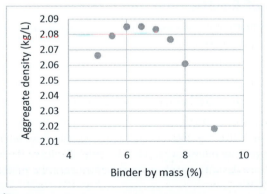

FIGURE 37.4 Aggregate Density

The density in kilogrammes per litre is equal to the specific gravity.

37.7.2 THE SUPERPAVE METHOD

The "superior performing asphalt pavement" method takes into account the expected environmental conditions. The design process has five distinct steps:

1. Selection of aggregate. This is based on the grading and shape/texture of the aggregate, and is specified for light, medium, and heavy traffic.
2. Selection of binder. This is based on a requirement for viscosity at the maximum and minimum expected pavement temperatures.
3. Design aggregate structure. Samples of asphalt mixture with different aggregate gradings and binder contents are compacted, using a gyratory compactor. The amount of compaction is varied, depending on expected traffic levels. VTM, VMA, and VFA are measured, and also the theoretical maximum density, and the mass of the aggregate passing a 75 μm sieve expressed as a percentage of the mass of binder. An adjustment is made to estimate these parameters for a mix with 4% VTM. The aggregate grading is then chosen, so all of them comply with specified limits.
4. Design binder content. Samples of the asphalt mixture are made at different binder contents, and subject to compaction tests. The mix with 4% porosity (VTM) is chosen and then checked to ensure the other volumetric parameters meet the specified limits.
5. Moisture susceptibility. Samples of the chosen asphalt mixture are vacuum saturated, and subject to cycles of freezing and thawing. The tensile strength is then compared with control specimens.

37.8 USE IN ROAD CONSTRUCTION

37.8.1 TYPES OF PAVEMENT

There are two main types of road construction. "Rigid pavement," such as concrete, can only be used on stable ground. "Flexible pavement," such as bituminous materials, as described below, can be used on ground where some movement is expected.

Figure 37.5 shows typical road construction types.

Wearing course
Base course
Base
Sub-base
Subgrade (e.g., limestone)

Wearing course
Base
Lower base (crack resistant)
Sub-base
Subgrade (e.g., limestone)

FIGURE 37.5 Pavement Types

37.8.2 THE WEARING COURSE

There are conflicting requirements for durability, requiring a high binder content, and skid resistance, requiring a high content of harsh aggregate. Care should, of course, be taken to ensure that no further openings for services will be excavated in a road after a surface of this type is laid. Some definitions of the different types of surfacing materials are:

- Hot rolled asphalt: a gap-graded mix consisting of a fines/bitumen mortar that, in the case of the wearing course mix, also contains added filler. The fines are generally natural sand, in contrast to the crushed material normally used in dense macadams. The mortar is mixed with a single sized coarse aggregate to provide the gap grading.
- Asphaltic concrete (US) or dense bitumen macadam (UK): an asphalt mixture in which the aggregate particles are continuously graded from the maximum down to filler, to form an interlocking structure. Asphaltic concrete differs from dense macadam in that it is slightly denser, and usually contains a higher quantity of a harder grade of bitumen, for example, 70 pen. In general, asphaltic concrete is designed using the Marshall method of mix design.
- Surface dressing: a sprayed bitumen binder with stone chippings rolled on. This is used for repair work.
- Slurry surfacings and micro asphalts: bitumen emulsions with selected aggregate combinations. Also used for repair.

  The wearing course is designed to prevent rutting, potholes (bond failure), and loss of skid resistance.

37.8.3 THE BASE COURSE

Since this is protected from the direct load of traffic and the weather, it is constructed with mixes with higher aggregate contents, and lower viscosity binders, than the wearing course. If it fails, it is often due to fatigue cracking.

37.8.4 THE BASE

Often known as the road base, this layer is the thickest, and is built with the cheapest materials. Typically, this will be made with a coated coarse aggregate and a low binder content. These mixes are equivalent to no-fines concrete, and are highly porous. Semi-dry concrete is also often used for the base, under bituminous surfacing. Failure of the surface layers may occur due to failure of lower courses (e.g., reflected cracking from Semi-dry concrete).

37.9 OTHER APPLICATIONS OF BINDERS

37.9.1 TANKING

This is the traditional treatment for the outside of basement walls. It consists of layers of mastic asphalt up to 30 mm (1.2 in.) total thickness, or layers of bitumen sheeting bonded with hot bitumen (Fig. 37.6). Failure is normally caused by damage during backfilling, or excessive backfill pressures causing extrusion. There is renewed interest in this system, as a means of extending the life of reinforced concrete in saline groundwater. Clients are demanding a 120 year life for infrastructure, and even with good concrete and deep cover, chlorides in the groundwater could cause corrosion of the reinforcement.

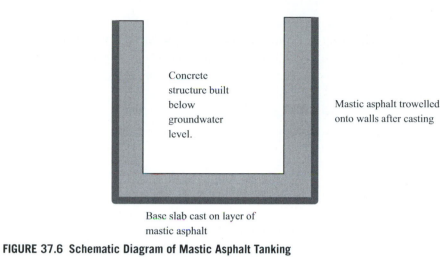

Concrete structure built below groundwater level.

Mastic asphalt trowelled onto walls after casting

Base slab cast on layer of mastic asphalt

FIGURE 37.6 Schematic Diagram of Mastic Asphalt Tanking

HDPE sheeting can be used as a barrier, but its long-term performance in groundwater is not known; however, mastic asphalt is known to have very good durability. Historic structures over a 1000 years old still have mastic asphalt in good condition.

37.9.2 ROOFING

Bituminous materials are often used for roofing. Such roofs often leak, but this is frequently due to damage from foot traffic, poor detailing at parapets and roof vents, or poor workmanship. Failure due to the material itself is rare. The main alternative systems are hot-trowelled mastic asphalt, or layers of felt bonded with hot bitumen.

37.10 CONCLUSIONS

- Bitumen is produced from oil refining; tar is obtained from coal or wood.
- The penetration test is used to measure the viscosity of a binder.
- Cutbacks and emulsions can be used cold, and will harden in a few days.
- Polymer modified bitumens are more elastic and less viscous.
- Asphalt mixtures contain aggregate that gives them higher strength to resist rutting.
- A road constructed with bituminous materials is a flexible pavement.
- In road construction, the lower courses contain less binder.

TUTORIAL QUESTIONS

1. A sample of asphaltic concrete is made with 7% binder by mass, and oven-aged. 5 kg of the loose material is vacuum saturated, and weighed submerged to give a mass of 3.08 kg after correction for the mass of the container and the cradle. 5 kg of the aggregate is also weighed submerged, to

give a mass of 3.12 kg after correction for the mass of the container and the cradle. 1.25 kg of the sample is compacted to give a cylinder 63.5 mm long and 101.6 mm diameter. Calculate VTM, VMA and VFA.

Solution:

For the loose material:

Absolute density $= 5 \times 1/(5 - 3.08) = 2.604$ kg/L   (8.4)

(The density of water $= 1$ kg/L)

For the aggregate:

Density $= 5 \times 1/(5 - 3.12) = 2.660$ kg/L   (8.4)

For the compacted sample:

Volume $= \pi \times 0.0635 \times 0.1016^2/4 = 0.514 \times 10^{-3}m^3 = 0.514$ L

Density $= 1.25/0.514 = 2.429$ kg/L

The porosity (VTM) $= 100 \times (2.604 - 2.429)/2.604 = 6.72\%$ (8.5)

The % of aggregate in the sample $= 100 - 7 = 93\%$

The mass of aggregate in the sample $= 0.93 \times 1.25 = 1.16$ kg

The volume of aggregate in the sample $= 1.16/2.660 = 0.437$ L

The total voids in the aggregate (VMA) $= 100 \times (0.514 - 0.437)/0.514 = 15.05\%$

The % of voids filled with bitumen (VFA) $= 100 \times (15.05 - 6.72)/15.05 = 55.39\%$

2. A sample of asphaltic concrete is made with 7% binder by mass, and oven-aged. 10 lb of the loose material is vacuum saturated, and weighed submerged to give a mass of 6.16 lb after correction for the mass of the container and the cradle. 10 lb of the aggregate is also weighed submerged to give a mass of 6.24 lb after correction for the mass of the container and the cradle. 2.75 lb of the sample is compacted to give a cylinder 2.5 in. long and 4 in. diameter. Calculate VTM, VMA and VFA.

Solution:

For the loose material:

Absolute density $= 10 \times 1686/(10 - 6.16) = 4390$ lb/yd^3 (8.4)

(The density of water $= 1686$ lb/yd^3)

For the aggregate:

Density $= 10 \times 1686/(10 - 6.24) = 4484$ lb/yd^3 (8.4)

For the compacted sample:

Volume $= \pi \times 2.5 \times 4^2/4 = 31.42$ in.^{3}

Density $= 2.75/31.42 = 0.0875$ lb/in.$^3 = 4083$ lb/yd^3

The porosity (VTM) $= 100 \times (4390 - 4083)/4390 = 7.0\%$ (8.5)

The % of aggregate in the sample $= 100 - 7 = 93\%$

The mass of aggregate in the sample $= 0.93 \times 2.75 = 2.56$ lb

The volume of aggregate in the sample $= 2.56/4484 = 5.71 \times 10^{-4}$ yd$^3 = 26.64$ in.^{3}

The total voids in the aggregate (VMA) $= 100 \times (31.42 - 26.64)/31.42 = 15.2\%$

The % of voids filled with bitumen (VFA) $= 100 \times (15.2 - 7)/15.2 = 53.95\%$

COMPOSITES

38

CHAPTER OUTLINE

38.1 INTRODUCTION

The word composite simply implies the use of more than one material. There are many composites in nature (e.g., wood is a fibrous composite), and artificial composites have been made for a long time (e.g., horse hair to reinforce plaster). The essential point about a successful composite is that each material should make up for weaknesses in others. The purpose of this chapter is to introduce some examples, and discuss what makes them successful.

38.2 REINFORCING BARS IN CONCRETE

This has proven to be a very successful composite because concrete is strong in compression, while steel is strong in tension, and steel is protected by the alkaline environment in the concrete. Reinforcing concrete with wood is less successful because wood is not so strong, and is preserved by acids and damaged by alkalis.

This type of composite is the most ordered (when compared to fibres) and, therefore, efficient but expensive to construct. The steel is placed precisely where it is required, so minimum quantities are used.

It is assumed that, unless it is prestressed, the concrete will crack before the steel takes a full design load.

38.3 FIBRE REINFORCEMENT IN CONCRETE

Fibres are introduced into the mixer, and are thus randomly distributed in the hardened concrete. Structurally, this is the least efficient method, although random reinforcement is best for some types of impact loading.

431

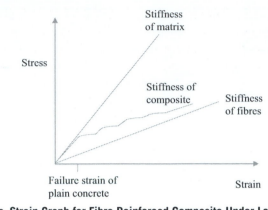

FIGURE 38.1 Typical Stress–Strain Graph for Fibre Reinforced Composite Under Load

Glass, polypropylene, and steel fibres are used. Glass fibres are damaged by alkalis, so they are coated when used in concrete (see Section 36.3.4). Steel fibres will cause rust staining, if they are not plated because some fibres will be at the surface. Polypropylene does not rust, and is not damaged by alkalis, but has the lowest modulus, and is thus not so effective.

The proportion length/diameter for the fibres is known as the aspect ratio. If this is too low (e.g., below about 160 for steel), the bond of the fibre to the concrete will be insufficient to make use of the tensile strength. In order to make steel fibres more efficient, shaped profiles are used to improve the bond.

Figure 38.1 shows the performance of the composite under load. This could be a uniaxial compressive load but, as noted in Section 3.3.2, the Poisson's ratio effect means that it effectively fails in lateral tension. The load initially increases until the first crack occurs in the concrete. If there are insufficient fibres present, the material will fail suddenly when this first crack occurs. However, if there are sufficient fibres, they will take the load, once the strain has increased to the point where they can carry it. As the load increases, more cracks will occur, and the process will be repeated. The ability of the fibres to prevent the cracking will depend on their stiffness.

Figure 38.2 shows an application for glass fibre mat in concrete. The mat is laid against the surface of a mould, and the paste penetrates through it, so the glass is just below the surface of the concrete (this sample was cast the other way up from that shown in the Fig. 38.2). A rifle bullet has been fired through the concrete, and the glass has significantly reduced dangerous fragmentation from the exit surface.

38.4 STEEL/CONCRETE COMPOSITE BRIDGE DECKS

Methods of using external steel members with concrete bridge decks are shown in Fig. 38.3. In the more common application, the concrete deck is cast on top of steel beams with shear connectors (this may be seen in Fig. 31.22). This is an effective composite because the deck is not continuous across the supports, so the concrete is always in compression, and the steel in tension. Corrosion in this type

FIGURE 38.2 Concrete Panel With Glass Mat in the Surface, Showing Damage From a Rifle Bullet

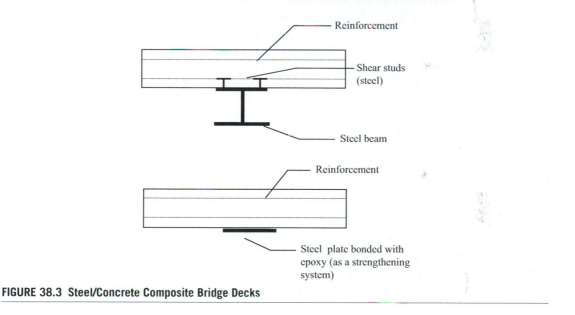

FIGURE 38.3 Steel/Concrete Composite Bridge Decks

of construction is discussed in Chapter 31, and is normally caused by saline water penetrating through the expansion joints that are needed above each support.

In the second application, glued plates are used to increase the strength of an existing bridge. This may be used if the reinforcement is insufficient, either due to deterioration of the bridge, or increased loading. The main difficulty with this system is the problem of lifting heavy steel plate into position, while keeping the surface clean, and applying the adhesive. For this reason, carbon fibre reinforced composites may be used in place of the steel. These are far lighter, but very expensive.

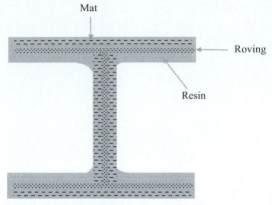

FIGURE 38.4 Section Through Fibre Reinforced Polymer Composite Beam

38.5 FIBRE REINFORCED PLASTICS

Polyester or epoxy resins are reinforced with glass, polymers such as aramid (aromatic polyamide) or carbon fibre. Fibreglass (see Section 36.3) is used for shuttering and cladding panels, where a finely detailed finish is needed. Fibreglass domes and other features are convenient to use because their relatively low weight makes them easy to install, and they also have good durability. Fibre reinforced plastics make very effective composites because they are lightweight (often because they can be made very thin), strong, and durable.

Figure 38.4 shows a composite beam. By using rovings (i.e., unidirectional fibres) that are positioned to carry the highest stresses, these beams can be very strong. They are also lightweight, and thus easy to install. However, they lack the impact resistance of steel, and are far more expensive.

Figure 38.5 shows a typical application of fibreglass in a house.

38.6 STRUCTURAL INSULATED PANELS

Structural insulated panels (Fig. 38.6) may be used to construct buildings without the need for a frame. This is an effective composite because if the panel is in bending, the foam can carry the shear load, and if it is in end loading, the foam will stop the timber panels from buckling. The strength depends on good adhesion between the foam and the panels. If the walls of a building are made from these panels, they have sufficient strength to support the roof. They are lightweight and, thus, quick and easy to assemble. However, the timber outer surface needs a protective coating that must be maintained, and due to their very light weight, they may provide poor sound insulation.

38.7 CONCLUSIONS

- Reinforcing bars in concrete make a good composite because the steel carries the tensile loads, and is protected by the alkaline environment.

Lead

Fibreglass

FIGURE 38.5 The Roof Section Over the Bay Window is Made out of Fibreglass to Look Like Lead

Compared with lead, it is cheaper and easier to install, and likely to require less maintenance. The flashing around the top of it is lead.

Oriented strand board or plywood

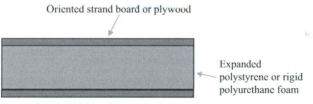

Expanded polystyrene or rigid polyurethane foam

FIGURE 38.6 Typical Structural Insulated Panel

- Glass fibre in concrete is a good composite structurally, but the glass must be protected from the alkali.
- Steel–concrete composite bridge decks are effective, provided the joints are well sealed to prevent salt ingress.
- Structural insulated panels are lightweight, structurally efficient composites, but need protection to make them durable.

ADHESIVES AND SEALANTS

39

39.1 INTRODUCTION

39.1.1 APPLICATIONS

Adhesives and sealants are often similar, and some materials may be used for both purposes. Sealants are generally used for the primary purpose of preventing moisture/air ingress, with adhesion being a secondary consideration.

Adhesives are used for a wide range of applications in construction, ranging from glued segmental bridge construction, in which massive precast sections are glued together, to fixing minor fittings to walls. Similarly, sealants may form the primary seal in tunnel construction or just a finish around a hand basin.

39.1.2 FAILURE OF ADHESIVES AND SEALANTS

The failure of adhesives and sealants is one of the major causes of the failure of structures to achieve their required performance. This failure is frequently not a problem with the material itself, but a problem with the specification, detailing, or installation. The main purpose of this chapter is, therefore, to give guidance on the specification and use of adhesives and sealants.

437

39.1.3 SAFETY

Many adhesives and sealants are toxic (e.g., containing isocyanates), and are readily absorbed through the skin. Many also give off toxic solvent vapours during setting. Always download and read the safety sheet before using these materials. Some materials that are relatively harmless in domestic applications/ quantities (e.g., epoxies) represent a considerable hazard in large quantities in industrial applications.

39.2 ADHESIVES

39.2.1 TYPES OF ADHESIVE

There are very large numbers of adhesives available. The broad categories are as follows:

- Mechanical adhesives. All of the original animal glues were of this type. They only work on porous materials – the adhesive penetrates into the pores and, on hardening, it forms a mechanical key.
- Solvent adhesives. These are used on materials that dissolve in solvents (e.g., plastics). The surfaces to be joined are softened in the solvent that subsequently evaporates to leave a continuous joint.
- Surface adhesives. These will adhere to surfaces that are neither porous, nor soluble (e.g., glass). Adhesives such as epoxies are of this type. They work by bonding in the same way as the bonds within a material.

39.2.2 SETTING PROCESSES

The different setting processes are:

- Adhesives that set on cooling (e.g., bitumens). These materials are heated for use and will fail if reheated.
- Adhesives that set by evaporation of water. These are clearly not durable in a wet environment, and will not set in contact with impermeable surfaces that prevent the evaporation.
- Adhesives that set by evaporation of solvents. These are more durable than the water based adhesives, but may also not set if evaporation is prevented.
- Adhesives that set by internal reaction (e.g., polymerisation of epoxies and polyesters).
- Anaerobic adhesives that set by exclusion of air (e.g. "superglue").
- Adhesives that set by reaction with water. Some polyurethane adhesives are stored in contact with a desiccant, and set by reaction with moisture in the air.

  Some adhesives set by a combination of two or more of these processes.

39.2.3 SURFACE PREPARATION

In order to adhere to a surface, the adhesive must wet it, and not form into droplets. Scoring of surfaces (a method often used on timber by joiners) is detrimental to a joint made with surface adhesives because it causes stress concentrations and air inclusions. It is only beneficial if traditional mechanical adhesives are used.

Obviously the strength of bond formed by an adhesive will be severely reduced, if the surfaces to be joined are loose or dirty. Water will also prevent the bond of many of them, but some, including some polyurethanes, will work well underwater. Surface treatments will destroy the bond, if they are impermeable (e.g., paint), and many other treatments, such as impregnated wood preservatives, will prevent some adhesives from working, and reduce the performance of others.

39.2.4 TYPES OF JOINT

An adhesive will ideally form a joint that is stronger than any form of mechanical fixing because it does not cause stress concentrations to the same extent.

Figure 39.1 shows different joint arrangements. For the contact joints, most adhesives are suitable. Joints of this type are normally clamped during setting. The clamping force must be sufficient to ensure contact, but not so high that the adhesive is forced out.

For gap filling applications, many adhesives are unsuitable. They either have insufficient viscosity to remain in place during setting, or have high shrinkage during setting. A pure epoxy resin will not fill gaps, so it is normally mixed with an inert filler (e.g., colloidal silica) to give it gap-filling properties. This will, however, reduce its strength for contact applications.

For fillet formation, even greater viscosity and stability is required. Inclusion of fibres may be beneficial.

39.2.5 MODES OF FAILURE OF ADHESIVES

Adhesive joints have a poor record for failures. Some of the reasons for these are as follows:

- Inadequate joint preparation.
- Failure of the substrate due to local stress concentrations (see 39.2.6).
- Bond failure due to moisture ingress, or exposure to ultraviolet light.
- Progressive failure due to eccentric loading.
- Failure of the adhesive due to inadequate mixing, or incorrect or prolonged storage.

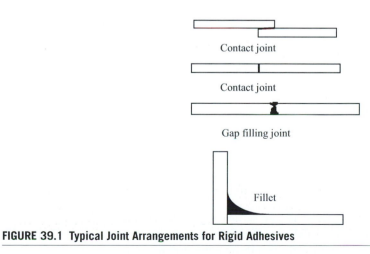

Contact joint

Contact joint

Gap filling joint

Fillet

FIGURE 39.1 Typical Joint Arrangements for Rigid Adhesives

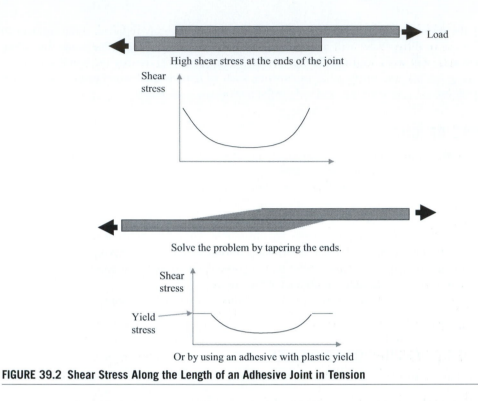

FIGURE 39.2 Shear Stress Along the Length of an Adhesive Joint in Tension

39.2.6 STRESS CONCENTRATIONS IN LAP JOINTS

An adhesive should not have a higher modulus than the substrates because this will cause stress concentrations at the ends of the joint that may cause progressive failure. Joining materials with different elastic moduli is difficult for this reason.

However, even if the substrates and the adhesive have the same elastic modulus, the shear stresses in the joint shown in Fig. 39.2 will be concentrated at the ends. This is caused by the elastic deformation of the substrates, and may be solved either by tapering the ends, as shown, or by using an adhesive with plastic yield which can redistribute the load.

It is often better to use a mechanical fixing, as well as the adhesive, for critical applications.

39.3 SEALANTS
39.3.1 TYPES OF SEALANT

Most sealants are of three main types:

Putties. These are the traditional materials for sealing to glass. They harden by surface oxidation, and subsequent slow loss of solvent.

Mastics. These generally do not harden. They are sufficiently viscous to prevent sagging, but offer little mechanical strength. They are often poured into joints. Mastic asphalts are described in Chapter 37.

Elastomeric sealants. These set to a tough but elastic condition by a number of different processes. Two pack types must be mixed on site, but one pack systems are more convenient because

FIGURE 39.3

A range of different sealants, ranging from a cheap caulk (left) that will set hard, to a high-performance polyurethane sealant (right) that will set under water.

they are supplied in cartridges ready for use (Fig. 39.3). Polyurethane sealants have very strong adhesive properties, as well as sealing. Silicones and polysulphides are the more common and cheaper materials.

39.3.2 SEALANT DETAILING

Figure 39.4 shows three different details that could be used for a typical joint between window frames. Shrinkage of the timber may cause the gap between the two frame sections to double in width. The simple joint in detail Fig. 39.4a will fail because it will experience a strain of 100%. This failure may be loss of bond with the timber, or rupture of the sealant itself. The detail in Fig. 39.4b has a greater width of filler, so the strain will be far lower, and the movement may be accommodated. However, if the sealant formed a bond with the back of the joint, this would cause severe local stresses, so a backing strip (that may be paper or plastic) is positioned to prevent it. The detail in Fig. 39.4c includes a drip, so failure of the filler may not permit moisture ingress. In all the details, it is only necessary to fill part of the joint. Filling the entire gap between the frames would be a waste of time and material, and might cause unwanted stresses in the frames.

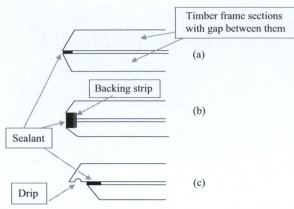

FIGURE 39.4 Joint Details for Timber Window Units

(a) Sealant in this will fail. (b) Sealant in this detail with a backing strip will be durable. (c) Note the "drip" in this detail.

39.4 CONCLUSIONS

- Inadequate detailing and joint preparation are the most common causes of failure of adhesives and sealants.
- The stiffness of the substrates and the adhesive must be considered when detailing a joint.
- If gap filling is required, an adhesive must have a high viscosity.
- Common sealants have poor adhesion to most materials.
- When designing a sealant joint, the expected movement of the substrates, and the resulting strain in the sealant, must be considered.

COMPARISON OF DIFFERENT MATERIALS

CHAPTER OUTLINE

40.1 INTRODUCTION

Materials are most often compared on the basis of cost, strength, environmental impact, safety issues, appearance, or durability. Clearly, these affect each-other and cannot be considered in isolation. The introduction of carbon trading raises the cost of materials with a high environmental impact, and materials with poor durability will have high long-term costs and environmental impact (from maintenance and replacement), and poor appearance.

This chapter provides an overview of methods of comparing strength, environmental impact, and health and safety of different materials. The durability has been considered in detail in the chapters on the individual materials.

Hazards from fire have not been considered in previous chapters because they are best seen on a comparative basis, and are discussed in Section 40.4.2.

40.2 COMPARING THE STRENGTH OF MATERIALS

When comparing the strengths of materials for use in structures, it is generally most relevant to consider the strength to weight ratios. Since the deflection of a structure is often more critical in design than the ultimate failure load, the modulus to weight ratio may be even more important. Table 40.1 shows a number of materials ranked in order of modulus to weight ratio. These are only typical values; all of the materials are produced in a wide range of different grades. For example, by curing under pressure and reinforcing with steel fibres, special cement mortars can be made with strengths above

443

Table 40.1 Typical Mechanical Properties of Different Materials

	Density (kg/m³)	Density (lb/yd³)	Ultimate Strength (MPa)	Ultimate Strength (ksi)	Modulus (GPa)	Modulus (ksi)	Strength/ density (MKS units)	Modulus/ density (MKS units)
Concrete (compressive)	2,300	3,887	40	5.8	20	2,900	0.017	0.009
Polypropylene	900	1,521	400	58	8	1,160	0.444	0.009
Polyester	1,380	2,332	1,100	159	14	2,030	0.797	0.010
Titanium	4,500	7,605	950	137	110	15,950	0.211	0.024
Mild steel	7,800	13,182	450	65	200	29,000	0.058	0.026
High tensile steel	7,800	13,182	1,600	232	200	29,000	0.205	0.026
Glass fibre	2,550	4,309	3,000	435	70	10,150	1.176	0.027
Aramid	1,450	2,450	3,000	435	130	18,850	2.069	0.090
Carbon fibre	1,900	3,211	3,000	435	400	58,000	1.579	0.211

200 MPa (29 ksi) in compression, and relatively high tensile strengths (see Section 29.7). These high strength materials offer the potential for advanced structural forms, such as very long span suspension bridges, but their high cost has severely restricted their use in most civil engineering.

40.3 COMPARING ENVIRONMENTAL IMPACT
40.3.1 TYPES OF IMPACT

The production and use of construction materials will cause a number of different impacts on the environment, from either the production process, or associated road transport, etc.:

- Smoke and dust emissions
- Contamination of groundwater
- Noise pollution
- Internal environment pollution (sick building syndrome)
- CO_2 and other gas emissions leading to climate change

Opinions may differ, regarding the relative importance of these impacts, and environmental legislation will often focus on local impacts. However, it is the opinion of the author that climate change is the more serious issue, and the reduction of the carbon footprint should be the highest priority in materials selection. While the anticipated global warming may not be occurring exactly as predicted, there is strong evidence that the risk of climate change is serious, and must be addressed.

The use of PFA (see Section 18.3.3) illustrates this conflict. PFA is produced in vast quantities from coal burning power stations. Some of the toxic elements from the coal are concentrated in the ash; however, if it is mixed with cement or lime, and made into concrete, they are bound, and leaching rates are very low. Any PFA that cannot be put to beneficial use, such as cement replacement, is disposed of in large impoundments at the power plant. There has been a prolonged discussion in many countries concerning classifying it as a hazardous waste. If it is classified as hazardous, it becomes very difficult to use it in concrete. The situation in the United States has been complicated by two major spills from impoundments (in 2008 in Tennessee, and in 2014 in North Carolina) that have had the effect of encouraging the argument that the material is hazardous. The effect of the uncertainty regarding the regulatory position has meant that beneficial use of PFA, that had been rising steadily year on year, fell after 2008. The "environmental lobby," who are seeking the classification as a hazardous waste, clearly hoped that this would make the power stations switch to gas burning, or possibly close as energy conservation reduced demand. The actual effect has been that at least 20 million tonnes of PFA, that could have been used beneficially, have been placed in the impoundments, and millions of tonnes of additional cement has been produced (with resulting CO_2 emissions, see Section 17.9) that could have been replaced with PFA. An example in the United Kingdom was the backfilling of old mine-workings beneath a wealthy city. This city has many fine old buildings, and a proposal to fill the workings beneath them with a cement-PFA grout was rejected because it was regarded as "waste disposal," so a pure cement grout was used. In reality, the leaching from a PFA grout would be lower because it would have a higher total cementitious content and lower permeability, so, as well as increasing greenhouse gas emissions, the risk of groundwater pollution was increased.

40.3.2 MEASUREMENT OF IMPACT

Choosing a method to measure impact to guide the choice of materials is difficult because it involves predicting the outcome of current environmental problems, and the introduction of new technologies. If a structure is to be built with a design life of 120 years, it can be argued that the full impact of use, maintenance, and demolition throughout the 120 years should be included. Alternatively, it may be justified to say that current environmental problems are so severe that the short-term impact must be given priority, in order to prevent disasters in the next 30 years. Currently, some steels are more readily recycled than other materials. However, it is questionable whether material choices should be based on the predicted impact of recycling them toward the middle of the twenty-second century.

40.3.3 CERTIFICATION SCHEMES AND STANDARDS

Some examples of these schemes are:

*   Leadership in Energy & Environmental Design (LEED) is a green building certification programme that recognises best-in-class building strategies and practices.
*   The Building Research Establishment Environmental Assessment Methodology (BREEAM) is a widely used method of assessing, rating, and certifying the sustainability of buildings.
*   The Civil Engineering Environmental Quality Assessment & Award Scheme (CEEQUAL) is an Assessment and Awards Scheme for improving sustainability in civil engineering, infrastructure, landscaping, and works in public spaces, based in the United Kingdom.
*   National and International standards and codes have also been developed to provide uniform assessments of the environmental impact of materials and designs.

In practice, it is not possible to consider all of the questions raised in the previous sections, so these schemes and standards will be used to give weightings to the different impacts, and assess and select the materials in a design. However, care should be taken to ensure that the default variables in the software are appropriate to the local conditions.

40.4 HEALTH AND SAFETY

40.4.1 SPECIFIC HAZARDS WHEN USING THE MATERIALS

Materials are not generally selected on the basis of health and safety. There are, however, considerations that may influence the choice, and two examples of these are:

- Self-compacting concrete (see Section 22.2.8) has the advantage that vibration is not needed. This avoids the risk of "white finger" for the operatives, and reduces the noise hazard from the vibrators.
- Welding steel is a hazardous process not only due to the heat generated, but also because the arc can damage eyesight both for welders and other operatives, or members of the public who may be close to it. On-site welding should therefore be minimised.

The safety data sheets of all materials should be downloaded before a decision is made to use them in any unusual applications.

40.4.2 HAZARD FROM FIRE

Fires on construction sites are, unfortunately, not uncommon. The possibility of fire during the use of a building is also a major consideration in design. Design for fire is always considered on the basis of the safety of the users during evacuation, rather than the preservation of the structure. The performance of different materials in fire is summarised in Table 40.2.

Figure 40.1 shows how unprotected steel will fail rapidly in a fire.

FIGURE 40.1 Steel Frame After a Fire

Table 40.2 Performance of Materials in Fire

	Flammability	Loss of Strength	Other Hazards	Design
Reinforced concrete	• Nonflammable • Concrete: loses free/bound water at 90–130°C • Lime will calcine at 450°C (see Section 7.9) • Limestone will calcine at 650°C • Steel: no reactions	• Low thermal conductivity of concrete protects steel and concrete • Failure is by spalling (tensile failure) exposing steel • Concrete lost all strength by 850°C • Steel by 550°C • Coefficients of thermal expansion similar, so bond OK	• If the concrete has very low permeability and is saturated, it may explode. Polypropylene fibres may be used to prevent this • Spalling concrete may fall • CaO (quicklime) formed in concrete will react violently with water • Possible fumes from additives	• Standards may specify minimum cover and minimum secondary reinforcement for given fire resistance • Lightweight or limestone aggregate is good • Siliceous aggregate is bad • High creep is good
Steel	Non-flammable but often painted	• High thermal conductivity leads to rapid failure • Strength lost by 550°C • Cold worked steel is worse	• High thermal expansion will disrupt structure • High thermal conductivity may ignite other areas • Welding is a common cause of fires	Must normally be protected with: intumescent paint, plaster, brick, wood, or water filling, etc.
Plastics	• Thermoplastics melt, and then burn • Thermosetting plastics char, and then burn • Great variations between types, but typical ignition at 400°C	• Thermoplastics may melt by 100°C • Thermosets OK to about 300°C • Generally high creep at high temperatures	• Toxic fumes • Melts and drops	• Additives may help. These give off small quantities of gases, such as chlorine, that locally displace the oxygen • Careful selection will reduce smoke hazard
Asphalt mixture	Burns, and volatiles will flash, but it is safer than pure bitumen/tar	Softens at low temperatures	• Melted binder sticks to skin • Hot walkways are skid hazard • Bitumen fire is spread by water • Fumes are carcinogenic	Maximise aggregate content and use chippings on surface

(Continued)

Table 40.2 Performance of Materials in Fire *(cont.)*

	Flammability	Loss of Strength	Other Hazards	Design
Glass	Non-flammable	• Sheet glass shatters due to differential thermal expansion • Toughened glass will break into small fragments that reduce hazard • Fibres OK to high temperatures	Explosive failure	• Hazard from toughened glass is less • Wire glass or, intumescent laminate can give resistance • Heat treated borosilicate glass is heat resisting • Fibers often used for fire resistance (e.g., fire blankets)
Timber	• Depends on species. • Protected by charcoal (up to 500°C)	• Little loss of strength due to heat (drying increases strength) • Low thermal conductivity protects it	• Low expansion (unlike steel) • Possible fumes from preservatives	• Sacrificial design • Retardants
Masonry	Non-flammable	May buckle due to heat on one side only	Stays hot for a long time	Details: expansion joints, fire stops in cavities

40.5 CONCLUSIONS

- Carbon fibre offers the best stiffness to density ratio of materials currently available for construction.
- When considering environmental impacts, local issues may be in conflict with global issues.
- There are a number of different schemes that enable designers to compare the environmental impact of different design options. However, a professional designer should always confirm that the assumptions in them are appropriate.
- Steel fails rapidly in fire, but timber performs remarkably well.

TUTORIAL QUESTIONS

1. A footbridge is to be constructed with the bottoms of the piers below the normal water level in a river. Describe the processes that will affect the long-term performance of the piers, if they are constructed with:
   a. Brickwork
   b. Reinforced concrete
   c. Steel
   d. Timber

Solution:

a. Brick
- Frost resistance of bricks and mortar is OK, if correctly selected.
- Possible abrasion/impact damage from debris/ice.
- Cracking likely due to lack of tensile strength, if not reinforced or massive.
- Attack on mortar, if the water is acidic.
- Sulphate resistance of mortar and bricks may be relevant.

b. Concrete
- Reinforcement corrosion from carbonation (salt possible, if the deck is salted, or the water saline)
- Frost attack, if not air entrained.
- Possible abrasion/impact damage.
- Possible sulphate attack.

c. Steel
- Corrosion likely due to impact damage to paint (from ice and debris), and difficulties with repainting.
- Weathering steel would help, if the water is not saline, and appearance is not critical.

d. Timber
- Wet rot will be a major cause of deterioration.
- Reduced, with durable timber or effective impregnation of preservative.
- Paint would be useless because it would soon fail.

2. Outline the relative merits of the following materials for the external cladding of a new library for a university.
   a. Brickwork.
   b. Precast concrete.
   c. *In situ* concrete.
   d. Glass with aluminium panels.
   e. Glass with plastic coated steel panels.

Solution:

a.
- Appearance: good, particularly if features are included.
- Durability: good.
- Insulation: OK, if cavity walls are used.
- Cost: high.

b.
- Appearance: could be good with exposed aggregate etc.
- Durability: could be good if cover is adequate, and w/c is low.
- Insulation: could contain insulating layer.
- Cost: average.

c.

- Appearance: poor, unless very carefully done.
- Durability: probably OK, but could have defects if site supervision is poor.
- Insulation: average. Difficult to put in insulating detail in site work.
- Cost: average-high.

d.

- Appearance: good, for those who like it.
- Durability: good, if aluminium anodised.
- Insulation: poor, but could have insulation behind it.
- Cost: high.

e.

- Appearance: poor.
- Durability: OK, provided it is protected from scratches, etc.
- Insulation: poor, but could be insulated.
- Cost: low.

NEW TECHNOLOGIES

CHAPTER OUTLINE

41.1 INTRODUCTION

During recent years, possibly the most important new technology in construction materials has been the introduction of self-compacting concrete, using high range superplasticisers, and viscosity modifiers. The impact of search engines, and the ever-increasing amounts of performance data for different materials, had also been very significant. These are now established technologies, and are discussed in other chapters. The purpose of this chapter is to discuss some ideas that are not covered elsewhere in the book because they are not yet fully developed and ready for use. The author has carried out research on some of the topics that are discussed in this chapter, and selected others based on papers he has seen while working as a journal editor, and running conferences. Some of the technologies are totally new, while others are traditional methods that are experiencing a revival of interest.

The term "smart material" has been applied to materials that react with their environment in some way. Both the photocatalytic admixtures, and the self-healing concrete described in Section 41.3 and 41.4 may be defined as smart materials.

This chapter only discusses areas where solutions to problems are being researched. There is a great need, in construction, for a reliable test to measure the potential durability of a structure. Durability tests are for concrete are discussed in Section 22.8, and measurement of the corrosion of steel is discussed in Section 31.5, but none of these tests give a real capability for a client's representative to test a new structure, and give a reliable estimate of how long it will last in its environment (this issue is also discussed in Section 14.2). The author is not aware of any really promising ideas that are being researched for this.

41.2 3D PRINTING

Small scale 3D printing using plastics is an established technology. In construction, 3D printing involves placing concrete using a robotic arm fitted to a gantry. The concrete mix design is critical because it must not slump, and must remain exactly where it is placed, without any shuttering. It must also be able to develop adequate strength without compaction, so it needs some of the characteristics of self-compacting concrete. Fibre reinforcement is used. Large numbers of thin layers are placed to develop the required shapes that may include curved cladding panels and architectural features. The printer can make items that cannot be cast using conventional processes. A further advantage is that it can make them as soon as the shape had been developed in a design package, without the need for drawings or formwork.

41.3 PHOTOCATALYTIC ADMIXTURES

Titanium dioxide is produced in very large quantities worldwide, and has a vast range of applications, ranging from the white pigment in toothpaste, to industrial catalysis. If it is added to concrete and exposed on the surface in sunlight, it acts as a catalyst to break down pollutants such as nitric oxide and sulphur oxides that can lead to acid rain and smog. If used on roads, it has the effect of reducing pollution from car exhausts. On buildings, it has the added benefit of breaking down the pollutants that adhere to the surface of concrete, and makes them "self-cleaning." However, titanium dioxide is expensive and, when mixed with concrete, most of it is wasted in the bulk of the mix, and not exposed on the surface – so methods are being investigated to impregnate it into the surface, rather than adding it to the mix.

41.4 SELF-HEALING CONCRETE

A number of systems have been investigated to make cracks in concrete heal when they form. Cracking is discussed in Section 23.3. All concrete will benefit from some autogenous healing of cracks. However, if the concrete is old or carbonated, or flow rates of water through the crack are high, this will be insufficient to heal a crack. Materials, such as microencapsulated sodium silicate, are being tested, that will be released when a crack forms, and promote the healing process. These would be most useful in water retaining structures.

41.5 ZERO CEMENT CONCRETE

This is not a new technology. The Romans had no cement, but made large amounts of concrete using pozzolans such as volcanic ash, and ground bricks and tiles mixed with lime. Ancient examples of similar concrete have also been found in Central America. Environmental concerns about carbon emissions from cement production have renewed interest in alternative ways of making low strength concrete, rather than just using normal cement with a high water/cement (w/c) ratio. If ashes, such as incinerator ash or PFA, are mixed with alkalis, such as kiln dust (a waste from cement or lime production), a blended powder with a very low cost and carbon footprint can be produced to replace cement in low strength concrete. Other blends, such as steel slag and waste gypsum, can be used to make a super-sulphated cement (see Section 18.2.6). The concretes in Figs 17.2, 17.4, and 27.7 are actually a zero cement supersulphated slag–gypsum mixes.

The concrete in Fig. 41.1 contains "red mud" that is a waste from bauxite processing, "terra rosa" that is a natural pozzolan, waste gypsum, and lime. The lime is a commercial product, but it only forms 10% of the cementitious content.

41.6 DURABILITY MODELLING

Construction clients are increasingly requiring design lives of more than a 100 years for structures such as road and rail bridges. The modelling of concrete durability is discussed in Section 25.9, and normally involves predicting chloride levels in the concrete at the cover depth where the steel is located. The theory of this modelling is well understood; however, the effect of age on the transport properties of concrete is largely unknown. Current predictions estimate that the diffusion coefficient, and permeability may reduce by a factor of more than 20, during a 100-year life. A small change in this factor leads to a substantial change in the predicted chloride levels and, thus, the probability of the reinforcement corroding. There is also a lack of data on adsorption. In Section 10.5, this was assumed to be linear (i.e., proportional to concentration), and the concept of the capacity factor was introduced; however, this is only an approximation, and there is a lack of data to give a more accurate estimate. New models based on extensive data sets will give accurate predictions of the life of structures, so designs can be improved.

FIGURE 41.1 Roller Compacted Zero Cement Concrete in Jamaica

41.7 HEMP LIME

Hemp lime is one of a number of technologies that include rammed earth construction, and exploit and extend the good thermal performance and low carbon footprint of traditional construction methods. Hemp lime is a composite of a lime-based binder, and the hemp as a plant based "aggregate." The hemp is very fast growing, and produces a crop in just 4 months. The resulting blocks are lightweight, and have a very low thermal conductivity. They can be used for underfloor or roof insulation, or even as an attractive internal finish. It is permeable to water vapour but, provided the outer surfaces are sealed with a render, this is seen as an advantage because it enables the structure to "breathe," thus preventing accumulation of moisture.

41.8 WOOD–GLASS EPOXY COMPOSITES

Fibre reinforced plastics are discussed in Section 38.5. If these are made with epoxies, they will adhere very effectively to the surface of most timber. They can then be used to make joints in timber that are more efficient than those discussed in Section 33.6 because they do not involve cutting into the timber in any way, causing loss of section. The fibre may also be bonded to whole timber sections to carry tensile loads. The composite has the additional major advantage of providing a durable waterproof finish, and preventing degradation of the wood.

This technology has been used extensively for the fabrication of very large laminated blades for wind turbines.

Figure 41.2 shows this technology in use in boat construction, where it has been common for many years. However, there has been little technology transfer to construction.

41.9 BAMBOO

Bamboo has been used in construction throughout history but, as noted in Section 33.9, its use in current construction is limited by the inefficient jointing systems used (see Figs 33.10 and 33.11). This can be overcome by reforming the bamboo into laminates that are square sections, but this means that

This area has bonded aramid (Kevlar) fibre to give additional resistance to punching shear. It is kept below the water line because it is not attractive.

This area has two layers of bonded woven glass rovings. They become completely transparent when wetted with epoxy.

FIGURE 41.2 Wood–Glass–Epoxy Composite Construction of a Boat

the intrinsically efficient hollow circular section of the stems is not used. The stems are not perfectly regular and straight, but 3D scanning can be used to take account of this, if joints are machine cut for each particular location. There is some debate regarding the best route to take to improve the method of jointing, but there is considerable potential for application of the wood–glass–epoxy composites mentioned in the previous section (although standard epoxy does not bond well to the outer surface of bamboo).

There is also a lack of a good stress grading system for bamboo. The machines of the type shown in Fig. 33.2 are clearly inappropriate.

41.10 CONCLUSIONS

- Construction is a deeply traditional industry. Our main method of construction (reinforced concrete) has changed little in over 100 years.
- There are a number of exciting new innovations currently in development.
- There have been a number of innovations, such as high-alumina cement and calcium chloride accelerators that have been used inappropriately, and led to major failures.
- It has been the aim of this book to help students identify good technologies, and avoid the inappropriate ones.

Tutorial Questions

Knowledge of material from other chapters may be assumed in some questions.

CHAPTER 1

(Answers to be in scientific notation.)

1. How many μm are there in 1 m?
2. How many Pa are there in 1 MPa?
3. How many Pa are there in 1 GPa?
4. How many ns are there in 1 s?
5. If 10^{17} is written in non-scientific notation, how many zeros are there after 1?
6. If 10^{-14} is written in non-scientific notation, how many zeros are there between the decimal point and 1?
7. How many g are there in 10^7 kg?
8. How many μs are there in 10^{-2} s?
9. The force required to make a mass m kg accelerate with an acceleration a is ma Newtons.
   a. What are the units of acceleration?
   b. What are the units of force when expressed in metres, kilogrammes, and seconds?
   c. If the force is 3.2×10^2 kN and the mass is 2×10^6 kg, what is the acceleration?
   d. If the force is 4.3×10^3 MN and the mass is 3 g, what is the acceleration?
   e. If the force is 2.7×10^7 N and the mass is 20 mg, what is the acceleration?
10. The force on a mass m kg due to gravity is mg Newtons where $g = 9.81$ m/s^2
    a. If the force is 3.5×10^3 kN, what is the mass?
    b. If the mass is 35,700 kg, what is the force?
    c. If the mass is 2×10^{13} g, what is the force?
11. The energy needed to apply a force F Newtons over a distance L metres is FL joules.
    a. What are the units of energy when expressed in metres, kilogrammes, and seconds?
    b. If the force is 3.5×10^8 N and the distance is 5 km, what is the energy?
    c. If the force is 2.7×10^5 kN and the distance is 20 mm, what is the energy?
    d. If the energy is 2×10^4 MJ and the distance is 2×10^7 m, what is the force?
12. The stress on an area A m^2 due to a force F Newtons is F/A Pascals.
    a. What are the units of stress when expressed in metres, kilogrammes, and seconds?
    b. If the force is 3 GN and the stress is 20 GPa, what is the area?
    c. If the force is 5×10^5 N and the stress is 10 kPa, what is the area?
    d. If the force is 2.2×10^3 N and the area is square with side 100 mm, what is the stress?
    e. If the force is 20 MN and the stress is 20 mPa on a square area, what is the length of the side?
    f. If the force is 3.2×10^7 N and the stress is 2×10^{15} Pa on a square area, what is the length of the side?

CHAPTER 2

1. The following observations are made when a 100 mm (3.94 in.) length of 10 mm (0.394 in.) diameter steel bar is loaded in tension:

Load (kN)	Extension (mm)	Load (kip)	Extension (in.)
0	0	0.00	0.0000
1.57	0.01	0.35	0.0004
4.71	0.03	1.06	0.0012
7.85	0.05	1.77	0.0020
9.42	0.06	2.12	0.0024
15.7	0.1	3.53	0.0039
18.84	0.12	4.24	0.0047
21.98	0.14	4.95	0.0055
23.55	0.15	5.30	0.0059
24.5	0.2	5.51	0.0079
25.5	0.3	5.74	0.0118
25.9	0.365	5.83	0.0144
26.5	0.5	5.96	0.0197
28.5	1	6.41	0.0394
31.5	1.9	7.09	0.0748

Calculate the following:
   a. The Young's modulus.
   b. The 0.2% proof stress.
   c. The yield stress.
   d. The ultimate stress.
   e. Explain which of these measurements is used for specification of steel, and why it is used.
2. A cube is made of a material that has no volume change when compressed (rubber comes close to this). What is its Poisson's ratio?
3. **a.** A concrete cylinder has a length of 300 mm (11.81 in.) and a diameter of 100 mm (3.94 in.). If the length when supporting a load of 4,700 kg (10,363 lb.) is 299.93 mm (11.807 in.), what is the Young's modulus?
   b. If the diameter of the cylinder becomes 100.004 mm (3.94016 in.) when the load is applied, what is the Poisson's ratio?
   c. If the characteristic strength of the cylinder is 35 MPa (5.075 ksi) and a factor of safety of 1.4 is used, what is the load, in tonnes (lb), that the cylinder will support?
4. A water tank is supported by four identical timber posts that all carry an equal load. Each post measures 50 mm (1.97 in.) by 100 mm (3.94 in.) in cross-section and is 1.0 m

(39.4 in.) long. When 1 m³ (1.31 yd³) of water is pumped into the tank, the posts get 0.08 mm (0.003 in.) shorter.

a. What is the Young's modulus of the timber in the direction of loading?

b. If the cross-sections measure 100.003 mm (3.940118 in.) by 50.00025 mm (1.97000985 in.) after loading, what are the relevant Poisson's ratios?

c. If 300 L (0.393 yd³) of water are now pumped out of the tank, what are the new dimensions of the posts?

(Assume that the strain remains elastic.)

5. A 6 m (19.68 ft.) long section of steel pipe has an outside diameter of 100 mm (3.937 in.) and an inside diameter of 90 mm (3.543 in.). The steel has a Young's modulus of 200 GPa (29,000 ksi) and a 0.2% proof stress of 340 MPa (49.3 ksi). Calculate the load in tonnes (lb) required to:

a. Increase the length to 6.006 m (19.6997 ft.).

b. Increase the length to the point where it has an irreversible plastic extension of 0.2%.

c. Reduce the length to 6.015 m (19.729 ft.) by partially unloading after applying the load in part (b).

6. The following observations are made when a 100 mm (3.94 in.) length of 10 mm (0.394 in.) diameter steel bar is loaded in tension:

Load (kN)	Extension (mm)	Load (kip)	Extension (in.)
0	0	0.00	0.0000
1.88	0.01	0.42	0.0004
5.65	0.03	1.27	0.0012
9.42	0.05	2.12	0.0020
11.3	0.06	2.54	0.0024
18.84	0.1	4.24	0.0039
22.61	0.12	5.09	0.0047
26.38	0.14	5.94	0.0055
28.26	0.15	6.36	0.0059
29.4	0.2	6.62	0.0079
30.6	0.3	6.89	0.0118
31.1	0.365	7.00	0.0144
31.8	0.5	7.16	0.0197
34.2	1	7.70	0.0394
37.8	1.9	8.51	0.0748

Calculate the following:

a. The Young's modulus.

b. The 0.2% proof stress.

c. The estimated yield stress.

    d. The ultimate stress.

    e. Explain which of these measurements is used for specification of steel, and why it is used.

7. a. A round steel bar with an initial diameter of 25 mm and length of 2 m is placed in tension supporting a load of 2000 kg. If the Young's modulus of the bar is 220 GPa, what is the length of the bar when supporting the load?

    b. If the Poisson's ratio of the steel is 0.4 what will the diameter of the bar be when supporting the load?

    c. If the 0.2% proof stress of the steel is 150 MPa what load would be required to give an irreversible extension of 0.2% in the bar?

    d. Describe what will happen if the load is increased beyond the 0.2% proof stress.

8. A 2 m long tie in a steel frame is made of circular hollow section steel, with an outer diameter of 60 mm and a wall thickness of 3 mm. The properties of the steel are as follows:

- Young's modulus: 200 GPa
- Yield stress: 300 MPa
- 0.2% proof stress: 320 MPa
- Poisson's ratio: 0.15

    a. What is the extension of the tie when the tensile load in it is 100 kN?

    b. What is the reduction in wall thickness when this load of 100 kN is applied?

    c. What load is required to produce an irreversible strain of 0.2%?

    d. If the load calculated in (c) is applied, and then reduced by half, what is the extension (while the reduced load is still applied)?

9. a. A flat steel bar measuring 25 mm by 210 mm by 3 m long is placed in tension supporting a load of 4 tonnes. If the Young's modulus of the steel is 210 GPa, what is the length of the bar when supporting the load?

    b. If the Poisson's ratio of the steel is 0.30, what are the dimensions of the steel section when supporting the load?

    c. If the 0.2% proof stress of the steel is 300 MPa, what load would be required to give an irreversible extension of 0.2%? (Give your answer in tonnes.)

    d. Describe what will happen if the stress is increased beyond the proof stress.

10. The following figure shows the load-displacement plot when a timber sample measuring $15 \times 16 \times 42$ mm is tested in compression on end (with the load applied to the 15×16 mm faces). The equation on the graph is for the best line fit shown.

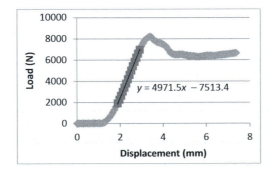

a. What is the approximate failure load for the sample in N?

b. What is the approximate failure stress in MPa?

c. What is the Young's modulus in MPa?

d. If a 100 mm square, 2 m tall column in a building is made from the same timber (with the grain in the same direction), and is subject to a load of 150 kN, what displacement will this cause at the top? (Assume it does not buckle.)

e. If the grain in the timber test sample had been in a different direction, what effect would this have had on the failure stress?

11. **a.** Describe the difference between load and stress.

b. A 4 in. diameter concrete cylinder that is 8 in. long is loaded on the end with 17920 lb. What is the stress in it?

c. The Young's modulus of the cylinder is 3000 ksi. What is its height after loading?

d. The Poisson's ratio of the cylinder is 0.17. What is its diameter after loading?

e. A column 6 ft. high, measuring 12 in. by 20 in. on plan, is made with the same concrete as in the cylinder, and supports a bridge. In order to prevent damage to the deck, the maximum permitted reduction in length of the column is 0.016 in. Assuming that the column is not reinforced, calculate the reduction in length of the column when it is supporting the full weight of a 89600 lb lorry and state whether the deck will fail.

CHAPTER 4

1. Two identical water tanks are supported 100 mm (3.94 in.) above a flat rigid floor on timbers that are 100 × 100 × 750 mm (3.94 × 3.94 × 29.5 in.) long. Each tank rests on the whole of one of the 100 mm wide edges of two timbers.

a. If the tanks are at the same level when empty, what is the difference in level when one tank has 2.5 m³ (3.275 yd³) more water in it than the other?

b. If 0.7 m³ (0.917 yd³) of hot water is added to the tank with less water in it, and the temperature of the timbers under it is raised by 25°C (45°F), what does the difference in level become?

c. Describe how the answer to part (a) would have been affected if the timbers had been (i) wet, or (ii) loaded parallel to the grain.

The properties of the timbers are as follows:

Young's modulus: 80 N/mm² (11,600 psi)

Coefficient of thermal expansion: 34 μstrain/°C (19 μstrain/°F)

2. The bricks in a wall have a specific heat of 900 J/kg/°C, (0.216 BTU/lb/°F), a thermal conductivity of 0.7 W/m/°C (0.414 BTU/hft.°F), and a density of 1700 kg/m³ (2873 lb/yd³). The wall is solid and is 215 mm (9 in.) thick, 3 m (10 ft.) high and 10 m (33 ft.) long.

a. If the wall is at 25°C (77°F) on the inside and 10°C (50°F) on the outside, what is the rate of heat loss through it in watts (BTU/hr)?

b. If the inside and outside temperatures are lowered by 5°C (9°F), what is the heat loss from the wall in joules (BTU)?

c. Describe how the thermal performance of walls is improved relative to this example in modern construction.

3. The four materials listed below are being considered for the cladding of a building:

Material	Proposed Thickness (mm)	Thermal Conductivity (W/m/°C)	Coefficient of Thermal Expansion (μstrain/°C)
Concrete	100	1.4	11
Steel	5	84	11
Aluminium	8	200	24
Brick	215	0.9	8

For each material:
a. Calculate the heat loss in watts through each m^2 of the cladding, if the temperature difference between the interior of the building and the outside air is 10°C.
b. State the assumptions you made in your calculation in section (a) and indicate whether they are realistic.
c. Calculate the spacing of the expansion joints in the cladding if the maximum acceptable expansion of a panel between joints is 2 mm for a 20°C temperature rise.
4. a. A 250 mm thick concrete wall is cast using a mix containing 350 kg/m^3 of cement. The shuttering on each side of the wall is 20 mm plywood. If the heat of hydration is being generated at a rate of 7 W/kg of cement, and is all being lost through the shutters, what is the temperature drop across them?
(The thermal conductivity of plywood = 0.15 W/m^2/°C)
b. Describe two methods for reducing the heat of hydration in a concrete mix.
5. A building has a 120 mm thick concrete floor slab with rooms above and below it. The slab is 5 m long and 4 m wide. The properties of the concrete are as follows:
- Thermal conductivity: 1.5 W/m/°C
- Density: 2300 kg/m^3
- Specific heat: 800 J/kg/°C
- Coefficient of thermal expansion: 8 μstrain/°C
a. What is the heat transmission through the floor in W if the temperature difference between the two rooms is 15°C?
b. What is the energy in joules that will be absorbed by the slab, if the average temperature of the rooms rises by 12°C on a hot day?
c. What is the increase in length of the slab when the temperature is increased by 10°C?
d. Describe two methods for reducing the heat transmission.

CHAPTER 5

1. What are the velocity and wavelength of ultrasound with a frequency of 5 kHz in a thin concrete element with density 2400 kg/m^3, a Young's modulus 50 GPa, and a Poisson's ratio of 0.15?
2. a. 4 L of an ideal gas is at a pressure of 1 atm and a temperature of 0°C.
   (i) If the volume is reduced to 3 L and the temperature rises to 20°C, what does the pressure become?
   (ii) If the molecular weight of the gas is 18, what is its mass?

b. A pulse of ultrasound takes 1.4×10^{-4}s to travel through a concrete wall 400 mm thick. If the Poisson's ratio of the concrete is 0.13, and the density is 2400 kg/m³, what is the Young's modulus?

3. A 200 mm thick concrete wall is tested using ultrasonics. The measured transit time is 50 μs.

a. Calculate the Young's modulus of the concrete, assuming a density of 2300 kg/m³ and a Poisson's ratio of 0.11.

What is the percentage error in the answer in (a) if:

b. The actual density is 2380 kg/m³?

c. The actual Poisson's ratio is 0.13?

d. The actual wall thickness is 220 mm?

e. Discuss the consequences of these percentage errors on the precautions that should be taken when carrying out ultrasonic testing.

4. a. A pulse of ultrasound takes 10^{-4}s to travel through a concrete wall 400 mm thick. If the Poisson's ratio of the concrete is 0.13 and the density is 2400 kg/m³, what is its Young's modulus?

b. A container of ideal gas at atmospheric pressure is heated from 20°C to 50°C. What is the final pressure?

c. 1 L of ideal gas at atmospheric pressure is compressed into 0.3 L at the same temperature. What is the final pressure?

d. A fixed volume of ideal gas at 30°C and 1 MPa is heated to 100°C. What is the final pressure?

CHAPTER 6

1. A 4 mm diameter copper wire has a resistivity of 1.9×10^{-8} Ωm and is carrying a current of 100 A.

a. What is the power loss per m length in the wire?

b. If the wire has a specific heat of 370 J/kg°C and no heat is lost from it, what is its temperature rise per minute? (Density of copper = 8900 kg/m³.)

c. If the coefficient of thermal expansion of the wire is 17×10^{-6}/°C, what is the percentage expansion after 3 min?

2. A current of 20 A is flowing first through a copper conductor, then through an aluminium one, and finally through a lead one. All of the conductors have a circular cross-section with a diameter of 4 mm. The properties of the conductors are as follows:

Material	Copper	Aluminium	Lead
Resistivity (Ωm)	1.7×10^{-8}	2.8×10^{-8}	1.8×10^{-7}
Density (kg/m³)	8,800	2,600	11,000
Specific heat (J/kg/°C)	390	880	200

For each metre length of each conductor calculate the following:

a. The resistance.

b. The voltage drop.

c. The power loss.

    d. The mass.

    e. The temperature rise in 10 min (Assume no heat loss).

3. a. State what is meant by the following terms in electrical circuits:

    (i) Current.

    (ii) Resistance.

    (iii) Resistivity.

  b. The maximum acceptable voltage drop on a 32 kV transmission line is 1%/km.

    (i) What cross-section area of copper wire is required for a power of 50 MW?

    (ii) How much would the wire between two pylons 100 m apart weigh?

    (iii) What would the weight be if aluminium wire was used?

Use the following:

	Copper	Aluminium
Resistivity (Ωm)	1.7×10^{-8}	2.8×10^{-8}
Density (kg/m³)	8900	7800

4. A 4-mm diameter copper wire is carrying a current of 20 A and has a resistivity 1.8×10^{-8} Ωm.

  a. What is the power loss per m length in the wire?

  b. What is the voltage drop per m length along the wire?

  c. If the wire is 20 m long and the supply voltage is 240 V, what is the percentage power loss?

5. a. Describe the difference between resistance and resistivity.

  b. A 3 mm diameter copper wire has a resistivity of 1.9×10^{-8} Ωm; what is its resistance per m length?

  c. If the wire is carrying a current of 100 A, what is the power loss per m length in the wire?

  d. If the wire has a specific heat of 370 J/kg°C and no heat is lost from it, what is its temperature rise per minute? (Density of copper = 8900 kg/m³.)

  e. If the coefficient of thermal expansion of the wire is 17×10^{-6}/°C, what is the percentage expansion after 3 min?

CHAPTER 7

1. Give the meaning of the following terms. Where possible, give examples:

  a. The atomic number of an element

  b. A molecule

  c. An exothermic reaction

  d. An acid

  e. pH

  f. Hard water

2. a. Calculate the volume of 3 mols of a gas at 20°C and a pressure of 3 bar.

  b. Explain what a catalyst is, and give an example of one.

  c. List three factors that will increase the rate of a chemical reaction.

  d. Give an example of how one of the factors from part (c) has a significant effect in construction.

3. a. State typical values for the chemical pH of the following:
   (i) A strong acid
   (ii) Concrete
   (iii) "Hard" water
   (iv) "Soft" water
   b. (i) Describe the process in which lime is produced from limestone.
   (ii) Describe the process in which lime is carbonated in concrete.
4. a. Describe three factors that affect the rate of a chemical reaction. Give examples of reactions in which each of the factors is significant.
   b. Describe the "lime cycle" and its relevance to construction processes.

CHAPTER 8

1. a. (i) A brick is fully saturated and has a density of 1800 kg/m^3. When it is dried it has a density of 1500 kg/m^3. What is its porosity?
   (ii) If the brick is cut in half, what proportion of the cut surface will be made up of pores?
   b. Grout is being pumped along a pipe. All of the grout that is more than 10 mm from the wall of the pipe is moving at 0.3 m/s, and the material in contact with the pipe wall is not moving. Assuming that the grout has a viscosity of 0.15 PaS, calculate the drag on each m^2 of the surface of the pipe.
2. a. A sample of wood is fully saturated and has a density of 800 kg/m^3. It is dried completely and the density is reduced to 600 kg/m^3. What is the porosity?
   b. On a cold morning, the air in a building is at 100% RH and the partial pressure of the water vapour in it is 0.01 bar. As the building warms up, the saturation pressure for water in the air in it is increased to 0.07 bar. Assuming that no water evaporates from surfaces in the building, and the partial pressure of the water vapour remains constant, what will the humidity be?
   c. Discuss how the conditions in the building will change as the temperature rises.
3. a. What would the change in temperature of a 2 m^3 block of concrete with a density of 2300 kg/m^3 and a specific heat of 850 J/kg/°C be, if it was heated by a 4 kW heater for 30 min? (Assume no heat loss.)
   b. If the floors of a building are constructed with a material with a high specific heat, what effect is this likely to have on the temperature of the rooms?
   c. (i) A brick is fully saturated and has a density of 1850 kg/m^3. When it is dried it has a density of 1550 kg/m^3. What is its porosity?
   (ii) If the brick is cut in half, what proportion of the cut surface will be made up of pores?

CHAPTER 9

1. A brick is 215 mm long, 65 mm high, and 102 mm wide, and has no frog or perforations. The brick is fully saturated with water and weighs 2.4 kg. The brick is then dried and weighs 2.17 kg. What is:
   a. The bulk dry density of the brick?
   b. The skeletal density of the solid material in it?

c. The porosity of the brick?

d. The brick is now placed in a high humidity environment, so 70% of its porosity is full of water. What will its mass be?

e. A brick wall is built without a damp proof course, and the bottom of it is exposed to water. If the typical pore radius in the bricks is 0.13 μm, what is the theoretical height to which the water will rise?

f. Why will it not rise to this height in practice?

(Surface tension of water, $s = 0.073$ N/m)

2. a. The flow through a concrete plug in a pipe is 2 mL per day. What is the flow in m^3/s?

b. A 300 mm diameter pipe is blocked with a concrete plug 100 mm thick. A 20 m head of water is applied to one side of the plug, and the other side is open to the atmosphere. The flow rate through the plug is 2 mL per day; what is the coefficient of permeability of the concrete?

c. The viscosity of the water is 10^{-3} PaS. What is the intrinsic permeability of the concrete?

d. Describe what happens when ordinary concrete is exposed to sulphates.

e. Describe two methods of preventing the effect of sulphate on concrete.

3. a. A river has been diverted to flow through a rectangular concrete culvert with external dimensions 4 m high and 2 m wide and with a wall thickness of 120 mm, contains water at an average pressure head of 20 m. If there is no water pressure on the outside of the culvert, and the coefficient of permeability of the concrete is 1.3×10^{-12} m/s, what is the flow in mL/day through the walls of each m length of culvert?

b. Describe the effect on the concrete if the water contains salt.

4. a. Why is concrete not waterproof?

b. A concrete wall 150 mm thick forms the side of a water tank with the outside of the wall open to the atmosphere. If the coefficient of permeability of the concrete is 6.5×10^{-12} m/s, calculate the velocity of the flow of water through the wall in m/s at a depth of 5 m below the water surface.

c. Calculate the rate of flow of water in $mL/m^2/s$ through the wall.

d. If the tank contains seawater with chlorides in it, what effect will they have on the concrete and the reinforcing steel in it?

e. If the water contains sulphates, what effect will they have?

CHAPTER 10

1. a. A 200 mm thick concrete wall is fully saturated with water with no pressure differential. On one side of it, the water contains salt at a concentration of 10% by mass, and on the other side the water is kept at zero concentration. The capacity factor for the salt is 0.5, the porosity of the concrete is 7%, and the apparent diffusion coefficient is 10^{-12} m^2/s. Describe the process that transports the salt through the concrete, and draw three graphs showing the changes in salt concentration across the thickness of the wall:

(i) Early in the transient phase.

(ii) Late in the transient phase.

(iii) In the steady state.

On each graph, show the concentration in solution and the concentration adsorbed onto the matrix.
   b. Calculate the total mass of salt in each m^2 of wall in the steady state.
   c. Calculate the mass of salt leaving each m^2 of the wall per second in the steady state.
2. a. What is the theoretical height to which water will rise in concrete due to capillary suction?
   b. A 150 mm concrete cube is placed in water, with the water level just above the base of the cube, and the top surface is kept dry by evaporation. The sides of the cube are sealed. What is the theoretical rate of flow of water up the cube in mL/s?
   c. If the water contains 8% salt, what will be the flux of salt up the cube in the steady state?
   d. If a second cube is tested in similar conditions, but with a constant flow of clean water across the top, what will the steady state flux now be?
   e. For each experiment, describe the following:
   The effect of changing the capacity factor of the concrete.
   The salt concentration in the cube in the steady state.
   The ultimate effect on the cube if the experiment is continued for a long time.
   For this question use the following data:
   typical diameter of pores in concrete = 0.015 μm
   surface tension of water = 0.073 N/m
   coefficient of permeability of concrete = 5×10^{-12} m/s
   porosity of concrete = 8%
   intrinsic diffusion coefficient of salt in water = 10^{-12} m²/s
3. a. State why the process of diffusion is important to the durability of concrete. The mix design of concrete elements exposed to chlorides is changed to include PFA. This increases the capacity factor from 1 to 5, and reduces the porosity from 10% to 8%. The density remains unchanged at 2300 kg/m³, and the intrinsic diffusion coefficient is 5×10^{-12} and does not change.
   b. Calculate apparent diffusion coefficient before and after the change is made.
   c. Describe the changes to the flux of chloride ions, both in the transient phase (shortly after exposure), and in the long term steady state.

CHAPTER 11

1. a. Describe the different situations in which ionising radiation might be encountered in construction.
   b. A concrete wall is being irradiated with 50 W of radiation. If the power is reduced to 25 W at a depth of 200 mm, what is the power at a depth of 300 mm?
2. A radioactive source has an initial power of 10 W and a power of 3 W after 1 year. What is its half-life, and what will the power be after 2 years?
3. A 5 mm steel plate reduces the power of incident radiation from 10 W to 0.1 W. What thickness of plate must be added to this to reduce the power to 0.001 W?
4. a. Describe three different types of ionising radiation.
   b. A concrete wall is being irradiated with 40 W of radiation. If the power is reduced to 15 W at a depth of 150 mm, what is the power at a depth of 200 mm.

CHAPTER 12

1. **a.** Explain why a single low result from testing a construction material should not normally result in the rejection of a batch of it.
   b. 5 bricks are tested for strength. If the average is 20 MPa and the standard deviation is 3 MPa, what is the strength below which 5% would be expected to lie?
   c. What is the probability of one strength in the set of five being below the 5% level?
   d. If three further sets of five are tested, what is the probability of all three sets having one each below the 5% level?

2. **a.** Why is the failure of a single concrete cube not necessarily a cause for concern about the quality of the structure in which the concrete was used?
   b. Concrete cubes are being tested and the mean strength is 40 MPa, and the standard deviation is 5 MPa. What is the strength below which 5% of the strengths will lie?
   c. Four concrete cubes are tested with probability of failure of 5%. What is the probability of:
      (i) One failure?
      (ii) No failures?

3. Concrete is delivered to site with a characteristic strength of 30 MPa and a percentage defect rate of 5%. If four cubes are made, what is the probability of:
   a. No failures.
   b. One failure.
   c. If four cubes are made each day, what is the probability of getting one failure per day on each of 3 consecutive days?
   d. If the observed standard deviation of the test results is 7 MPa, what is the mean strength?
   e. State three causes for variations in test results.

4. Concrete is delivered to site with a characteristic strength of 30 MPa, and a percentage defect rate of 5%. If 5 cubes are made, what is the probability of:
   a. No failures?
   b. One failure?
   c. If five cubes are made each day, what is the probability of getting one failure per day on each of 3 consecutive days?
   d. If the observed standard deviation of the test results is 6 MPa, what is the mean strength?

CHAPTER 13

1. After 1 year of production with satisfactory results, an increasing number of cube failures are reported from a concrete batching plant.
   a. Describe how you would determine whether the trend is significant.
   b. Describe the various possible causes for the change, and how you would identify them.

2. **a.** Explain why cube failures from the concrete delivered to a site may not be significant.
   b. Describe the normal method used to determine whether the failures are significant.
   c. List the different causes, which might be responsible for cube failures, and describe how the relevant one may be identified when significant failures occur on site. Include a brief description of any test methods you suggest should be used.

CHAPTER 14

1. **a.** What is the difference between a material that carries a British Standard "kitemark," and one that is supplied as complying with the relevant British Standard but does not carry a kitemark.

   b. What are the main implications of the introduction of European Standards for specifiers and purchasers of materials in the United Kingdom?

2. **a.** Describe what is meant by the terms "Quality Assurance" and "Quality Control," and state the difference between them.

   b. Discuss the differences between an agrément certificate and a National Standard.

3. Explain the meaning of the following terms:

   a. British Standard

   b. ASTM Standard

   c. ISO Standard

   d. Euro Norm

   e. Agrément Certificate

   f. Quality assurance

CHAPTER 16

1. Discuss the merits and weaknesses of current research that is being carried out into new materials and methods for concrete. Have the major failures, such as widespread corrosion from carbonation, HAC, and calcium chloride been caused by poor research? Discuss.

2. With the decline in coal burning power generation in the United Kingdom, there is likely to be a shortage of PFA. Describe a 3-year research programme to evaluate imported PFA from Eastern Europe to assess its suitability for use in concrete in highway structures. You may assume that you have a good supply of the material, and that you may obtain access to structures built with it in the country where it is produced.

3. A slag arising from the production of a nonferrous metal has been found to exhibit cementitious properties, and is being proposed for use in concrete. You have been awarded a 6-month contract to evaluate the material for use in mass concrete foundations. Describe the major parts of your programme.

4. When designing a large concrete structure, you find a substantial source of secondary (recycled) aggregate close to the site, and are considering it in the concrete in order to reduce costs.

   a. Describe how you would carry out a literature search to find information on the performance of concrete made with the aggregate. Describe the type of literature you would expect to find indicating the relative merits of each type.

   b. Describe two experiments that you would propose to determine the durability of the concrete, and give reasons why the two you have chosen are the most suitable tests for this purpose. These may be either laboratory or *in situ* tests.

5. When designing a large concrete structure, you find a substantial source of mineral waste close to the site, and are considering it as aggregate in the concrete. Describe how you would assess the material to see if it was suitable. Justify your choices. You may assume that you have about 1 year, full laboratory facilities, and access to some old concrete structures built with the aggregate.

CHAPTER 18

1. **a.** Describe four different types of cement, and give typical applications for them.
    b. List two different types of cement replacement material, and describe how they are produced and used, and what properties they give to concrete.
2. **a.** Describe the different types of cement used to make concrete, and how they may be identified from an analysis of dust samples drilled from a concrete structure.
    b. A 50 mm thick concrete barrier has pure water on one side and 80 kg/m³ salt solution on the other side. D_i is 2×10^{-12} and the porosity is 15%. What is the flow in kg/m²/s in the steady state? What is the velocity of the ions in solution 20 mm from the pure water side?
3. **a.** Identify the types of cement which have:
    (i) A high C_3S content
    (ii) A low C_3A content
    (iii) A low C_4AF content
    For each one, explain why they are made in this way.
    b. What are pozzolanic materials, and what are the effects of the pozzolanic reaction in concrete?
    c. Identify two pozzolanic materials that are used in concrete production, and describe the properties of concrete made with them, and applications where these properties are used.
4. **a.** How are the proportions of the cement compounds adjusted to give the following properties:
    1. Low heat
    2. Rapid hardening
    3. Sulphate resistance
    4. White cement
    b. Describe two cement replacement materials, and describe how they may be used, and what properties they give to concrete.
5. State the full names of the following materials, and describe their production and their uses in concrete construction:
    a. PFA
    b. GGBS
    c. CSF
    d. $CaCO_3$
6. **a.** Calculate the compound compositions for Portland cements with the following oxide analyses.

Sample	A	B	C
C	63	64.15	62
S	20	21.87	22.5
A	6	5.35	3
F	3	3.62	4.2
$\bar{S}$	2.5	2.53	4

b. Calculate the total heat generated by 1 m³ of concrete made with each of the cements, with a cement content of 300 kg/m³.

c. Outline the main consequences of the heat of hydration in concrete.

You may use the data in the following table:

Compound	Heat of Hydration (J/g)
C_3S	502
C_2S	251
C_3A	837
C_4AF	419

CHAPTER 19

1. The following results are obtained after sieving three different samples of aggregate:

	Mass Retained in Sieve (g)		
Sieve	Sample A	Sample B	Sample C
20 mm	0	0	0
9.5 mm	150	0	1100
4.75 mm	200	0	0
2.36 mm	170	100	0
1.18 mm	200	200	0
600 μm	150	200	150
300 μm	200	100	200
150 μm	50	100	100
Base	20	35	20

a. Plot the grading curves

b. Identify which is (i) sand, (ii) gap-graded aggregate, and (iii) all in aggregate.

c. Discuss the problems that occur with aggregate shrinkage, and how they may be avoided by using standard tests.

2. a. The following results are obtained after sieving three different samples of aggregate:

Sieve	Mass Retained in Sieve (g)		
	Sample A	Sample B	Sample C
20 mm	0	0	0
9.5 mm	1200	0	300
4.75 mm	0	0	100
2.36 mm	0	80	200
1.18 mm	0	100	300
600 μm	150	200	200
300 μm	200	200	100
150 μm	100	50	20
Base	54	35	16

  (i) Plot the grading curves.
  (ii) Identify which samples are (i) gap graded aggregate, and (ii) all in aggregate, (iii) sand.
 b. Describe two different major sources of aggregate for concrete.

CHAPTER 20

(For these questions, assume the specific gravity of the unhydrated cement is 3.15 kg/L, and the for the hydrated cement it is 2.15 kg/L.)

1. Define the following for cementitious materials:
  a. Porosity
  b. Specific gravity
  c. A concrete sample is made with the following proportions:
    Water/cement ratio: 0.45
    Cement content: 350 kg/m^3
    Aggregate content: 1800 kg/m^3
  The sample is cured and dried and the density after drying is 2200 kg/m^3. What is the porosity?
2. a. A concrete sample is made with a w/c ratio of 0.53 and a cement content of 370 kg/m^3 and an aggregate content of 2000 kg/m^3. A cube of the concrete is dried and the weight loss on drying is 4.5%. What percentage of the cement has hydrated, and what is the porosity?
  b. Describe the effects of the heat of hydration of cement in concrete.
  c. Describe the different methods that may be used to reduce the heat of hydration and the effect it has on the concrete.
3. a. Explain why a cement with a high w/c ratio will have poor durability.
  b. 5 kg of cement are mixed with 2.5 kg of water. The resulting paste sample is cured and dried, and its final mass is 6.2 kg. Calculate the porosity.

4. 10 kg of cement are mixed with 3 kg of water. The resulting paste sample is cured and dried, and its final mass is 12 kg. Calculate the porosity.

CHAPTER 21

To use United States method for these questions, assume:

- Fineness modulus of fine aggregate: 2.8
- Dry-rodded density of coarse aggregate: 1600 kg/m³ (2700 lb/yd³)
- Absolute density of aggregate: 2600 kg/m³ (4400 lb/yd³) (for question 1)

The cylinder strength should be taken as the characteristic strength given.

1. Calculate the mix quantities for a 0.2 m³ trial concrete mix to the following specification:
- Characteristic strength: 30 MPa (4350 psi) at 28 days
- No. of test samples: 5
- Cement: OPC
- Percent defectives: 5%
- Target slump: 100 mm (4 in.)
- Coarse aggregate: 10 mm (3/8 in.) crushed
- Fine aggregate: crushed, 40% passing a 600 μm sieve
2. Calculate the mix quantities for a 0.25m³ trial mix to the following specification:
- Characteristic strength: 40 MPa (5800 psi) at 28 days
- No. of test samples: 3
- Cement: OPC
- Proportion defectives: 5%
- Slump: 100 mm (4 in.)
- Coarse aggregate: 20 mm (3/4 in.) crushed
- Fine aggregate: crushed, 60% passing a 600 μm sieve
- Aggregate SSD density: 2500 kg/m³ (4225 lb/yd³)

CHAPTER 22

1. a. Describe the precautions that should be taken when carrying out a slump test on site.
 b. Rheological tests are carried out on three different concrete mixes, and the results are as follows:

	Shear Stress (Relative Units)		
Shear Rate (Relative Units)	**Control Mix**	**Mix A**	**Mix B**
0	0.5	0.3	0.4
0.2	0.62	0.41	0.5
0.4	0.73	0.52	0.6
0.6	0.83	0.61	0.62
0.8	0.91	0.7	0.64
1.0	1.0	0.8	0.66

Discuss the conclusions that may be drawn about the performance of these mixes in the following:

(i) A slump test

(ii) A concrete pump

(iii) Placement from a skip

2. **a.** What is the difference between air entrained concrete and foamed concrete?

b. When the air content of a sample of concrete is measured using a pressure type air meter, the following observations are made:

- Volume of concrete: 15 L
- Initial pressure: open to atmosphere
- Pressure change: 1.3 atm
- Volume change: 160 mL

What is the percentage of entrained air?

c. Describe the applications and correct methods of use of air entrained concrete.

CHAPTER 23

1. Describe the different types of nonstructural cracks that might occur in a concrete road bridge within the first year after casting. Describe the causes of the cracking and the methods that may be used to prevent it. Indicate which parts of the bridge might be affected by each type of crack.

2. Define the following terms in concrete construction:

a. Segregation

b. Honeycombing

c. Bleeding

d. Plastic settlement

e. Plastic cracking

f. Describe two different methods that could be used to rectify plastic cracking.

g. Cracks have been observed in a suspended concrete floor slab in a new residential development. Outline a programme of investigation to determine the cause of the cracks. Justify your choices.

CHAPTER 24

1. **a.** Describe the consequences of using concrete with insufficient workability in construction.

b. Describe two ways in which the workability of a concrete mix may be increased, and state the advantages and disadvantages of each.

c. Describe the precautions that should be taken when carrying out slump tests on site to ensure that an accurate result is obtained.

2. **a.** Describe what calcium chloride is used for in concrete, and give examples of where it should and should not be used.

b. Discuss the difficulties that occur when frost resistance is required in ready-mixed concrete that must be batched at a plant that is a substantial distance from the site. Your discussion should include a description of methods that may be used to overcome the difficulties.

c. Discuss the different methods that may be used to achieve a high early strength of concrete. Describe the circumstances under which each method should be used.

3. How may admixtures be used to achieve the following in concrete?
 a. High slump
 b. Frost resistance
 c. Low heat of hydration
 d. High ultimate strength
 e. High early strength
 f. Reduced cost
 g. Reduced plastic cracking
 h. Reduced bleeding
 i. Improved durability
 j. Improved pumping

4. State how a typical concrete mix may be modified by the use of admixtures to achieve the following, and in each case describe the effect on the durability of the mix:
 a. High workability
 b. High strength
 c. Low cost
 d. Self-compacting concrete

CHAPTER 25

1. What would the influence of the following be on the durability of the concrete in a structure:
 a. Decreasing chloride diffusion rates.
 b. Decreasing the capacity factor for chlorides.
 c. Decreasing the hydroxyl ion content.
 d. Applying a positive voltage to the steel.
 e. Applying a negative voltage to the steel.
 f. Increasing the pore sizes in the concrete.

2. Describe the processes that will limit the life of a reinforced concrete beam in a road bridge. Indicate how materials selection and construction practice may be used to extend the life.

3. a. Describe the processes that take place during the carbonation of reinforced concrete.
 b. If a structure is constructed with 20 mm of cover to the steel, and the carbonation depth is 10 mm after 5 years, when would it be expected to reach the steel?
 c. Describe the effect of applying a negative voltage to the steel, after the carbonation has started to reach it.
 d. What will happen in (c) above if chlorides are present?

4. a. State, with reasons, which one or more of the four main transport processes for ions in concrete will be most significant in the following circumstances. In each case, also describe the effects that will be caused by the ions in the concrete.
 (i) The parapet of a road bridge subject to frequent applications of deicing salt.
 (ii) A concrete slab placed directly over hardcore in moist conditions.
 (iii) External cladding panels on a building.

       (iv) A tank containing seawater.

       (v) A bridge beam under a direct current power rail.

   b. Describe how chloride transport in concrete may be influenced by materials selection.

CHAPTER 26

1. You are advised that segregation has occurred in a concrete pour.
   a. If you inspect the concrete when the shutters are removed, what would you expect to see?
   b. What are the effects of segregation on the quality of the structure?
   c. What changes should be made to subsequent pours to avoid it?
2. a. Explain the consequences of the following in reinforced concrete construction:
      (i) Too much cover to the reinforcement.
      (ii) Insufficient cover to the reinforcement.
   b. Describe the two types of curing that are required in concrete construction. Outline the effects of not using them.
   c. Describe the difference between a shutter oil and a release agent.
   d. Describe the difference between air entrained concrete, and foamed concrete.
3. Explain why the following materials have been used in concrete, and describe the consequences of their use.
   a. High-alumina cement
   b. Calcium chloride accelerator
   c. Reactive siliceous aggregate
   d. Sea water for mixing
4. a. Why is vibration important when concrete is placed?
   b. Describe the correct methods for vibrating concrete.
   c. State the two distinct functions of curing in concrete construction.
   d. Outline the effects of not curing concrete.
   e. Describe how curing is carried out.

CHAPTER 27

1. Describe a programme to investigate the structural condition of an estate of houses with precast concrete frames that are 40 years old.
2. Very large variations in the results from strength and slump tests are observed from a particular site. Describe the possible causes of this. State which of these causes would indicate the presence of poor quality concrete in the construction, and outline the methods that could be used to detect it.
3. Describe a programme of investigation to determine the condition of the reinforced concrete in a road bridge. You should give brief descriptions of the tests that you would plan to use.

4. Describe four methods that are used to measure the strength of concrete either in the laboratory, or on an existing structure. For each method, give typical uses and discuss the limitations of the method.

CHAPTER 28

1. **a.** What are the requirements of a masonry mortar when used for the external elevation of a house?
   b. Describe how these requirements may be met with good site practice and materials selection.
2. You are specifying materials to be used in the masonry mortar for a large housing project. The work is to be carried out in an area where there is a good supply of a natural pozzolanic material. Describe how you would decide on the specification.

CHAPTER 29

1. A substantial concrete retaining wall is to be built at the top of a beach at a coastal resort. Outline the methods that could be used in the design and specification of the wall to give it:
   a. Good durability.
   b. An improved appearance.
2. **a.** Describe the advantages and disadvantages of using high strength concrete in a 10-story office block.
   b. Describe the different methods that may be used to give an exposed concrete wall an attractive appearance.

CHAPTER 30

1. **a.** Explain why the grain size of steel has a significant effect on its mechanical properties.
   b. Describe two different methods for controlling grain size.
   c. For each of the two methods in (b), describe the effect of subsequent welding.
2. Explain the following terms when applied to structural steels. For each one also describe their effect on the steel:
   a. Lattice dislocation.
   b. Grain boundary.
   c. Work hardening.
   d. Quenching.
   e. Annealing.
3. **a.** Define the terms "0.2% proof stress" and "yield stress," and state the relative advantages of using each of them to specify the strength of steel.
   b. Describe the effect of increasing the carbon content in iron compounds. Your description should include the effect on the microstructure.
   c. Describe the effect of cold working steel, and indicate how the microscopic structure is affected, and how this causes the changes.

CHAPTER 31

1. **a.** Describe the procedure for determining the condition of the reinforcement in a concrete structure using linear polarisation.
   b. The polarisation resistance and rest potential of a steel sample are measured at 1600 Ω and −300 mV.
      (i) What is the corrosion current?
      (ii) If the rest potential falls to −400 mV, what does the corrosion current become?
   (Assume values for the Tafel constants $B_1 = B_2 = 0.13$ V.)
2. The Tafel constants for a corroding steel sample in concrete are $B_1 = B_2 = 0.16$ V.
   a. If the polarisation resistance is measured as 1700 Ω, what is the corrosion current?
   b. If the rest potential subsequently falls from −350 mV to −600 mV, and the cathode conditions do not change, what is the new corrosion current?
   c. Discuss the relative advantages and disadvantages of linear polarisation and rest potential measurements on reinforcing steel in concrete.
3. **a.** Contrast the different methods that may be used to assess the rate of deterioration of steel reinforcement in a concrete structure.
   b. Why is the word "linear" used in the term "linear polarisation?"
   c. The polarisation resistance and rest potential of a corroding steel sample are measured at 1300 Ω and −350 mV.
      (i) What is the corrosion current?
      (ii) If the rest potential falls to −500 mV, what does the corrosion current become?
   (Assume values for the Tafel constants $B_1 = B_2 = 0.12$ V.)
4. A steel sample is cast into a concrete structure and is kept electrically insulated from the reinforcing cage. By applying a voltage relative to the reinforcing cage, the following observations are made:

Voltage Relative to Reference Electrode at Concrete Surface (mV)	Current Through Sample and Reinforcing Cage (μA)
−380	35
−390	0
−400	−35
−410	−70
−420	−105

   a. Assuming B = 26 mV, calculate the corrosion current in the bar.
   b. Assuming the mass and charge of a Fe^{++} ion are 9.3×10^{-26} kg and 3.2×10^{-19} C, calculate the rate of loss of mass from the bar.
   c. Give the name of the test described above, and discuss its relative merits when compared to other electrical methods for measuring corrosion of steel in concrete.

CHAPTER 32

1. Describe the processes that take place when each of the following are cooled from a liquid state to a solid state:
   a. A mixture of two metals that are completely insoluble in one another.
   b. An alloy of two metals that are partially soluble in one another.
   c. An alloy of two metals that are completely soluble in one another.
   d. Discuss the use of two different nonferrous metals (or alloys) in construction, and explain why they are used in preference to ferrous metals.
2. Some properties of some nonferrous metals are as follows:

Property	Lead	Zinc	Copper	Aluminium
Standard electrode potential (V)	−0.12	−0.76	+0.34	−1.7
Relative density	11.3	7.1	8.7	2.7
Melting point (°C)	327	419	1083	659
Elastic modulus (GPa)	16.2	90	130	70.5
Tensile strength (MPa)	18	37	210	45
Thermal conductivity (W/m°C)	35	113	300	200

Describe a major application of each metal that takes advantage of one of the properties listed, and describe how the property makes the metal suitable for the application. Also describe the influence of the other properties that affect the performance of the metal in the application.

CHAPTER 33

1. A client is considering the use of tropical hardwood for the front facade of a public building.
   a. Briefly discuss this choice of materials in the context of current public concern about the destruction of tropical forests.
   b. Describe the processes that will affect the long-term performance of the wood.
   c. Indicate how materials selection and detailing may be used to improve this long-term performance.

CHAPTER 34

1. a. Describe what is meant by the term "efflorescence" on brickwork, and discuss how the problem may be avoided.
   b. Describe the methods that should be used to achieve good appearance and durability of brickwork.

CHAPTER 35

1. a. What is meant by the following:
   (i) Organic materials.
   (ii) Polymeric materials.
   (iii) Thermoplastics.
   (iv) Thermosetting plastics.
   b. What are the processes that limit the long-term performance of polymeric materials in construction?
   c. Identify two thermosetting resins and describe how good site practice may maximise their long-term performance.

CHAPTER 36

1. Describe the properties of the following materials and explain how these properties are used in construction.
   a. Toughened glass
   b. Laminated glass
   c. Low-emissivity glass
   d. Tinted glass
   e. Clear acrylic sheet
   f. Clear polycarbonate sheet

CHAPTER 37

1. Discuss the factors that contribute to the relative durability of concrete and bituminous road pavement construction materials.
2. a. Define the following terms:
   (i) Binder
   (ii) Bitumen
   (iii) Tar
   (iv) Asphalt
   (v) Mastic
   (vi) Macadam
   b. Describe the components of a typical flexible road construction.
   c. Describe the common failure mechanisms for the components.

CHAPTER 38

1. Give three examples of composites that are used in construction, and discuss why they are better than equivalent components made from a single material.

CHAPTER 39

1. **a.** Describe the common failure mechanisms for adhesive joints.
   b. Explain why detailing of joints with sealant is important, and give an example of correct practice. Illustrate your answer with a sketch.

CHAPTER 40

1. Discuss the factors that should be taken into consideration when choosing one of the following materials for the frame of a small industrial building:
   a. Timber
   b. Steel
   c. Precast concrete
   d. Brick structural walls with steel roof structure
2. Describe the processes that will cause loss of long-term performance of each of the following materials, if used to construct a pedestrian access bridge to a residential building.
   a. Reinforced concrete
   b. Steel
   c. Timber
   d. Masonry

Subject Index